Special Relativity, Tensors, and Energy Tensor

With Worked Problems

Other Related Titles from World Scientific

A Tale of Two Twins: The Langevin Experiment of a Traveler to a Star
by Lucien Gilles Benguigui
ISBN: 978-981-121-909-2

General Relativity: A First Examination
Second Edition
by Marvin Blecher
ISBN: 978-981-122-043-2
ISBN: 978-981-122-108-8 (pbk)

Interacting Gravitational, Electromagnetic, Neutrino and Other Waves: In the Context of Einstein's General Theory of Relativity
by Anzhong Wang
ISBN: 978-981-121-148-5

Loop Quantum Gravity for Everyone
by Rodolfo Gambini and Jorge Pullin
ISBN: 978-981-121-195-9

Special Relativity, Tensors, and Energy Tensor

With Worked Problems

Somnath Datta

Formerly, Professor of Physics,
National Council of Educational Research and Training,
New Delhi, India

NEW JERSEY • LONDON • SINGAPORE • BEIJING • SHANGHAI • HONG KONG • TAIPEI • CHENNAI • TOKYO

Published by

World Scientific Publishing Co. Pte. Ltd.
5 Toh Tuck Link, Singapore 596224
USA office: 27 Warren Street, Suite 401-402, Hackensack, NJ 07601
UK office: 57 Shelton Street, Covent Garden, London WC2H 9HE

Library of Congress Cataloging-in-Publication Data
Names: Datta, Somnath, 1939– author.
Title: Special relativity, tensors, and energy tensor : with worked problems / Somnath Datta.
Other titles: Introduction to special theory of relativity.
Description: Hackensack, New Jersey : World Scientific, [2021] | "This book is actually an expanded version of my first book Introduction to special theory of relativity published by Allied Publishers in 1998"--Preface. | Includes bibliographical references and index.
Identifiers: LCCN 2020053102 | ISBN 9789811228117 (hardcover)
Subjects: LCSH: Special relativity (Physics) | Electrodynamics. | Electromagnetism. | Space and time. | Relativistic mechanics. | Continuum mechanics.
Classification: LCC QC173.65 .D383 2021 | DDC 530.11--dc23
LC record available at https://lccn.loc.gov/2020053102

British Library Cataloguing-in-Publication Data
A catalogue record for this book is available from the British Library.

For any available supplementary material, please visit
https://www.worldscientific.com/worldscibooks/10.1142/12036#t=suppl

Desk Editor: Ng Kah Fee

Typeset by Stallion Press
Email: enquiries@stallionpress.com

This book is dedicated
to
my parents

Snehalata Dutta
Gobindo Lall Dutta

Preface

This book is actually an expanded version of my first book *Introduction to Special Theory of Relativity* published by Allied Publishers in 1998. The reason I wanted to write another book on the same subject is that I published several articles and papers over the last two decades in the journal *Physics Education* (on-line since 2013) brought out by the Indian Association of Physics Teachers. These articles were intended to give students and teachers a better grip of some key concepts associated with Relativity, both Special and General. I have incorporated these articles in this book by merging them within its chapters and sections. The reader will find the original articles in the Bibliography. Their references are shown in the next paragraph in square brackets [ ]. My treatments of these concepts are supposed to give additional strength to this book, and constitute a new feature of this book.

I mention some of these features here: An elaborate introduction to tensors and stress tensors; Maxwell's stress tensor and conservation of momentum in electromagnetic fields [17]; graphical construction of Lorentz transformation, of time dilation, length contraction and simultaneity paradox [41]; magnetism as a relativistic effect [35]; Minkowski's equation of motion illustrated by a mathematical treatment of relativistic rocket [27]; energy tensor for the electromagnetic field, and of a system of incoherent charged dust creating its own electromagnetic field [39].

Einstein's two original papers that gave the world the Special Theory of Relativity, and how the well-known phenomenon of electromagnetic induction led him to this theory, have been detailed in my article *1905 Relativity Papers of Einstein* [40]. How Minkowski's further pursuit resulted in a four-dimensional geometrical world-view of events, and subsequently of the dynamical variables of mechanics, and of electrodynamics, has been detailed in another of my article *Minkowski's Space-Time* [41]. These two supplementary resource materials are not a part of this book, but they can be downloaded from my website, cited at the end of this Preface.

This book is comprised of four parts divided into twelve chapters. Part I starts with a story of Special and General Relativity in brief, and is intended for a layman (Chapter 1). I have outlined the basic tenets of Special Relativity (Chapter 2), followed by Lorentz transformation (Chapter 3) and relativistic mechanics (Chapter 4). I have worked out many innovative problems and exercises. Some of them were designed to remove any doubt centred on relativity paradoxes, like simultaneity, time dilation, and length contraction. In fact my derivation of Lorentz transformation is itself an exercise in this direction. The detailed worked-out problems on Lorentz transformation and relativistic mechanics are intended to strengthen the reader's understanding. It is hoped that an undergraduate student studying physics will not only understand and enjoy this part thoroughly but also derive all essential knowledge about Special Relativity from it.

The rest of this book is intended for a reader seeking advanced knowledge, in particular the covariant language, in which advanced texts in Classical Electrodynamics are written. This is where its wider reach is discovered and the journey to General Relativity begins.

The entire Part II (Chapters 5 and 6) is an introduction to tensors, with special emphasis on the stress tensor. The reader will discover to his amazement and enlightenment that empty space loaded with an electromagnetic field has stresses developing within it, just like a beam loaded with bricks, or any other ordinary matter, like solids and fluids, subjected to external forces. The special name for this stress is *Maxwell's stress tensor*, and constitutes an important ladder to the energy tensor which I have explained in Part IV.

Part III takes the reader into the four-dimensional world of *Minkowski's space–time*. The introduction to tensors initiated in Part II, where it was restricted to three dimensions, now begins to give rich dividends in four dimensions, with terms like contravariant, covariant, raising and lowering of indices, and an introduction to the all-important metric tensor (Chapter 7).

But what is the need for this esoteric trip to four dimensions? Because without it, it would be impossible to look at *relativistic mechanics* in a holistic manner. Without it how would one explain that energy and momentum together form one unit, the 4-momentum (for which I have coined a new name *En-Mentum*, to remind the reader of its four components and their sequence), and transforms as one unit under Lorentz Transformation; or 4-force (*Pow-Force*)? I have shown Lorentz transformation of a 4-vector, and specialized it to En-Mentum and Pow-Force and explained their significance and corollaries. Then I worked out *Minkowski's equation of motion* (EoM), and conservation of energy and momentum as a single law of physics (Chapter 8). The Minkowski EoM is best illustrated by its application to *relativistic rocket* (Chapter 9), and the Lorentz transformation of 4-force by *magnetism as a relativistic effect* (Chapter 10). The Principle of Covariance, for which the covariant equations of electrodynamics stand as a shining example, is the culmination of this trip (Chapter 11).

Pat IV, titled *Physics of a Relativistic Continua*, has only one very important chapter, namely Chapter 12, dedicated to the *energy tensor*. We take a brief look at the non-relativistic EoM of a perfect fluid, known as Euler's equation, and extend the lessons to relativistic perfect fluid, which, in combination with the energy conservation of electromagnetic field, completes the *energy tensor*.

Writing the former edition of this book, and its expansion to this version was a major challenge with manifold obstacles in the way. I owed it to the educational values inculcated at the hallowed precincts of the University of Illinois at Urbana-Champaign, and to the professional ethics of my great teachers, to face these challenges with determination in order to "present a true account" of the gifts I have inherited from my Alma Mater. My books on *Special Relativity* and *Mechanics* are a presentation of this account in a humble way. With deepest reverence and respect I recall some of my mentors: Professor Dillon Mapother (who was instrumental in my changing over from civil engineering to physics), Professor James H. Smith (from whom I learnt *Special Theory of Relativity*), Professor O. Hanson, Professor G. C. McVittie (legendary authority in General Relativity who taught me an introductory course on this subject), Professor Yavin, and Professor Peter Axel, to name a few.

The memory of Professor James Allen, my research advisor, is etched permanently on my mind. He stood by me during some of my most trying times. To me he was the most shining example of love and kindness.

There were two other persons who left their indelible footprints on the seashore of my life: Dr. Pratap Chandra Chunder, former Minister of Human Resource Development, Government of India, and Dr. Rais Ahmed, former Director of the NCERT.

Fig. 1. Alma Mater Statue at the University of Illinios

The entire typesetting of this book, up to its last details, plotting of graphs and drawings were done by me, under the operating system *Linux Mint 17.2.* I used

- The document preparation system LaTeX 2_{ε} for typesetting texts and equations, however difficult and complex they might appear before our eyes,
- *Kile 2.1* for making the typesetting and editing operations easy,
- *Gnuplot* for preparing plots and graphs of mathematical equations, however difficult they may be,
- *Maxima* for evaluating most difficult integrals, which are impossible to perform manually,
- *Xfig* for the drawings and integrating all plotted graphs in the drawings, and
- *GIMP* for integrating .jpg images into this book and into my previous book *Mechanics.*

I bow my head in humble respect for those who gave their precious times to write these programs for the students and teachers of the world, and distributed them free of cost.

I got my first tutorials in *Linux*, LaTeX and *Xfig* from my first son-in-law Michael Murphy way back in 1995, and guidance in the use of *Gnuplot* and *Maxima* from my second son-in-law G. R. Santhosh.

There were times when I was at my wit's end, whether in understanding certain aspects of Relativity or in overcoming typesetting problems under LaTeX 2_ε. I rushed to Professor A. V. Gopala Rao, himself a relativist of great repute, who gave his time liberally to help me out of the difficulties.

Professor Ashok Singal, Department of Astrophysics at Physical Research Lab, Ahmedabad, read my manuscript patiently and corrected the errors and mistakes that came to his notice. Without his watchful eyes some embarrassing errors would have gone into my book undetected.

The final formatting of my manuscript to fit into the specified page size, has been done so meticulously by Mr. Kah-Fee Ng, Senior Editor, World Scientific Publishing Company. My special thanks to him.

I shall mention a few more names: My former students Lakshmi Narayanan and B. Rajeswari who helped me in publishing this book; Mr. K. S. Venkatesh, CEO, QDP Technologies from whom I received all help to sort out my computer problems.

I am grateful to Dr. H. Basavana Gowda, Cardiologist and Principal JSS Medical College, Mysuru, Dr. B. S. Ramesh, Radiation Oncologist at HCG Hospital, Bangalore, Dr. K. G. Srinivas, Medical Oncologist at Bharat Hospital and Institute of Oncology, Mysuru for their medical guidance during my most difficult and trying times. I am indebted to Dr. Shankara Narayana Jois, my Yoga Guru for guiding me to a healthier life and to Dr. K. V. Ravishankar of Usha Kiran Eye Hospital, Mysuru for helping my wife and me protect and maintain our vision as we age.

My granddaughter Barsha Manjari Kush has been the inspiration behind all my creative works in physics and music. My wife Aloka, and my two daughters Anuradha and Madhusmita propped up my sinking spirits when I lost hopes of completing my multi-dimensional projects.

I have been enjoying *Fuller Fund Membership* of the *American Association of Physics Teachers* since October 2001. This allowed me access to the *American Journal of Physics*, crucial for writing papers in *Classical Electrodynamics* and *Special Relativity*. I thank Prof. Rogers Fuller, Associate Director of Membership, AAPT, and Harold Q and Charlotte Mae Fuller for this precious gift.

I shall conclude by citing my website

http://sites.google.com/site/physicsforpleasure

where the readers will find many of my physics experiments, papers and articles (published or unpublished), and my music projects.

I conclude by wishing the readers an enjoyable experience in going through this book.

Somnath Datta
Mysuru, November 30, 2020

Contents

Symbols Used in This Book

Symbol	*Meaning*	*Page*
"… "	"event"	21
Δt, $\Delta\tau$	improper/proper time between two events	59
$\mathbf{u} = c\boldsymbol{\beta}$, $\mathbf{v} = c\boldsymbol{\nu}$	boost velocity, particle velocity	81
$\mathbf{v} = c\beta$	particle velocity	95
f, ν	frequency of light	62, 118
γ, Γ	boost, dynamic Lorentz factors	39, 86
$\hat{\Omega}$	Lorentz transformation matrix	53
m, m_0, m_o	relativistic mass, rest mass	95
$\boldsymbol{a}, \mathbf{p}, \mathbf{F}, p$	acceleration, momentum, force, pressure	88, 95, 96, 165
$\mathbf{n}$	unit vector in the direction of boost, photon propagation	54, 241
$E, \mathcal{E}$	energy	103, 239
ρ, ρ_o, $\mathbf{J}$	charge density (proper), electric current density	296
J^{μ}	current density 4-vector	298
$\mathbf{E}$, E, $\mathbf{B}$, B	electric, magnetic fields and their strengths	Chapters 6, 11
ε_0, μ_0	permittivity, permeability of free space	Chapters 6, 11

(*Continued*)

(*Continued*)

Symbol	*Meaning*	*Page*
$\boldsymbol{g}$, $\mathbf{P}$	(field, particle) momentum density	190
E^3, M^4	Euclidean space, Minkowski Space–Time	209, 211
$\vec{\mathbf{A}}, A^\mu, \nabla_\mu, \mathbf{A}, \boldsymbol{\nabla}$	4-vector, 4-gradient, 3-vector, 3-gradient	223, 224
$\vec{\mathbf{e}}_\mu, \mathbf{e}_i$	unit 4-vector, unit 3-vector	237, 146
$\widehat{\mathbf{T}}, \widehat{\mathcal{T}}, \hat{\mathbf{p}}$	3-tensor, 3-stress tensor, pressure tensor	147, 156, 165
$\widehat{\mathcal{T}}^{(\mathrm{e})}, \widehat{\mathcal{T}}^{(\mathrm{m})}, \widehat{\mathcal{T}}^{(\mathrm{em})}$	Maxwell's stress tensor for $\mathbf{E}, \mathbf{B}$, em fields	174, 181, 190
$\vec{\vec{\mathcal{T}}}$	Maxwell's 4-stress tensor	322
$\vec{\mathbf{V}}, V^\mu, \vec{\mathcal{A}}$	4-velocity, 4-acceleration	237, 238
U^μ	4-velocity of a fluid particle	298, 333
$\vec{\mathcal{P}}, \mathcal{P}^\mu$	4-momentum (En-Mentum)	239
$\vec{\mathcal{F}}, \mathcal{F}^\mu, \vec{\mathcal{K}}$	Minkowski 4-force, convective 4-force	242, 243
Π, ϖ, w	power, power density, field energy density	242, 321, 323
$\vec{\boldsymbol{f}}, f^\mu$	Vol 4-force density	321
$\boldsymbol{g}$, σ	momentum density, mass density	190, 324, 337
$\mathbf{S}$	energy flux density, Poynting's vector	323
$F^{\mu\nu}$, $\mathfrak{F}^{\mu\nu}$	electromagnetic field tensor & its dual	299, 301
$\vec{\vec{\mathbf{M}}}, M^{\mu\nu}, E^{\mu\nu}$	Maxwell's energy 4-tensor, Einstein tensor	16, 325
$\mathcal{D}^{\mu\nu}$	energy tensor for incoherent dust	333
$T^{\mu\nu}$	energy tensor for a closed system	336

Abbreviations Used in This Book

Short form	*Long form*	*Short form*	*Long form*
		I.R.F.	*Instantaneous rest frame*
LT	*Lorentz transformation*	EoM	*equation of motion*
em	*electromagnetic*	fld	*field*
mch	*mechanical*	mat	*matter*

Conversion Factors

u = atomic unit for mass; eV = electron volt; MeV = mega electron volts; Å = angstrom

Mass	1 u	$= 1.66 \times 10^{-27}$ kg	1 kg	$= 6.02 \times 10^{26}$ u
Mass	1 u	$= 932.0\,\text{MeV}/c^2$	$1\,\text{MeV}/c^2$	$= 1.074 \times 10^{-3}$ u
Length	1 Å	$= 10^{-10}$ m		
Length	1 lt-sec	$= 3 \times 10^{8}$ m	1 lt-year	$= 9.46 \times 10^{12}$ m
Energy	1 MeV	$= 1.602 \times 10^{-13}$ J	1 J	$= 6.242 \times 10^{12}$ MeV

Universal Constants

Symbol	*Name*	*Value in SI units*
c	speed of light in empty space	3×10^{8}m/s
h	Planck's constant	6.63×10^{-34} J.s
$\mp e$	electron/proton charge	$\mp 1.6 \times 10^{-19}$ C
m_o	electron mass	$5.49 \times 10^{-4}\,\text{u} = 0.511\,\text{MeV}/c^2 = 9.11 \times 10^{-31}$ kg
m_p	proton mass	$1.0073\,\text{u} = 938.8\,\text{MeV}/c^2 = 1.67 \times 10^{-27}$ kg
m_H	hydrogen atom mass	1.0078 u
ε_0	permittivity of free space	$8.85 \times 10^{-12}\,\text{C}^2/\text{N.m}^2$
$\frac{1}{4\pi\epsilon_0}$	$= 1/(4\pi \times 8.85 \times 10^{-12})$	$9 \times 10^{9}\,\text{N.m}^2/\text{C}^2$
μ_0	permeability of free space	$4\pi \times 10^{-7}$
$\dfrac{1}{\mu_0\varepsilon_0}$	$= 9 \times 10^{16} = c^2 \quad \Rightarrow \quad c =$	$\dfrac{1}{\sqrt{\mu_0\varepsilon_0}} = 3 \times 10^{8}$ m/s

Relationship between the Energy $\mathcal{E}$ and the Wavelength λ of a photon

$$\mathcal{E} = h\nu = \frac{hc}{\lambda} \text{ J},$$
$$\mathcal{E}(\text{eV}) \times \lambda(\text{Å}) = 1.24 \times 10^4 \quad \text{eV} \cdot \text{Å}.$$

New name *En-Mentum* for the conventional name *4-momentum*. See Sec. 8.4 for the justification.[a]

[a]Data have been taken from Ref. [42].

Part I
Einsteinian Relativity

Chapter 1

What Is Relativity?

1.1. Influence of the Human Value System on the Evolution of Physical Theories

The physical laws, the way we understand them, are man-made constructs of mathematical models, capable of discerning order and pattern in a myriad manifestations of the insentient, impersonal and impercipient nature. They are a code of conduct, believed to be obeyed by all children of nature — particles and objects of various hues and forms, forces and fields — in their mutual interactions, propagation and dynamical behaviour — as they weave a web of evolution through the vast expanse of space and time. It is not unnatural, therefore, that these codes of the impercipient world — being codification by the sentient human mind — should borrow from the ambience of human society, concepts, models, imageries and even value systems.

For example, the principle of least action — which asserts that all dynamical processes of change in the universe are motivated by a propensity to extremize the "action integral" — which is the Lagrangian function summed over time — is an adaptation from human society of man's own drive for optimizing the fruits of his efforts and the value for his money. As a corollary, it is the most sensible thing for an insentient electron or a missile to reach its target by that shortest route which also takes into consideration the compulsions of a prevailing ambient electric or gravitational field, just as it is the most sensible thing for a sentient human to reach his destination by that shortest route which also takes into consideration the various compulsions of his mission, places of special interest as well as lurking dangers in an unfamiliar terrain.

As another example, the facility with which many of us handle, and even enjoy, trigonometrical exercises, lends a natural instinct to extend geometrical models for unfathoming mysteries of the physical nature. And what else can be a simpler geometrical object than a straight line of a measured length and orientation — more descriptively, a directed line segment — popularly known as a vector? Naturally, therefore, all of classical physics is dominated by the vector model of various physical quantities — like force, velocity, momentum, electric and magnetic fields and others — each one of which is modelled as the segment of a straight line whose changing length and direction tell us all about the dynamical nature of the prototype. Understanding, as well the application, of physical laws thus greatly simplify to the construction of straight lines and triangles, and interpreting them with the help of trigonometrical formulas.

With the passage of time, human society has changed and its value systems have transformed, leaving their mark on the evolution of physical theories. The emergence of democracy as a universally accepted social order, for instance, also contained the seeds of two powerful streams of physical theories that have occupied the centre stage since the turn of the 20th century. One of them, called quantum mechanics, starts with Einstein's interpretation of photo-electric effect in terms of light quanta and de Broglie's hypothesis of wave-particle duality. In effect, they served to remove any fundamental distinction between the world of radiation and the world of matter — treating both on the same footing, endowing both with wave and corpuscular attributes. The other stream of thought, called the theory of relativity proclaims democracy of all frames of reference. It starts with an outright rejection of the prevailing notion of an absolute frame, just as democracy starts with an outright rejection of an absolute ruler. Relativity teaches us to look upon all frames of reference as equivalent, just as democracy teaches us to treat different frames of opinion with equal respect.

1.2. Rejection of Absolute Frame

A frame of reference is a set of permanent benchmarks with respect to which the coordinates of all moving objects are defined. To concretize the idea, a frame of reference is often viewed in classical physics as a set of three mutually perpendicular rigid bars extending up to infinity (Fig. 1.1).

These bars serve as the X-, Y- and Z-axes with respect to which the coordinates (x, y, z) of a moving object are measured. Time rates of change of these coordinates lead to velocity, momentum, acceleration and a host of

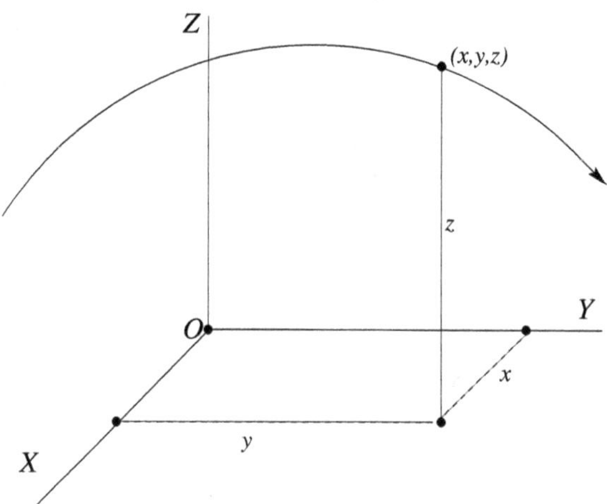

Fig. 1.1. Coordinates of a projectile.

other derived physical quantities. Systematic observation, followed by careful study and analysis of the collected data, often leads to the speculation of certain types of temporal, spatial and interconnecting relations among such physical quantities, which are enunciated as physical laws. From its very nature, therefore, a law is presumed to be valid in one particular frame namely the one from which, or with respect to which, the observations were made.

The apparent diurnal motion of stars and planets across the sky, for instance, led to the earlier Aristotelian speculation of a geocentric model in which the earth was taken to be a "fixed frame" of reference at the centre of the universe. In the 15th century, Nicolaus Copernicus disbanded the geocentric model in favour of a heliocentric universe in which the sun is the centre of a fixed frame of reference with the entire clan of "fixed stars" forming a backdrop of reference points against which the earth, the moon and the other planets and their satellites moved. From the astronomical data of these motions, collected by Tycho Brahe and mathematically analyzed by Johannes Kepler, Newton arrived at the Universal Law of Gravitation.

Thus, at least till the end of the 19th century, man thought only in terms of one supreme standard of frame of reference, the so-called Absolute Frame (often intuitively identified with the sun and the "fixed stars") which also served as the standard of absolute rest. This also defined "motion" in an absolute sense, to be one in which the coordinates of an object, as measured in the absolute frame, are seen to be changing with time. Theory of relativity brings about a complete and radical overthrow of that attitude by declaring that such a notion of Absolute Frame, absolute motion, absolute rest are only a myth, which can never be substantiated by any experiment. Relativity asserts that all motions are relative, all states of rest are also relative — relative to the particular frame of reference we have fixed for the convenience of our observation and calculations. The so-called Absolute Frame does not exist, because there is no special attribute, special quality that can be experimentally measured, and then recognized to be an exclusive preserve of any particular frame. The laws of physics work equally well in all frames of reference, provided we write them in the correct mathematical language.

1.3. Relativity Principle: A Rudimentary Form

We can explain this in the following way. Suppose, either by subjective judgement or by established convention, one particular frame has long been looked upon as the standard, which, for fixing ideas, we take as the one riveted on the surface of the earth. A team of experimenters had discovered the laws of physics by performing a chain of experiments in this frame and analyzing the table of collected data. The same team now decides to repeat these experiments inside a mobile laboratory set up on a train which is coasting smoothly along a straight line with uniform velocity. (Imagine the journey to be so smooth that not even a sound or jerk comes from the rails.) Relativity foretells that the results of this second series of experiments will be identical to those in the first, that the new set of data tables will replicate the older one. This implies that the same laws of physics are at work when observed from the ground as when observed from the train. This also means that there is no experimental way of discriminating between the two frames, or of establishing any special feature shared by one of them — say the ground frame — and not by the other. This being the case, the experimenters have no objective means of deciding which one of the two frames is moving in an absolute sense and which one is stationary. They can, however, "look out", see the trees, the mileposts, the hills and the rivers passing them by, and thereby conclude that their lab is moving relative to the ground. Alternatively, they can also believe that the ground — along with the hills and the trees — is moving while his lab is stationary. Relativity considers both these viewpoints equally valid and prohibits any objective judgement about who is moving and who is stationary, because motion is always and intrinsically relative.

We can now summarize the above experience as a limited relativity principle in the following not-so-precise form. Two frames of references that are "non-rotating"[a] and moving with uniform velocity with respect to each other are equivalent and indistinguishable in all respects. This statement forms the core of the special theory of relativity.

1.4. Inertial Forces

If we take this viewpoint somewhat far, we can even elevate the discarded geocentric view of the universe from the dustbin of history to a status of equality with the heliocentric view. To be more precise, let us define the geocentric frame of reference as three mutually perpendicular rigid bars (serving as X-, Y-, Z-axes) riveted to the body of the earth and stretching up to infinity. The spirit of relativity would seem to suggest that this geocentric frame and the heliocentric one are equivalent. The universe revolving around the earth, and the earth revolving around the sun, are two apparently different, but legitimate, modes of description of the same schemes of nature at work.

The reader will probably react to the above suggestions with disbelief. The universe with the stars and mighty galaxies all going around us! And to expect a modern scientific mind of the 20th century (or, 21st century) to place such a nonsense on par with the respectable and well-founded theory of the earth's absolute rotation around its axis! Isn't the fact that the earth is slightly thicker around the equator — making it look more like an oblate spheroid than a sphere — enough evidence that the earth is absolutely rotating? The necessary flattening force on the earth — known as the centrifugal force — comes only because the earth is turning. How can it originate if the earth is believed to be stationary and, instead, the universe is made to revolve around?

The centrifugal force, cited in the above paragraph as an evidence of the earth's rotation, is an example of a class of *inertial forces* which — as every student of mechanics knows — needs to be "invented" in order to make Newton's second law of motion valid in a general accelerating frame (within the class of which we shall include also a rotating frame of reference). Such inertial forces cannot be real, because they do not originate from any material source (the way gravity forces and electro-magnetic forces do). Being a

[a]Fix three gyroscopes with their axes perpendicular to each other. If the axes remain fixed in directions, then the frame of reference is non-rotating.

parentless, illegitimate child, an inertial force is often called a "fictitious force". Anyone who analyzes the motion from a "non-accelerating" frame does not see any inertial force at all.

Inertial forces are daily encountered by every commuter when the train or the bus he rides suddenly starts or suddenly halts. At such moments he feels a jerky backward thrust or a forward pull — which are examples of inertial forces. An observer who watches the motion from the ground would argue that the sudden backward or forward movement of the body of the commuter is not due to a real force, but due to inertia inherent in the commuter's body, in conformity with Newton's first law of motion.

As a general rule, an inertial force $-m\boldsymbol{a}$ is "imagined" to be acting on an object of mass m, when viewed from a frame of reference which is moving with acceleration $\boldsymbol{a}$. Anyone trying to stand, or walk on a merry-go-round experiences two kinds of inertial force, namely a centrifugal force and a Coriolis force. The first one of them is velocity independent, whereas the second one is strictly proportional to the velocity of the walker. These two forces are exactly analogous to the forces exerted on a particle carrying electrical charge e by an electric field $\mathbf{E}$ (force $= e\mathbf{E}$), and a magnetic field $\mathbf{B}$ (force $= e\mathbf{v} \times \mathbf{B}$, where $\mathbf{v}$ is the velocity) respectively. If a platform is rotating about an axis with an angular velocity $\boldsymbol{\omega}$, then an object of mass m at a distance r from the axis experiences a centrifugal force directed outwards from the axis and having magnitude $m\omega^2 r$; whereas the Coriolis force is given by $m\mathbf{v} \times 2\boldsymbol{\omega}$. A direct evidence of the existence of Coriolis force is provided by a Foucault pendulum, which is nothing but an ordinary pendulum with a rather heavy bob (so as to be relatively unaffected by air friction) and suspended by a very long thread from the ceiling of a very tall building. One such Foucault pendulum, which is in public display is the lobby of the UN headquarters in New York, is suspended from a 75-feet high ceiling.

The plane of oscillation of this pendulum (or, in principle, any pendulum suspended from a fixed support in any earthly physics lab) will not be confined to a fixed vertical plane. Instead it will turn slowly with a period of rotation equal to $1/\sin\lambda$ days where λ is the latitude of the location on the earth.

Thus, we find that the slight bulging at the equatorial plane of the earth and the slow precession of the plane of oscillation of a Foucault pendulum show the existence of centrifugal and Coriolis forces at any place on the earth. They, in turn, provide the irrefutable evidence that the earth is turning around its axis.

There is another absurd implication of the geocentric view which the reader may not have missed. A galaxy which is, say, one billion light years away, will be orbiting a circular path of 2π billion light years in just one day — suggesting an incredible speed of $730\pi \times 10^9 c$, where c is the speed of light. No physical theory will allow such a nonsense.

The above example helps to underscore certain complexities and subtleties associated with the concept of frame of reference in the complete theory of relativity. It is not legitimate to think of a frame of reference as a non-rotating set of rigid bars stretching up to infinity. A frame of reference in relativity is a mathematical construct, sometimes lacking a complete visual picture. Two different frames of reference are related to each other by means of mathematical transformation equations satisfying certain conditions. With the choice narrowed down to legitimate frames of reference satisfying required conditions, the fundamental credo of relativity is still equal status for them all. We do not intend to pursue this argument further for the fear of straying away from our main objective.

Fortunately, some of the above considerations do not fog the clarity of the special theory of relativity with which this book is primarily concerned. In this special theory, which Albert Einstein enunciated in 1905 while working as a clerk in a patent office in Zurich, does make a distinction between a class of "privileged" frames called the inertial frames and the non-inertial ones. Without going into proper definition right now, let us accept naively that an *inertial frame* of reference is the archetype of non-accelerating, non-rotating frames. *They are the ones in which the inertial forces are absent*, so that Newton's first law of motion (often called the law of inertia) is strictly valid. In other words, an inertial frame of reference is one with respect to which a point particle will continue to move along a straight line with uniform speed, so long as it is free from external forces.

The relativity principle, enshrined in the special theory of relativity proclaims that *all inertial frames of reference are equivalent.* This statement is a refinement of the relativity principle stated at the end of Sec. 1.3.

1.5. Principle of Equivalence

Even though our limited mission is an exposition of the special theory of relativity, it will be helpful to realize the central theme of relativity shared by both the general and the special theory.

Einstein was not satisfied with the limited pronouncement of the relativity principle in special relativity, in which the inertial frames have been given a special status. In order to remove all distinctions between inertial and non-inertial frames, one has to appreciate the features by which one distinguishes a non-inertial frame from an inertial one. This feature is the appearance of the *inertial forces* in *non-inertial frames*, and its absence in the inertial frames, as already mentioned.

Einstein himself considered the following thought experiment.[b] Suppose, while you are inside an elevator, someone cuts the supporting cable. The elevator will be falling freely in the earth's gravitational field. But so will you, with the same acceleration $\boldsymbol{g}$ of the elevator so that you will be floating inside. Seen in another way, there will be an inertial force $-m\boldsymbol{g}$ acting on you (see Sec. 1.4), which together with the gravitational force $m\boldsymbol{g}$ of the earth will result in zero force on you. As a consequence, you will be weightless, just like an astronaut inside his space lab orbiting around the earth.

On the other hand, suppose your elevator is moved upwards with a constant acceleration $\boldsymbol{g}$ (so that the acceleration of the elevator is $-\boldsymbol{g}$). Then you will feel twice as heavy. Because now the induced inertia force will be $m\boldsymbol{g}$ which, along with the existing gravitational force $m\boldsymbol{g}$, will result in a total force of $2m\boldsymbol{g}$.

Thus, the inertial force has the remarkable property that it can get mixed up with the gravitational force to cause either a total or partial cancellation of the same, or an enhancement, or even change in the direction and magnitude of the same.

As a preamble to the General Theory of Relativity, therefore, Einstein proposed his famous *Principle of Equivalence.*[c] According to this principle, *the inertial forces are equivalent to gravitational forces.* One can generate the force of gravity of arbitrary direction and magnitude by accelerating or rotating his spaceship, or frame of reference suitably. One can, conversely, destroy the existing force of gravity at will by letting his frame of reference fall freely in the existing gravitational field. There is no experiment whatsoever — either in electrodynamics, or in optics, or in any other discipline of physical science — which can differentiate between the effect of "true" gravity force (produced by the earth, or the sun) and the effect of inertial forces induced due to acceleration of his frame. In other words, at least the

[b]See Ref. [2].

[c]Equivalence Principle, in its "weak form" and "strong form", needs to be understood for a study of Relativity, in particular the General Theory of Relativity. Read what the masters have written on this [3–5].

local effects of inertial forces are exactly the same as that of the true gravity forces, so much so that the inertial forces are also a kind of gravity forces.

A trivial example of the equivalence principle is the prediction that light bends downwards under gravity. Consider the same elevator which has a hole A on its eastern wall. The elevator is accelerating with an acceleration g upwards in early morning, when a ray of light, progressing along a horizontal straight line, enters through A and falls on a mark B on the western wall. Let the time of flight of the light ray from A to B be t. In this time, the elevator has moved upwards by a distance $s = \frac{1}{2}gt^2$. Therefore, the mark B must be the same distance s below a corresponding horizontal line AH drawn inside the elevator. Since the accelerating elevator in gravity free space is equivalent to a stationary elevator under gravity, light must fall by the same distance in a gravitational field g. In other words, the trajectory of a photon deviates from its straight line path by bending towards a gravitating mass. Such bending of a ray of light in a "true" gravitational field can be confirmed by measuring the deflection angle of a light ray when it grazes the periphery of the sun during a solar eclipse from its normal value. Experiments have confirmed this effect.

In view of the equivalence principle, the concept of inertial frame — which is central in the theory of relativity — needs to be redefined. We can now define an *inertial frame* to be a frame of reference with respect to which all inertial and gravity forces are absent. In the vicinity of a gravitating mass (e.g. the earth) such a frame can be realized inside a non-rotating box which is falling freely in the given gravitational field. Einstein's freely falling elevator and earth orbiting space-labs (which are also freely falling under gravity) provide best examples.

1.6. Tidal Forces

With the proclamation of the Equivalence Principle, gravity seems to disappear into thin air. We can create and generate the force of gravity at will by changing the frame of reference. Moreover, seen from a truly inertial frame, there is no gravity at all. What is then the fate of the Universal Law of Gravitation discovered by Newton, the triumph of which had been heralded by all the planets and satellites sailing in the sky? Does that also vanish into thin air?

It is obvious that, since inertial forces and gravity forces are intimately intermixed, and since the inertial forces are off-springs of non-inertial frames, a truly generalized theory of relativity (i.e. the one that treats all frames of reference on equal footing) must also be a theory of gravitation. Indeed, Einstein's General Theory of Relativity — which he published in 1916 — is also the most modern theory of gravitation. It was the towering achievement of the genius of Einstein to isolate "real" gravity (e.g. the one that is generated by a massive object, like the sun) from piles of "spurious" ones (i.e. the inertial forces posing as gravity).

The *local* effects of both gravities being the same (both can be created, or destroyed by suitably selecting frames of reference; both have the same effect on material particles and light), can there be some *global effects* by which the "real" gravity can be isolated?

Real gravity manifests itself in the phenomenon of *tide*, which is a global effect.[d] This effect can be seen in the deformation of an extended massive object falling freely under gravity.

It should be easy to visualize the deformation of a spherical mass $\mathcal{S}$ of radius r which has been dropped from a height h above the surface of the Earth and falls vertically towards the Earth's centre, as shown in Fig. 1.2(a). The particle A, being nearer to the earth than C experiences a larger gravitational force than C and falls faster than C. The particle B, being further, experiences a lesser gravitational force and falls slower. As a consequence C lags behind A, B lags behind C, and the diameter AB slowly elongates from r to $r + \delta$.

The particles E and F fall with almost the same acceleration as C, along the radial lines EO and FO, joining to the centre of the Earth. However, these lines come closer as $\mathcal{S}$ comes closer to Earth. Therefore, the diameter EF contracts slowly from r to $r - \epsilon$.

The net result is that a massive spherical ball which had been dropped from a height h above the surface of the Earth becomes an ellipsoid. An observer who sits inside a "rigid box" (which actually gets deformed by the tidal forces) shown by a rectangular frame, finds deformation of the originally spherical mass into an ellipsoid as the box falls from P to Q.

If $\mathcal{S}$ is an entirely incoherent assembly of particles, falling toward a gravitating centre along an ellipse or a circle, its different parts will accelerate and fall in their own ways. As a consequence they will disperse, fall apart, and $\mathcal{S}$ will hardly look like one body after some time. However, extended objects, like comets, are not entirely incoherent, because

[d]The phenomenon of tide as the signature of real gravity is discussed by Taylor and Wheeler [6].

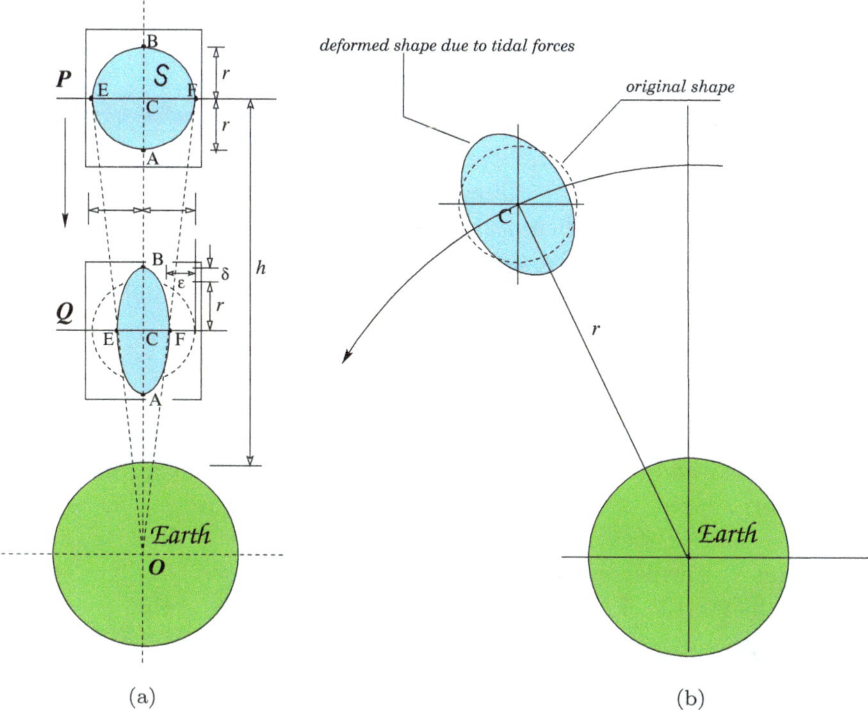

Fig. 1.2. Tidal deformation of a spherical object in free fall.

the internal gravitational pull among different parts acts like a bond. But they become distorted.

The same distortion will occur if the massive spherical ball had been a satellite of the earth, orbiting the earth in a circular orbit, as shown in Fig. 1.2(b). The mathematical analysis of this effect is not as simple as in the case of vertical free fall, but not so complicated either.[e]

This distortion, due to differential accelerations of different parts of an extended body, is what we call *tide*. The regions of elongation, around A and B are the locations of *high tide*. The regions of contraction around E and F are the locations of *low tide*. The real gravity is characterized by tidal forces.

---

[e]See, for instance, [7]. Or, get the mathematical analysis with diagrams, in [8]. For a layman's view, see [9].

1.7. The Scheme of General Relativity

Acceleration induced (spurious) gravity can be *transferred away* everywhere by suitably selecting a "global" inertial frame. "Real" gravity can be transformed away *locally*, but *not globally*. There does not exist a global inertial frame in the presence of a gravitating body like the sun or the earth. Perhaps the following examples will clarify.

Imagine a frame of reference S freely falling under the gravitational pull of the earth (and at the same time going in a circular orbit) as shown in Fig. 1.3. From this frame, observe the motion of two balls A and B, both freely falling. Of the two, A is very near the origin of S, say within a "small" radius R (which has to be defined properly), whereas B is far away. In the frame S, A will be either stationary, or moving along a straight line with uniform velocity (at least for a limited span of time, depending on the initial velocity and initial location of A), whereas, B will be moving with a non-uniform speed. This means that the law of inertia is seen to be valid for over a limited region around the origin of S, and that this region, i.e. the radius R, shrinks smaller and smaller as time passes. The frame S is only locally inertial, locally with respect to both space and time.

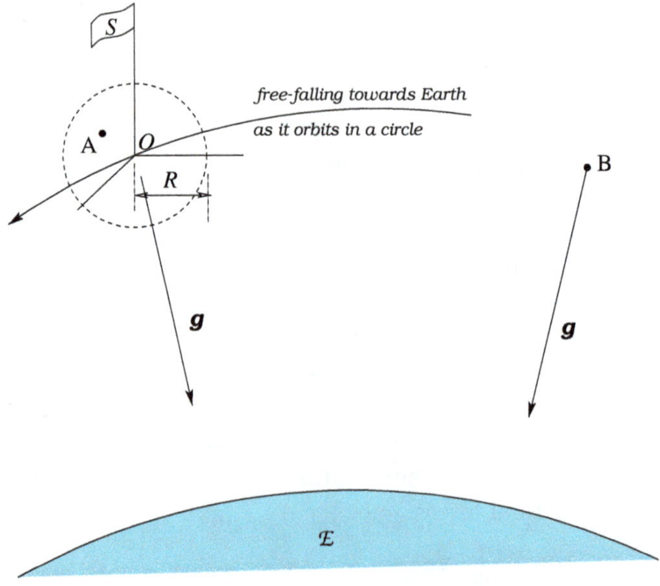

Fig. 1.3. Local inertial frame S.

A *local inertial frame* is analogous to Cartesian axes on the surface of a sphere. If the sphere is sufficiently large, like the earth, any city or town can be considered to be built on a surface which is flat locally, so that the map of the city can be drawn on a flat sheet of paper with Cartesian axes running in the W–E and S–N directions. On the other hand, it will not be possible to draw the map of the entire Asian continent, for instance, on a flat sheet of paper.

The surface of the earth is locally flat, so that we can draw straight lines locally. But if we stretch two parallel straight lines too far, they will ultimately cross. In the same way, if we consider two balls, originally floating stationary near each other inside the "Einstein's elevator" that is falling freely vertically downward, they will come closer to each other with the passage of time and will actually meet each other if the elevator is allowed to fall all the way to the centre of the earth.

Einstein conjectured that if we consider the *world line* of an object moving in a gravitational field (world lines are trajectories of particles, photons, in a four-dimensional world, called *space–time* in which time also is a coordinate axis, discussed in Sec. 7.1), that world line will be the *straightest possible path* in a *curved* four-dimensional space–time.

The straightest possible lines on a curved surface are called *geodesics*. Geodesics on the surface of our globe are also called great circles (e.g. the equator, the meridian circles). By analogy, Einstein proposed the famous *geodesic hypothesis*, according to which all freely falling objects, like the planets, satellites, moving in the gravitational fields of the sun or the earth, trace out geodesics (i.e. straightest lines) in a four-dimensional space–time. The path of a starlight progressing along a geodesic bends as it grazes the sun's periphery, the effect discussed in Sec. 1.5. This bending of "straight lines" is the manifestation of curvature in the space–time.

Coming back to the example of vertical free-fall of the particles E and F, alluded in Fig. 1.2(a), the geodesic lines of these two particles come closer to each other. This reminds us of the geodesic lines drawn on the surface of the earth meeting at some point. The geodesic lines that cross the equator perpendicularly converge at the North and South poles. This happens because the surface of the earth is curved.

In the same way, the four-dimensional space–time is curved. There is a *relative acceleration* between two objects both of which are falling freely (the particles E and F in the above example). *This relative acceleration, when seen in the four-dimensional space–time, constitutes the curvature of the space–time.* "Gravitation is a manifestation of space–time curvature, and that curvature shows up in the deviation of one geodesic from a nearby geodesic (relative acceleration of test particles)."[f]

Einstein constructed the *Curvature Tensor*, reshaped it through mathematical steps and identities into what is known as *Einstein Tensor* $E^{\mu\nu}$, and wrote the Field Equation of Gravitation in the esoteric form, known as *Einstein Equation*:

$$E^{\mu\nu} = -\frac{8\pi G}{c^4} T^{\mu\nu}, \tag{1.1}$$

in which the source term $T^{\mu\nu}$ on the right-hand side is the *energy tensor*. This term is a generalization of mass density ρ used in the Newtonian field equation of gravitation written in the form of the Poisson's equation

$$\nabla^2 \phi = -4\pi G \rho. \tag{1.2}$$

Note that the energy tensor $T^{\mu\nu}$ replaces mass density ρ, because of mass–energy equivalence, and Einstein tensor $E^{\mu\nu}$ replaces $\nabla^2\phi$ where ϕ is the gravitational potential. The gravitational constant $G = 6.67 \times 10^{-11}$ m^3/s^2 kg is the common factor in both equations.

We have discussed tensors and energy tensor in details in this book, but stopped short of the grand finale of celebrating Einstein's equation, because we are not yet prepared for that great journey.

How far are the predictions of Einstein's theory from those of Newton?

In the non-relativistic limit, i.e. when the source of the gravitational field is pure stationary matter, Einstein equation (1.1) converges to the Newtonian equation (1.2).

In Newtonian theory gravity is an action-at-a-distance force generated by massive objects, e.g. the sun. The path of a test particle is determined by solving Newton's second law of motion, which is a second-order differential equation. The second integral of this Equation of Motion (EoM) predicts an elliptic path that a planet must follow under the gravitational force of the sun. No such force acts on a massless photon, which must follow a straight path.

In Einstein's theory, all freely falling particles (i.e. particles falling under gravity) including a photon, follow straightest lines, or geodesics in the four-dimensional space–time. The path of a planet in the gravitational field of a star, e.g. the sun, when projected on the three-dimensional physical space, shows up as an ellipse, as in the Newtonian case. However, this ellipse has a slow precession rate, i.e. its major axis turns slowly in its plane about the sun.

---

[f]See [5, pp. 17–18 in Chapter 1, pp. 218–219 in Chapter 8, pp. 265–271 in Chapter 11].

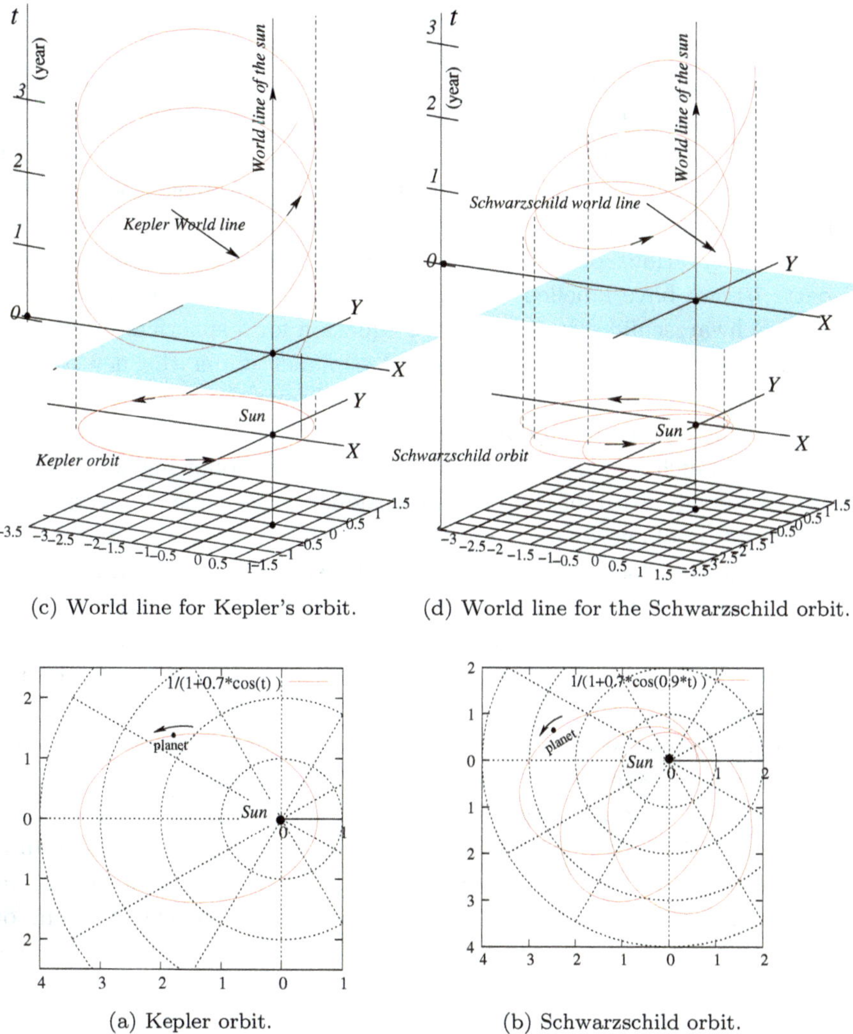

(c) World line for Kepler's orbit.

(d) World line for the Schwarzschild orbit.

(a) Kepler orbit.

(b) Schwarzschild orbit.

Fig. 1.4. The world line of a planet in the gravitational field of the sun (a) Kepler's orbit, (b) Schwarzschild orbit.

We have shown the precession of a hypothetical planet and its world line in Fig. 1.4. The effects shown in the diagrams are highly exaggerated to make an impression on the reader. The elliptic orbit shown in the diagram has an eccentricity of 0.7, whereas the maximum eccentricity of orbit pertains to Mars, having an eccentricity of 0.093 (so that all planets move in what appear to be circular orbits). The major axis of the orbit in the diagram shows an angular displacement of almost $45°$ in one revolution (i.e. in one planet year), compared to about $43''$ per century in the case of mercury (which has the maximum precession rate).

Keeping these exaggerations in mind let us look at Fig. 1.4. In part (a), we have shown the Newtonian orbit, labelled as Kepler orbit, because Johannes Kepler had discovered the elliptic orbits of planets through detailed observations of their positions in the sky long before Newton. In part (b), we have labelled the precessing orbit as Schwarzschild orbit, because Schwarzschild solved Einstein's equation for a spherically symmetric source, and obtained the geodesic of a planet under the new theory of gravitation. In parts (c) and (d), we have drawn the world lines of the planet moving along the said orbits.

Without much ado let us now summarize in the following words. The nature of the curvature of the four-dimensional space–time is governed by the distribution of energy and momentum (in the case of the earth and the stars, the matter itself acting as energy). Einstein's field equation, which establishes the relation between the distribution of energy–momentum and curvature, is actually equivalent to Newton's formula of universal gravitation when the bodies producing curvature of space–time (like the earth, the sun) are stationary. Therefore, Einstein's General Relativity theory does not invalidate Newton's theory. It provides an alternative approach to gravitation through the path of geometrodynamics, an approach in which gravity is just a manifestation of the intrinsic geometry of space–time. Even though this approach has nothing in common with Newton's universal law of gravitation, its ultimate predictions match those of Newton in a non-relativistic situation. However, Einstein's theory leads to bending of light, black holes and many other phenomena that Newton's theory cannot foresee.

1.8. Conclusion

Even though the last few sections digressed into a domain that has no direct application in this book, they may have placed the reader in a better perspective. Special Relativity hinges on the concept of inertial frames. The reader may have appreciated that this inertial frame is just an idealization, which does not exist globally. If anything, that was the one of the lessons conveyed by the last section.

There is another justification. The fundamental creed of relativity is equivalence of frames of reference. Such equivalence beckons like a mirage as long as one is confined within the bounds of special relativity. To explain what equivalence truly means, one has to consider equivalence in its entirety, i.e. one has also to consider accelerating frames. However, the moment one considers accelerating frames, gravity enters by the back door. Thus, gravity and equivalence of frames are inseparable.

The mention of relativity conjures up vision of a paradoxical world where space odyssey rejuvenates the youth, speeding clocks tick slowly and matter is transmuted into energy. How much of such mind-boggling stories truth and how many fiction?

In the following chapters, we shall follow the relativity postulate to a logical end to seek answers to some of these questions. We shall discover that this innocent looking postulate (equivalence of frames of reference) contains seeds of epoch making consequences. One may not miss its philosophical message — that an abiding faith in certain values, when carried through trials and tribulations to its logical end, can lead to a world of miraculous discoveries.

We quote the following lines from Professor S. Chandrasekhar [10] "It is an incredible fact that what the human mind, at its deepest and most profound, perceives as beautiful, finds its realization in external nature." To which we add: What the human mind perceives as just and equitable finds its realization in the man-made structure of the physical laws.

Chapter 2

Einstein's Postulates, Their Paradoxes, and How to Resolve Them

2.1. Event Point in Space–Time

Let us start with the following example. A rocket which was fired from the ground exploded in the atmosphere. This explosion is an *event*. Events such as this, and more varied than this, play a central role in the concept structure of relativity. We shall use *double quote* $\cdots$ *double unquote* to indicate an event. For example, we shall say that "the rocket exploded in the atmosphere" is an *event*, to be denoted by a symbol, say, "Θ".

An observer S in Delhi can pin-point the location and timing of the event "Θ" by stating that it occurred 200 km west, 250 km north, at a height of 60 km and at exactly 20 hours IST. To convey this information compactly, we could imagine a set of X-, Y-, Z-axes whose origin is in Delhi, such that the X-axis is directed eastward, the Y-axis northward and the Z-axis vertically upward. We shall designate this set of axes, or, the *reference frame* defined by this set of axes, by the symbol S. We shall often use the same symbol to mean either the observer or his frame of reference. With respect to S, the event "Θ" is completely identified by specifying the four numbers, namely, −200, 250, 60, 20 in this particular order. (The first number −200 means that the x coordinate of the event is 200 km in the negative X-direction.) These four members, when arranged in this particular order, are called the coordinates of "Θ" with respect to S. We write "Θ" $= (-200, 250, 60, 20)$ with respect to S.

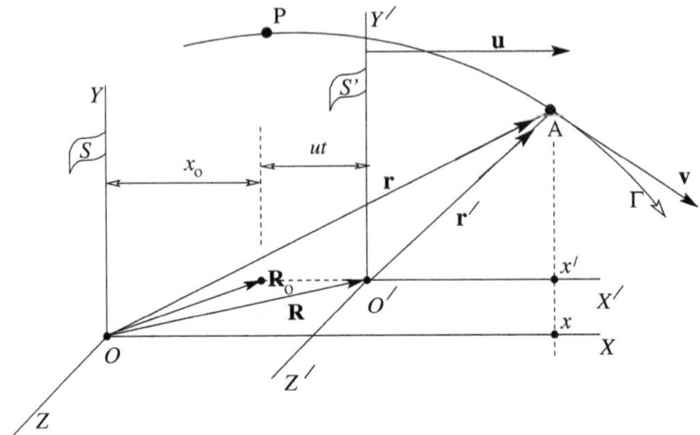

Fig. 2.1. Coordinates of a particle with respect to frames S and S'.

Conversely, *every set of four numbers, arranged in the order* (x, y, z, t), *will be considered to constitute an* "event". (There is one anomaly in the above coordinates, namely x, y, z have the dimension of length, whereas t has the dimension of time. This anomaly will be removed in Sec. 3.1 by multiplying t with the velocity of light, so that all the four coordinates of an event will have the dimension of length.)

Now consider two frames of reference S and S', of which S is the Absolute Frame of reference conceived by Newton in which his laws of motion were assumed to be valid, and S' is another frame of reference (the earth for example) which is moving in the X-direction with velocity $\mathbf{u}$, as shown in Fig. 2.1.

Note that we have tagged the frames by means of flags, a convention we shall follow in the rest of this book. We imagine a set of three mutually perpendicular rigid bars, serving as the frames of reference — XYZ for S and $X'Y'Z'$ for S'. We have taken the X-axis of S, and the X'-axis of S' to be parallel and directed along the velocity $\mathbf{u}$ of the earth. Written component wise, $\mathbf{u} = (u, 0, 0)$ with respect to S.

Now imagine a particle P of mass m moving in space. Let A be a point on the trajectory of the particle. "The particle arrives at A" is an event. We call it the event "Θ_A". The coordinates of this event are as follows:

$$\text{“}\Theta_A\text{”} = \begin{cases} (x, y, z, t) & \text{in } S, \\ (x', y', z', t') & \text{in } S'. \end{cases}$$

For the configuration shown in Fig. 2.1, the coordinates of the origin O' of S', with respect to S, are $\mathbf{R}_o = (x_o, y_o, z_o)$ at $t = 0$, and $\mathbf{R} = (x_o + ut, y_o, z_o) = \mathbf{R}_0 + \mathbf{u}t$ at time t. It is obvious that

$$\begin{cases} x' = x - x_o - ut, \\ y' = y - y_o, \\ z' = z - z_o, \\ t' = t. \end{cases} \tag{2.1}$$

Written compactly,

$$\begin{aligned} \mathbf{r}' &= \mathbf{r} - \mathbf{R} = \mathbf{r} - \mathbf{R}_0 - \mathbf{u}t, \\ t' &= t. \end{aligned} \tag{2.2}$$

Equation (2.1) and its equivalent form (2.2) constitute the familiar *Galilean Transformation* (GT) with the inclusion of the time coordinate.

2.2. Inertial Frames

2.2.1. *Newton's equation of motion*

Much of physics is based on mechanics and it deals with motion of objects. The physical quantities associated with motion are displacement, velocity, acceleration, etc. These quantities must be specified with respect to some reference frame. When we say that the velocity of a steamer is 25 km/hr, this implies that the steamer recedes from a point fixed on the bank of the river at the rate of 25 km/hr. Here, the river bank constitutes a reference frame. We can imagine a boat on the river sailing in the direction of the steamer at a speed of 10 km/hr. This boat constitutes another frame. The velocity of the steamer with respect to this second frame will be 15 km/hr. Therefore, it is meaningless to talk about the laws of motion without having fixed before our mind a particular reference frame. What reference frame, then, did Newton fix before his mind when he enunciated the laws of motion?

Newton presumed the existence of some *Absolute Frame*, as mentioned in the Sec. 1.2. He identified it with the frame of the 'fixed stars'. In our discussion, we shall tentatively identify this so-called Absolute Frame (AF) as one fixed with respect to the sun. Now let us state the first two laws of motion, which, by assumption, are valid in this AF. The first law, also called the *law of inertia*, states that the acceleration $\boldsymbol{a}$ of a particle is zero in the absence of an external force.

The second law states that $\boldsymbol{a}$ is not zero if the particle is acted on by an external force $\mathbf{F}$, in which case $\mathbf{F}$ equals mass times the acceleration of the particle:

$$\mathbf{F} = m\boldsymbol{a}. \tag{2.3}$$

Another Newtonian assumption is that the mass m of the particle is *absolute.* It does not change with the velocity of the particle and it is the same to all observers.

Should the above two equations be valid on the earth which is moving with respect to the Sun?

Strictly speaking, the answer is 'no'. Not because the earth is just moving, but because the earth is rotating about its own axis and is also going around the sun in a circular motion. However, for the time being let us ignore the spinning motion and the orbital motion around the sun, the effects of which are relatively small. Let us assume, for simplicity, that the earth is moving, without spinning, along a straight line with a uniform speed of 30 km/s with respect to the sun.

Let the velocity of the particle P be $\mathbf{v}$ with respect to S, and $\mathbf{v}'$ with respect to E. Component wise,

$$\begin{cases} \mathbf{v} = \dfrac{d\mathbf{r}}{dt} = \left(\dfrac{dx}{dt}, \dfrac{dy}{dt}, \dfrac{dz}{dt}\right) & \text{with respect to } S, \\[2ex] \mathbf{v}' = \dfrac{d\mathbf{r}'}{dt'} = \left(\dfrac{dx'}{dt'}, \dfrac{dy'}{dt'}, \dfrac{dz'}{dt'}\right) & \text{with respect to } S'. \end{cases} \tag{2.4}$$

Note from Eq. (2.1) that $dt' = dt$, $dx' = dx - u\,dt$, $dy' = dy$, $dz' = dz$. Therefore,

$$\begin{aligned} \mathbf{v}' &= \left(\frac{dx - u\,dt}{dt}, \frac{dy}{dt}, \frac{dz}{dt}\right), \\ \text{or} \quad \mathbf{v}' &= \mathbf{v} - \mathbf{u}, \end{aligned} \tag{2.5}$$

which is the transformation equation for velocity.

Let $\boldsymbol{a}$ and $\boldsymbol{a}'$ denote the acceleration of P with respect to S and E, respectively. Then

$$\begin{aligned} \boldsymbol{a} &= \frac{d\mathbf{v}}{dt}; \\ \boldsymbol{a}' &= \frac{d\mathbf{v}'}{dt} = \frac{d(\mathbf{v} - \mathbf{u})}{dt} = \frac{d\mathbf{v}}{dt} = \boldsymbol{a}, \end{aligned} \tag{2.6}$$

since $\mathbf{u}$ is a constant vector. We notice that even though velocity transforms under the GT, *acceleration remains invariant.*

By our assumption, Newton's second law of motion, as given by Eq. (2.3) is exactly valid in the Absolute Frame S. Also, the measure of the external force, as for example determined by the reading on a spring balance, should be the same in E as in S. Hence, from Eq. (2.6)

$$\mathbf{F} = m\boldsymbol{a}'. \tag{2.7}$$

Thus, Newton's second law of motion is valid in E, if it is valid in S.

As a special case let $\mathbf{F} = \mathbf{0}$. This would imply $\boldsymbol{a} = \mathbf{0}$, as well as $\boldsymbol{a}' = \mathbf{0}$. This is Newton's first law of motion, which is therefore valid in E, if it is valid in S.

In summary, if the E frame is imagined to be a non-rotating frame moving with uniform velocity with respect to the S frame, then Newton's first and second laws of motion — which are postulated to hold in S — also hold in E. These laws are valid in any frame S' with similar properties (i.e. non-rotating and moving with uniform velocity with respect to the AF). By induction, since these laws are valid in S', they are valid in any other non-rotating frame S'' moving with uniform velocity with respect to S'. By this process, we obtain an infinity of frames of reference which are non-rotating and moving with uniform velocities with respect to one another, and with respect to the AF — and Newton's laws of motion are valid in them all.

A frame of reference in which the law of inertia (i.e. Newton's first law of motion) holds is called an *inertial frame* (IF) — as we have discussed at some length in the previous chapter. We recognize that there is an infinite number of IFs. Let S be any one IF (not necessarily the AF) and let S' be another. If, at $t = 0$, the origin of S' is at the coordinates $\mathbf{r}_0 = (x_0, y_0, z_0)$ and moving with velocity $\mathbf{u}$ with respect to S, then the Galilean transformation from S to S' gives us the following relations:

$$\begin{aligned} \mathbf{r}' &= \mathbf{r} - \mathbf{r}_0 - \mathbf{u}t, \\ t' &= t, \\ \mathbf{v}' &= \mathbf{v} - \mathbf{u}, \\ \boldsymbol{a}' &= \boldsymbol{a}. \end{aligned} \tag{2.8}$$

All IFs will measure the same acceleration of a moving object.

In the above we have considered only Newton's first and second law of motion. What about the third law? The third law of motion is a corollary of the law of conservation of momentum. We shall now discuss the invariance of the laws of conservation of momentum and energy under GT.

2.2.2. *Conservation of energy and momentum*

Conservation of energy and momentum are among the foundational principles of physics. We shall show that these laws are valid in all inertial frames by examining a two-body collision.

Figure 2.2 shows a view of two particles A and B engaging in a collision, as viewed from some inertial frame S. The collision — a term which is often used to mean a passing interaction between two particles — may result in the creation of new particles after the original ones have encountered each other. Therefore, for the sake of generality, we are considering two different particles C and D emerging from the scene of collision.

In Newtonian physics mass is conserved, so that

$$m_A + m_B = m_C + m_D. \tag{2.9}$$

Let us postulate that the total momentum and the total kinetic energy of an isolated system (i.e. a system which is not influenced by anything from outside) be each conserved in an elastic collision, when viewed from an inertial frame S. This gives the following two relations:

(i) **Conservation of Momentum:**

$$m_A\mathbf{v}_A + m_B\mathbf{v}_B = m_C\mathbf{v}_C + m_D\mathbf{v}_D. \tag{2.10}$$

(ii) **Conservation of kinetic energy:**

$$\frac{1}{2}m_A v_A^2 + \frac{1}{2}m_B v_B^2 = \frac{1}{2}m_C v_C^2 + \frac{1}{2}m_D v_D^2. \tag{2.11}$$

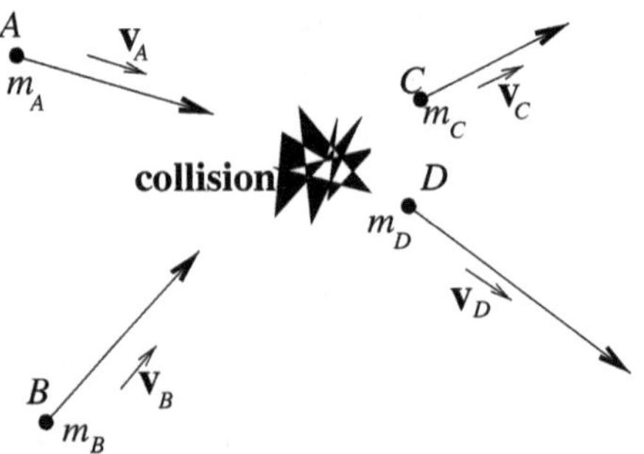

Fig. 2.2. Collision of two particles.

We shall now use Eq. (2.8) to rewrite the above equations in terms of velocities, as measured in the frame S'. Letting $\mathbf{v}'_A, \mathbf{v}'_B, \mathbf{v}'_C, \mathbf{v}'_D$ be the velocities of the respective particles in the frame S', we have

$$m_A(\mathbf{v}'_A + \mathbf{u}) + m_B(\mathbf{v}'_B + \mathbf{u}) = m_C(\mathbf{v}'_C + \mathbf{u}) + m_D(\mathbf{v}'_D + \mathbf{u}).$$

Using Eq. (2.9), the above equation reduces to the following form:

$$m_A\mathbf{v}'_A + m_B\mathbf{v}'_B = m_C\mathbf{v}'_C + m_D\mathbf{v}'_D. \tag{2.12}$$

This shows that the total momentum, as measured in the frame S', is the same after a collision, as it is before the collision. We now convert Eq. (2.11) along parallel lines:

$$\frac{1}{2}m_A(\mathbf{v}'_A + \mathbf{u})^2 + \frac{1}{2}m_B(\mathbf{v}'_B + \mathbf{u})^2 = \frac{1}{2}m_C(\mathbf{v}'_C + \mathbf{u})^2 + \frac{1}{2}m_D(\mathbf{v}'_D + \mathbf{u})^2,$$

i.e.

$$\frac{1}{2}m_A v'^2_A + \frac{1}{2}m_B v'^2_B + \mathbf{u}.(m_A\mathbf{v}'_A + m_B\mathbf{v}'_B) + \frac{1}{2}u^2(m_A + m_B)$$
$$= \frac{1}{2}m_C v'^2_C + \frac{1}{2}m_D v'^2_D + \mathbf{u}.(m_C\mathbf{v}'_C + m_D\mathbf{v}'_D) + \frac{1}{2}u^2(m_C + m_D).$$

Using (2.9) and (2.12) the above equation reduces to the desired form:

$$\frac{1}{2}m_A v'^2_A + \frac{1}{2}m_B v'^2_B = \frac{1}{2}m_C v'^2_C + \frac{1}{2}m_D v'^2_D. \tag{2.13}$$

Equation (2.13) now validates conservation of the kinetic energy in the frame S'.

In summary, if the laws of conservation of linear momentum and kinetic energy are established in one inertial frame, then they are established in all inertial frames, as a consequence of the Galilean transformation. We can extend the same principle to other physical quantities, like angular momentum and total energy (i.e. the sum of the kinetic energy and the potential energy) and establish that they are valid in all inertial frames.

2.2.3. *Equivalence of inertial frames*

The conclusions just reached in the previous sections lead to equal status of all inertial frames — at least within the limited scope in which we have examined them. They clearly tell us that all inertial frames are equivalent with respect to the experiments and laws of mechanics. A coin tossed inside a smoothly cruising jet plane will fall exactly in the same way as it will fall on the ground. Considered from a more general perspective, every IF has all the properties that characterize the hypothetical AF.

This is the *Newtonian principle of relativity.* In essence it says that *identical mechanical experiments performed in different inertial frames will yield identical results.* The laws of conservation of mechanical momentum, mechanical energy and the laws of motion which are presumed to be valid in the AF are seen to be valid in all IFs, as a consequence of the GT. Therefore, there is no experiment, at least in the domain of mechanics, by which this hypothetical AF, even if it exists, can be identified.

2.3. Historical Background

2.3.1. *Search for the absolute frame*

If it is impossible to tell, by any means whatsoever, which frame is AF and which frame is not, then why should there be any reason for hypothesizing an AF at all?

With the formulation of the laws of electricity and magnetism, however, the need for this AF became evident. Clerk Maxwell rationalized the phenomena of electricity and magnetism into a set of equations known as Maxwell's equation. Maxwell's equations lead to wave equations for the electric and magnetic fields, showing a characteristic wave speed c which, in vacuum, equals 3×10^8 m/s, the same as the speed of light. This means that if you change the charge–current configuration somewhere in space, the electric and magnetic fields will change everywhere in space, but the field will change earlier in the near region and later in the far region. The messenger that carries the command for change from a near region to a far region is the electromagnetic wave and it propagates with the speed of light, somewhat in the same manner that ripples propagate the information of a disturbance on the surface of a pond with a much slower velocity. Radio waves, visible light, X-rays are all such electromagnetic waves, lying in different band zones of the frequency spectrum.

Before the formulation of the theory of relativity it was generally believed by physicists that electromagnetic waves were similar to mechanical waves — like sound, seismic waves, ripples on the surface of a pond. Each of these examples is associated with a medium that carries the wave. A disturbance in the mechanical configuration occurs somewhere in the medium, and this information is sent outwards by the elastic properties of the medium in the form of a wave. Water surface is the medium for ripples, air for sound, earth for seismic waves.

It was generally believed that the "mechanical medium" that carries the disturbance called electromagnetic wave, or light, is *aether*. Many physicists, including Maxwell himself, dabbled with the hypothetical properties of aether. Aether was thought to be a fluid that pervaded all space, penetrated all materials, had some extraordinary properties, like perfect elasticity (so that no energy is extracted out of light when it propagates through it) and extremely high modulus of rigidity (so that light waves, oscillating at very high frequencies, of the order of 10^{16} Hz, could propagate through it).

One now gets a clue for identifying the AF. There must be some frame of reference, say S_0, in which Maxwell's equations are valid *exactly*. However, if they are valid exactly in S_0, then they cannot be exactly valid in some other frame S, because any GT applied to these equations will destroy their forms (equations of electrodynamics involve first-order derivatives whereas those of Newton involve second-order derivatives like $\frac{d^2\mathbf{r}}{dt^2}$). Therefore, people were inclined to believe that all inertial frames were *not* equal, that among all the IFs there did exist one privileged frame, which alone was entitled to claim the equations of electrodynamics, and had, therefore, an *absolute* character. That frame of reference must be the long-cherished AF.

It was therefore speculated that aether was *at rest* in the AF, and light, which propagates in aether, must have an absolute speed c in all directions with respect to this aether.

How to identify this AF? The answer should not be difficult even for a lay reader. In any other IF, which is moving, say in the X-direction with velocity u with respect to the AF, the speed of light in the $+X$ direction will be $c - u$, and in the $-X$-direction will be $c + u$. The velocity of light in different directions should in fact be different in this new IF. Among all the IFs there is one, and only one, IF in which the speed of light is same in all directions and that frame alone is to be identified as the AF.

An experiment can be devised to measure the difference in the velocities of light in two different directions on the surface of the earth. This will give us immediate information about the velocity of the earth relative to the AF. A large number of ingenious experiments, of which the Michelson and Morley's experiment is most well known, have been performed to measure this difference. Contrary to everybody's expectations no difference has ever been found. All experiments on the velocity of light have unmistakably shown that light propagates with the *same constant speed c in all directions in vacuum*, at all times and in all seasons, in the frame of reference of the earth.

The Michelson–Morley experiment attempts to determine the velocity of the earth relative to aether. Alternatively, since aether is assumed to be at rest in the Absolute Frame, the outcome of the experiment should determine the velocity v of the earth relative to the AF. It was found from this experiment that v is zero.

In summary, it will be sufficient to say that the theory of relativity is founded on the premise that *there is no aether and no Absolute Frame.*

We shall now give a brief account of the Michelson–Morley experiment.

2.3.2. *Michelson–Morley experiment*

The Michelson–Morley experiment (to be abbreviated as MM experiment in the following) utilizes Michelson interferometer. Figure 2.3(a) describes the basic set up of the apparatus. Here S is a source of a monochromatic light. A collimator takes a parallel beam out of this source. This beam is split into two components ϕ_1 and ϕ_2 by the partially silvered mirror A. The component ϕ_1, which is transmitted through A towards B, gets reflected back at the mirror B, and then comes back to A. The other component ϕ_2 is reflected upwards at A, goes to the mirror C, and is reflected back to A. These two beams, after return to A, get partially reflected again and partially transmitted again at A, so that a fraction of each one of them, say half of ϕ_1 and half of ϕ_2, will now recombine and proceed along the path

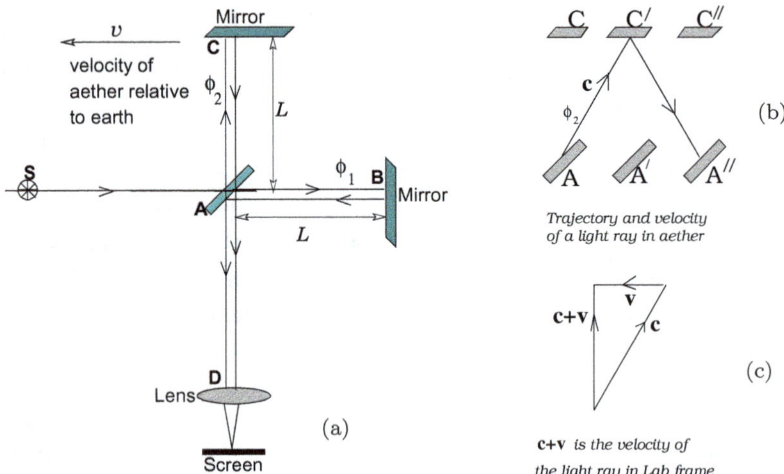

Fig. 2.3. Schematic arrangement of Michelson–Morley experiment.

AD as a single beam. This recombination of two fractions of what used to be a single beam earlier, after they have travelled through two different path lengths, causes what is known as "optical interference". The lens will focus these interfering beams onto a screen.

Suppose the earth is moving, with respect to aether, with velocity $\mathbf{v}$ in the direction of the line AB, which we take as the X-axis. This means that aether is moving in the negative X-direction with speed v with respect to the earth. Since light travels in aether with speed c, the speed of light in the Lab frame as it travels along the $+X$-axis will be $c - v$, and along the $-X$-axis will be $c + v$. If T_1 is the time required for the beam ϕ_1 to go from A to B and then travel back from B to A, then

$$T_1 = \frac{L}{c-v} + \frac{L}{c+v} = \frac{2L}{c}\left\{1 - \frac{v^2}{c^2}\right\}^{-1} \simeq \frac{2L}{c}\left\{1 + \frac{v^2}{c^2}\right\}.$$

Now consider the second beam ϕ_2 following the path ACA, which it covers in time T_2. To an observer at rest in aether, this beam must follow the slanted path $AC'A''$, shown in Fig. 2.3(b), in order for it to reach the mirror A which has moved to the point A'' during the same time T_2. If $\mathbf{v}$ is the velocity of aether with reference to the lab and $\mathbf{c}$ is the velocity of light with reference to aether along AC', then the velocity of the beam ϕ_2 in the Lab frame along the upward path AC is the vector sum $\mathbf{c} + \mathbf{v}$. The magnitude of $\mathbf{c} + \mathbf{v}$ is

$$u_1 = |\mathbf{c} + \mathbf{v}| = \sqrt{c^2 - v^2} = c\left\{1 - \frac{v^2}{c^2}\right\}^{1/2}.$$

In the same way, one computes the speed of the beam in the Lab frame down the path CA to be

$$u_2 = u_1 = c\left\{1 - \frac{v^2}{c^2}\right\}^{1/2}.$$

The time taken by the beam ϕ_2 to cover the path ACA is therefore

$$T_2 = \frac{L}{u_1} + \frac{L}{u_2} = \frac{2L}{c}\left\{1 - \frac{v^2}{c^2}\right\}^{-1/2} \simeq \frac{2L}{c}\left\{1 + \frac{1}{2}\frac{v^2}{c^2}\right\}.$$

Therefore, there is a time difference

$$\delta T = T_1 - T_2 = \frac{Lv^2}{c^3}$$

between the times of arrival of the two beams, when they recombine at A to cause an interference fringe pattern. The order of the fringe at the centre of the screen is

$$n = \frac{c\delta T}{\lambda} = \frac{Lv^2}{\lambda c^2},$$

where λ is the wavelength of the light beam used. As the earth goes round the sun in a circular orbit, its velocity relative to the AF keeps changing direction. Six months later, the direction of the velocity of the earth, which was in the $+X$-direction at the start of the experiment, will now change into the $+Y$-direction, so that the aether wind will now blow in the $-Y$-direction in the reference frame of the earth. The roles of the paths ABA and ACA in the above experiment will now get interchanged, so that now

$$T_1' = \text{time taken to cover ABA} = \frac{2L}{c}\left[1 + \frac{v^2}{2c^2}\right],$$

$$T_2' = \text{time taken to cover ACA} = \frac{2L}{c}\left[1 + \frac{v^2}{c^2}\right].$$

Hence, the new fringe order at the centre of the screen will be

$$n' = -n.$$

Therefore, in six months the interference pattern will shift through

$$N = n - n' = 2n = \frac{2Lv^2}{\lambda c^2} \text{ fringes.}$$

In the actual experiment no such fringe shift was observed.

2.4. Postulates of Special Relativity

The null result of the MM experiment demolishes the notion of Absolute Frame. Special Theory of Relativity therefore starts with the premise that all IFs are equal — not merely with respect to the laws of mechanics, but for the whole of physics. We have, however, seen that the laws of electrodynamics do not appear to be the same in all IFs. The answer to this paradox lies in recognizing the fallacy of the Galilean Transformation. Before showing how the paradox is resolved satisfactorily, it will be desirable to enunciate the two fundamental postulates of Special Relativity on which its entire concept structure rests.

Postulate 1 (Relativity Postulate). *All inertial frames are equivalent in the sense that identical experiments performed in different inertial frames will yield identical results.*

We mentioned earlier Maxwell's equations lead to a characteristic speed c of light (by light we shall mean any electromagnetic wave). These equations also show that light emitted by a moving source (for example, every charged particle under acceleration radiates light) propagates in all directions with equal speed c, and c is independent of the velocity v of the source. This is a key notion on which further progress of our theory depends. Hence the following postulate.

Postulate 2 (Source-independence of the speed of light). *Light propagates without any medium with a speed c whose value is same in all directions, and is independent of the velocity of the light-emitting source.*

By combining the above two postulates we get a very important corollary. Consider a light emitting source L which is moving with respect to a frame S. There is some comoving frame S_0 in which L is at rest, and light emitted by it propagates at speed c. Since propagation speed does not depend on the velocity of L (postulate 2), and since this characteristic speed should be same in all frames of reference (postulate 1), propagation speed c in S_0 (in which L is at rest) should be same as in S (in which L is moving). Hence the following corollary.

Corollary 2A. *The speed of propagation of light is given by* $c = 3\times 10^8\, m/s$ *in all inertial frames, independent of the motion of the light emitting source.*

2.5. Relativity of Simultaneity

Let us imagine two inertial frames S and S'. S' is moving relative to S with velocity u in the direction of the X-axis, which is taken parallel to the X'-axis of S' (Fig. 2.4(a)). Times t in S and t' in S' are measured from the instant when the origins O and O' of the two frames just pass each other. At that very moment a sharp flash of light is emitted from a source L which is fixed to the origin of S', so that L is stationary in S', but moving with speed u in S. According to Corollary 2A, light will be propagating radially with speed c with respect to S, as well as S'. This means that a spherical wavefront Σ, diverging from the origin O of S with speed c and having radius ct, will contain this light flash at the instant t. Similarly, another spherical wavefront Σ', diverging from the origin O' of S' with speed c and having radius ct' at the instant t', will contain this same light flash. If the clocks of the observers S and S' are assumed to tick at the same rate, then at the instant $t = t'$, the same flash of light is simultaneously contained

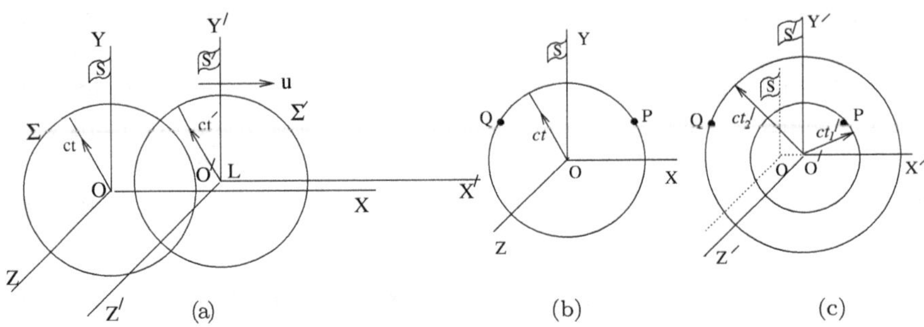

Fig. 2.4. Simultaneity paradox.

in two different wavefronts Σ and Σ' which have the same radius $ct = ct'$. This is absurd. What has gone wrong?

We went wrong by believing in one universal time which, as we shall find, does not fit with the postulates of relativity. Consider the same light flash as discussed in the previous paragraph. Imagine two points P and Q on a sphere of radius *ct* in frame S (Fig. 2.4(b)). Two events, e.g. "light reaches P" and "light reaches Q" which we designate as "Θ_P" and "Θ_Q" respectively, both occur at the *same* time t and are, therefore, *simultaneous* in S. Their coordinates are

$$\left.\begin{aligned} \text{“}\Theta_P\text{”} &= (x_1, y_1, z_1, t) \\ \text{“}\Theta_Q\text{”} &= (x_2, y_2, z_2, t) \end{aligned}\right\} \quad \text{in } S.$$

However, in this time t, O' has been displaced by the distance ut to the right of O. Consequently, P and Q no longer lie on the same sphere with centre at O' (Fig. 2.4(c)). P will be nearer to O' than Q.

Therefore, the event "Θ_P" occurs earlier than the event "Θ_Q", according to S'. The coordinates of the same events with reference to S' will be

$$\left.\begin{aligned} \text{“}\Theta_P\text{”} &= (x'_1, y'_1, z'_1, t'_1) \\ \text{“}\Theta_Q\text{”} &= (x'_2, y'_2, z'_2, t'_2) \end{aligned}\right\} \quad \text{in } S',$$

and $t'_1 < t'_2$. We therefore see that the relativity principles are in contradiction with the Newtonian concept of universal time. Time assumes a relative character in relativity. In particular, we have just discovered the following important rule.

Rule 1. *Two events which are simultaneous with respect to an observer S cannot be simultaneous with respect to another observer S' who is moving* *relative to S (unless the direction of motion happens to be perpendicular to the straight line joining the spatial locations of the events).*

One can advance arguments to show similar discrepancy with respect to distance measurement also. Time and distance are both divested of absoluteness in relativity. They give different measures to different observers who are moving with respect to each another.

Relativity rejects many of the intuitive notions of Newtonian physics. One of the first casualties is Galilean transformation, as the following exercise will illustrate.

Consider an event "Φ" namely "reception of the light flash at some point P". Let the coordinates of this event be (x, y, z, t) in the frame S. This means that the light ray has covered a distance $\sqrt{x^2+y^2+z^2}$ in time t, as measured in S. Therefore,

$$x^2 + y^2 + z^2 - c^2t^2 = 0. \tag{2.14}$$

This is an equation of a family of concentric spheres, representing wavefronts $W_1, W_2, W_3, \ldots$, corresponding to different times $t_1, t_2, t_3, \ldots$, as seen from S, and as shown in Fig. 2.5(a). In order to obtain the equations of these wavefronts in the frame S', we shall have to transform the above equation with the help of Eq. (2.1). This gives the desired equation:

$$(x' + ut)^2 + y'^2 + z'^2 = c^2t'^2. \tag{2.15}$$

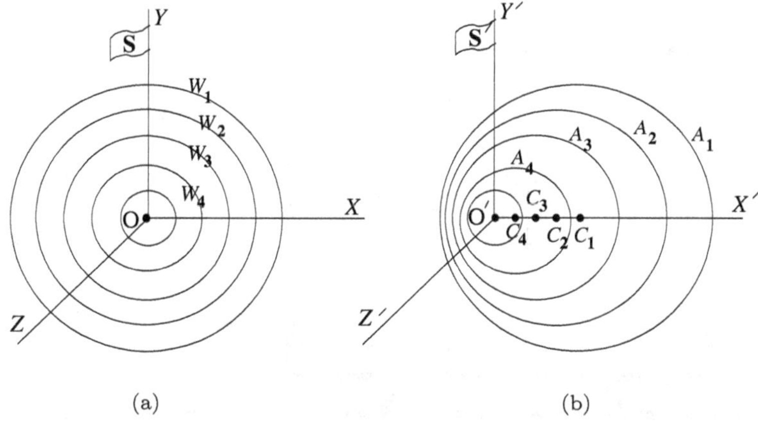

Fig. 2.5. Wavefront series seen from S and S'.

Equation (2.15) describes a series of spheres $A_1, A_2, A_3, \ldots$ whose centres $C_1, C_2, C_3, \ldots$ are located along the negative X-axis, as shown in Fig. 2.5(b). However, Corollary 2A requires that both S and S' should see concentric spherical wavefronts with centres fixed at their respective origins. Therefore, a new set of transformation equations between (x, y, z, t) and (x', y', z', t') is required to replace the GT. This new transformation should satisfy the requirement that if the coordinates of the event "Φ" with respect to S satisfy (2.14), then the coordinates of the same event "Φ" with respect to S' must satisfy a similar equation, namely

$$x'^2 + y'^2 + z'^2 - c^2t'^2 = 0. \tag{2.16}$$

As a prelude to the new transformation rule (to be called Lorentz transformation), we shall consider two paradoxical and important consequences of our postulates, namely time dilation and length contraction, in the following sections.

2.6. Time Dilation

Imagine the Michelson interference experiment being performed on a train which is moving with velocity **v** with respect to the platform as shown in Fig. 2.6(a). Let "α" represent the event that the ray ϕ_2 — after coming from the source S — is "reflected upwards at the mirror A". Let "β" represent the event that this ray — after bouncing downward from the overhead mirror — is "received back at the mirror A".

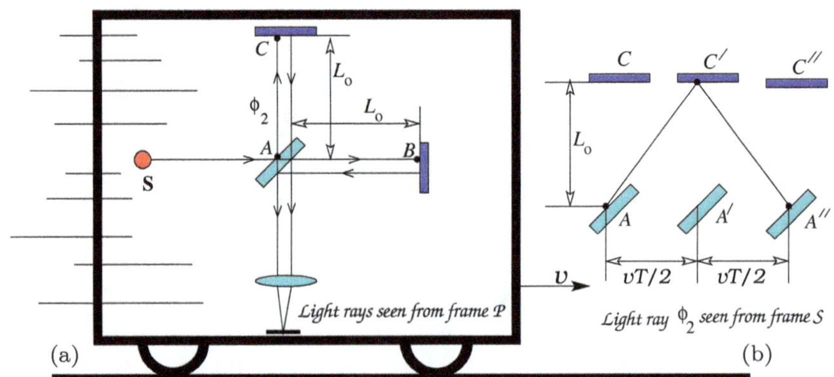

Fig. 2.6. Coordinates of a particle with respect to frames S and E.

An observer sitting in the train — call him Mr P (P for passenger) — marks the path followed by this ray. We shall denote his frame of reference — i.e. the frame fixed on the floor of the train — by the same letter P. Let the length of the arm AC, as measured in the frame P, be L_o. Since light propagates with speed c in P, the time of flight T_0 of the ray ϕ_2 in covering the round trip ACA, and as measured in the frame P, is

$$T_0 = \frac{2L_0}{c}. \tag{2.17}$$

Another observer Mr S (S for Station Master), standing on the railway station, watches the same pair of events "α" and "β". He finds them taking place at two different points A and A'' on the track (Fig. 2.6(b)). Light takes a longer route $AC'A''$ in the frame S, and therefore, travelling with the same speed c (according to Corollary 2A), must take a longer time T (as measured in S) in covering this route. Assuming that the length of the vertical arm of the MM apparatus (i.e. the perpendicular distance between C' and the line AA'') measures out to be the same length L_0 in S as in P (we shall justify this statement later), the length of the path traversed by light, as seen from the frame S, is $2\sqrt{L_0^2 + \{\frac{vT}{2}\}^2}$, which should now equal cT. Therefore,

$$T = \frac{2}{c}\sqrt{L_0^2 + \left\{\frac{vT}{2}\right\}^2}. \tag{2.18}$$

Using Eqs. (2.17) and (2.18), one obtains a relation between T and T_0:

$$T^2 = \frac{4}{c^2}\left\{L_0^2 + \frac{v^2T^2}{4}\right\} = T_0^2 + \frac{v^2}{c^2}T^2,$$

so that

$$T_0 = \sqrt{1 - \frac{v^2}{c^2}}T. \tag{2.19}$$

Note that T_0 and T are the time intervals, as measured in P and S respectively, between the same pair of events "α" and "β".

The reader may conclude from the above discussions that "moving clocks go slow". Mr P is moving and Mr S is stationary! Therefore, Mr P's clock registers smaller time than Mr S's clock. However, such arguments are wrong and contradict the very spirit of the relativity principle.

The fallacy in the above statement lies in that if S can claim that P's clock is going slow because P is moving (relative to S), then P can also claim that S's clock is going slow because S is moving (relative to P). As we have stressed in Chapter 1, motion is relative. And though it may sound strange, relativity postulates find both P and S to be correct. Each one's clock is going slow with respect to the other.

Let us resolve the paradox implied in the last sentence. We define *proper time* between any two events to be the time interval between these events, as measured in one particular frame of reference S_0 in which both the events happen to take place at the same spatial location. What we mean by this is that if two events have coordinates (x_1, y_1, z_1, t_1) and (x_2, y_2, z_2, t_2) in S_0, such that $x_1 = x_2, y_1 = y_2, z_1 = z_2$, then we call $T_0 = t_2 - t_1$ the proper time between the events. It should be remembered, however, that for a particular pair of events there may or may not exist a frame in which both the events will occur at the same location, and hence, there may or may not exist a "proper time" between these events.

In the above example, the events "α" and "β" occur at the *same* floor location A in P's frame of reference, whereas they take place at two different locations, namely A and A'', along the railway track, as seen from S's frame of reference. Therefore, P measures proper time between "α" and "β" whereas S measures "improper" time. *Please note that we are using the terms "proper" and "improper" not to mean right and wrong.* Proper time is just a nomenclature, a definition. "Improper time" is any time interval that does not satisfy that definition. It has been assumed that *both observers P and S are using standard clocks for measuring time intervals, and, therefore, both measurements, i.e. "proper time" and "improper time", are correct measurements, but with respect to different observers.* It should be understood that the term "proper time" has no meaning except with reference to a particular *pair* of events.

The correct conclusion from the result of the above calculation is that the proper time interval T_0 between a given pair of events "α" and "β" is always less than the corresponding "improper" time interval T between them.

It will be convenient to introduce at this stage the *Lorentz factor*, γ which we define as follows:

$$\gamma \equiv \frac{1}{\sqrt{1-\beta^2}}, \quad \text{where } \beta \equiv \frac{v}{c}. \tag{2.20}$$

Since the Lorentz factor γ defined above is associated with the operation boost (see the meaning in Sec. 3.1), we shall in future call it *boost Lorentz factor*, in order to distinguish it from the *dynamic Lorentz factor* Γ to be introduced later through Eq. (4.15), in Chapter 4

Note that

$$\gamma \geq 1, \tag{2.21}$$

and the following identities which can be proved easily:

$$\frac{\beta^2}{1-\beta^2} = \gamma^2\beta^2, \quad 1 - \frac{1}{\gamma^2} = \beta^2, \quad \frac{1}{\gamma} = \gamma\left(1-\beta^2\right), \quad \gamma^2 - 1 = \gamma^2\beta^2. \tag{2.22}$$

Using the Lorentz factor, we shall summarize the time dilation formula (2.19) in the form of the following very important rule:

Rule 2. *Let there be a frame of reference S_0 with respect to which two events "α" and "β" occur at the* same *spatial location. Let S be another frame of reference which is moving with uniform velocity $\mathbf{v}$ with respect to S_0. If T_0 and T be the time intervals, as measured in S_0 and S, respectively, between "α" and "β" (so that T_0 is the proper time between the events), then*

$$T_0 = \sqrt{1-\beta^2}T = \frac{1}{\gamma}T. \tag{2.23}$$

2.7. Length Contraction

In a similar vein, we can define *proper length* to be the length of an object in its *rest frame*, i.e. the length measured in that particular frame in which it is at rest.

Let us once again examine the MM experiment being conducted on the train as was shown in Fig. 2.6. This apparatus is stationary in the frame of reference of the train. Therefore, the length L_0 of the arms AB and AC, when measured with meter sticks laid on the floor of the train (or held stationary on board the train), are the proper lengths of these arms. The same lengths measured by meter sticks laid on the ground may be called "improper" lengths of these segments. Instead of using meter sticks, an alternative and better means of the length measurement will be with the help of light beams, which we shall now employ for conceptualizing the length paradox.

Let us, therefore, assume that both P and S measure the length of the arm AB using the time of flight of a light ray from A to B and then back from B to A. Let these measurements be L_0 to Mr P, and L to Mr S.

We shall establish a relationship between L_0 and L with the help of relationship (2.19) between proper time and "improper" time. For this purpose, we shall identify three significant events along the journey route ABA of the ray of light. These events are:

- "Θ_{A1}" = "the light ray passes through the half silvered plate A" (on its way towards the mirror B).
- "Θ_B" = "the ray is reflected back at the mirror B".
- "Θ_{A2}" = "the ray returns to the half silvered plate A" (after being reflected at the mirror B).

As light travels with velocity c in the reference frame of the train, Mr P measures the same propagation time T_1' for the light ray to go from A to B, i.e. between the events "Θ_{A1}" and "Θ_B", and also to return from B to A, i.e. between the events "Θ_B" and "Θ_{A2}". The total time of the round trip flight is then $T_0 = 2T_1'$. The total length travelled during this time is $2L_0$. Therefore, $2L_0 = 2cT_1' = cT_0$. Hence,

$$L_0 = \frac{cT_0}{2}. \tag{2.24}$$

Mr S observes the same phenomena from his own reference frame, the platform. According to his watch, the light ray takes time T_1 for its forward trip, i.e. between the events "Θ_{A1}" and "Θ_B" during which time the mirrors move from the locations A and B to A' and B', and a different time T_2 for the return trip (i.e. between the events "Θ_B" and "Θ_{A2}" during which time the mirrors move from A' and B' to A'' and B''), as suggested in Fig. 2.7.

It should be clear from these diagrams that $cT_1 = L + vT_1$, and $cT_2 = L - vT_2$, so that $T_1 = \frac{L}{c-v}$; $T_2 = \frac{L}{c+v}$.

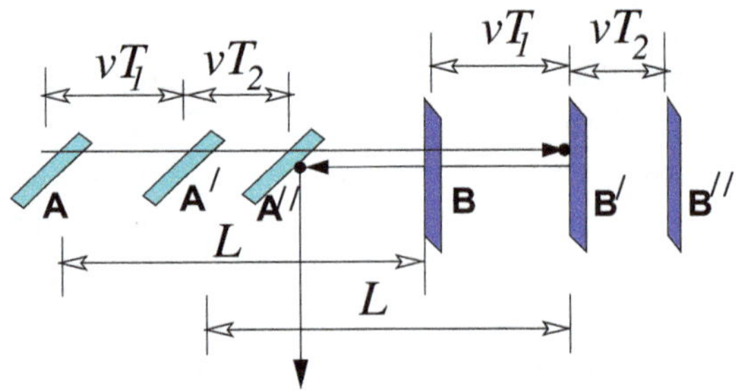

Fig. 2.7. The ray ϕ_1 traced by Mr S.

The total time for the round trip according to Mr P is then $T = T_1 + T_2 = \frac{2Lc}{c^2 - v^2}$. Therefore,

$$L = \left(\frac{c^2 - v^2}{2c}\right) T = \frac{cT}{2}\left(1 - \frac{v^2}{c^2}\right). \tag{2.25}$$

Note that T_0, appearing in Eq. (2.24), is the "proper time", and T appearing in Eq. (2.25) is an "improper time", between the events "Θ_{A1}" and "Θ_{A2}". Connecting Eqs. (2.25) and (2.24) with the help of Eq. (2.23) one now obtains in a straightforward way the required relationship between L and L_0, which we have written below as Eq. (2.26) at the conclusion of the following important rule.

Rule 3. *Let L_0 be the proper length of a rod (as measured in its rest frame S_0). Let the rod be moving longitudinally with velocity $\mathbf{v}$ with respect to another frame S (i.e. $\mathbf{v}$ is parallel to the axis of the rod), and let the (improper) length L of the rod, as measured in the frame S, be L. The relation between the two is given by the formula:*

$$L_0 = \frac{1}{\sqrt{1 - \frac{v^2}{c^2}}} L = \gamma L. \tag{2.26}$$

Compare this with the relation between the proper time and improper time as given in Eq. (2.23)

Since $\gamma \geq 1$, it is seen from Eq. (2.26) that Mr S measures a smaller value for the longitudinal dimension (i.e. the dimension which is parallel to its direction of motion) than Mr P. He therefore concludes that *when an object moves, its longitudinal dimension contracts.*

It will be in order to suggest an operational model of length measurement both for subsequent reference as well as for elucidation of the meaning of Eq. (2.26). The length of a moving stick can be measured by taking its shadow-graph on a photographic plate laid on the floor (or on a table) of the laboratory as the stick shoots past it. We have illustrated this in Fig. 2.8(a). An overhead array of flashguns has been provided for casting shadow on the plate. The end guns G_1 and G_2 are triggered *simultaneously*, so that the shadows P_1 and P_2 of the endpoints A_1 and A_2 of the stick are also etched simultaneously on the plate. (Note that these etching events are simultaneous in the Lab frame only.) The distance L between the marks P_1 and P_2 (which are permanent marks on the table) measured subsequently using any standard meter stick will give the laboratory measure of the length of the moving stick. It is this L which is related to the

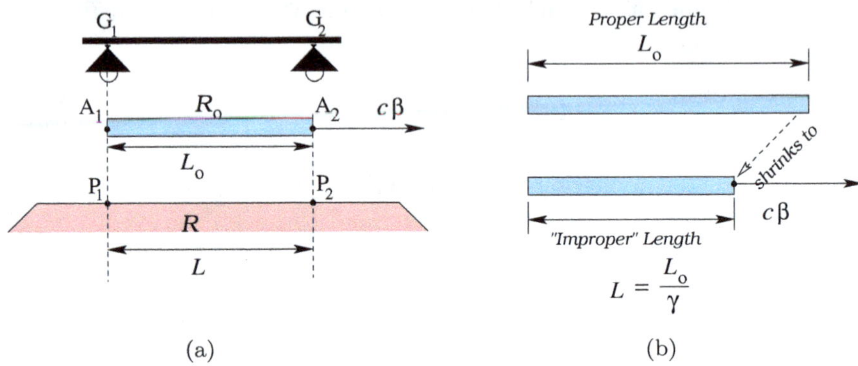

Fig. 2.8. Measurement of longitudinal length.

"proper length" L_0 of the stick through Eq. (2.26). (We have obtained this relationship using Lorentz transformation in Sec. 3.3).

We shall adopt the following mathematical description for the outcome of the above length measurement experiment.

Theorem 2.1. *Let R and R_0 be two straight rods sliding past each other longitudinally with a relative speed $c\beta$. If the segment $\overline{\mathrm{P_1P_2}}$ of R coincides with the segment $\overline{\mathrm{A_1A_2}}$ of R_0,* when viewed from R *(so that R is stationary and R_0 is seen to be moving), and if the* proper lengths *of $\overline{\mathrm{P_1P_2}}$ and $\overline{\mathrm{A_1A_2}}$ are L and L_0, respectively, then*

$$L = \frac{1}{\gamma} L_0. \tag{2.27}$$

A popular way of saying the same thing is: *a rod of proper length L_0 shrinks to a smaller length $L = \frac{1}{\gamma} L_0$, when moving with velocity βc.* We have portrayed this concept graphically in Fig. 2.8(b).

Theorem 2.1 will play a very important role as the mathematical model of length measurement in the subsequent development of the formalism.

In Sec. 3.6, in Chapter 3 we have reinforced the concepts of simultaneity, time dilation, length contraction through Theorems 3.1–3.3. We have shown through actual calculations:

- the exact time difference between two events in a second frame, when the *same* events are simultaneous in the first one;

- the relation between the two different lengths ℓ' and ℓ'' measured in a given frame of reference, when they match the *same* length ℓ measured in another frame of reference.
- the relation between time intervals T_1' and T_2' in a given frame of reference when the corresponding time intervals are the *same* in another one.

The reader should understand the logic presented there in order to get a better grip of these exotic concepts.

We shall now take up transverse length and suggest a reason why the transverse dimension of an object (i.e. the dimension measured *perpendicular* to the direction of its motion) *should not* change due to the motion.

Imagine two identical and parallel rods L and R, each equipped with markers at their ends (Fig. 2.9). Let A and B be the markers of L, and C and D the markers of R. When at rest, the marker A coincides with C, and the marker B with D, thereby confirming that the proper lengths of the rods are equal. Now let the rods move transversely towards each other. As they zip past each other, the markers mark the opposite rods. If the length of the rod L, as seen from the rest frame of R, remains unchanged, then the marks made by A and B will fall on C and D, respectively. Otherwise, only one set of markers, say, A and B of the rod L, will be able to make marks on the body of the other rod R, implying thereby that the rod L has become shorter due to its motion relative to R. An observer in the rest frame of R would then conclude that the length of the rod L has shrunk due to its transverse motion.

An observer in the rest frame of L can also examine the marks. There cannot be disagreement between the two observers about the fact that these

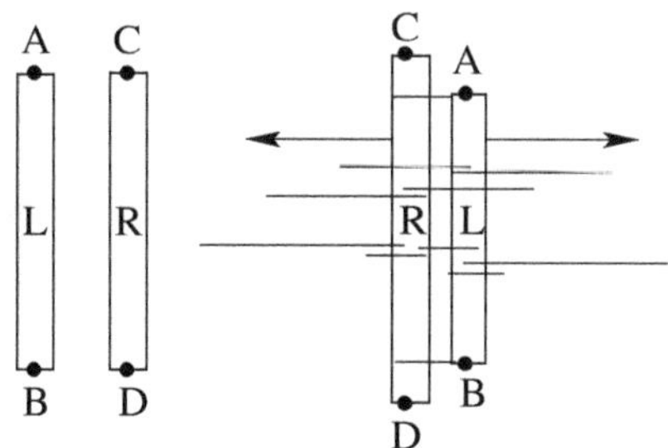

Fig. 2.9. Measurement of transverse length.

marks lie within the body of R. Therefore, this observer would conclude that the length of the rod R (which is in motion relative to L) has expanded due to its transverse motion.

Thus two different observers observe two different effects of speed on the transverse dimension of a rod (i.e. expansion in one case and contraction in the other). This contradicts postulate 1. Hence, the conclusion that transverse dimensions cannot change due to motion.

We shall therefore record the above finding in the following rule.

Rule 4. *When a rod moves transversely, as seen from a frame S, its length L as measured in S equals its proper length L_0.*

We shall supplement rules 3 and 4 with two important corollaries. We illustrate them with the help of Fig. 2.10. It shows an object A of arbitrary shape. S_0 is the rest frame of the object (we assume that every part of A is at rest in S_0)

Corollary 4A. *Let the rest frame S_0 be moving with velocity $\mathbf{v}$ along the X-axis with respect to another frame S. Let the dimensions of the object along the X-, Y- and Z-directions be $\delta x_0, \delta y_0, \delta z_0$ in S_0 and $\delta x, \delta y, \delta z$ in S. Then*

$$\delta x = \frac{1}{\gamma}\delta x_0; \quad \delta y = \delta y_0; \quad \delta z = \delta z_0. \tag{2.28}$$

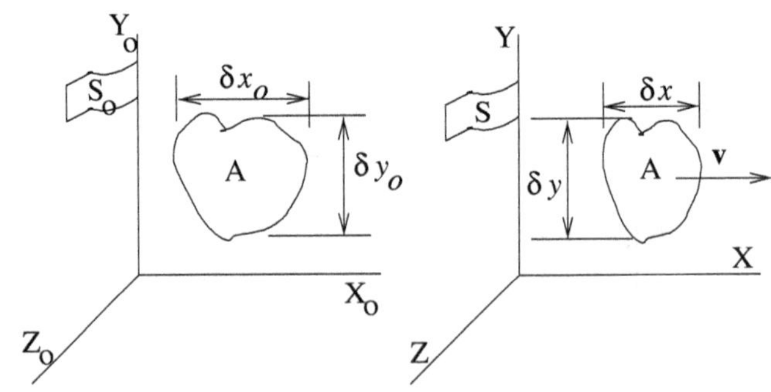

Fig. 2.10. Volume transformation.

Corollary 4B. *If δV_0 is the proper volume of the object (i.e. the volume measured in S_0), then its volume, as measured in S is*

$$\delta V = \frac{1}{\gamma}\delta V_0. \tag{2.29}$$

With the foresight gained through the discussions presented in this section, we shall now search for the correct transformation equation connecting the coordinates of a given event in two different inertial frames of reference.

Chapter 3

Lorentz Transformation

3.1. Lorentz Transformation I: Special Case

We shall obtain the transformation equations between the coordinates (x, y, z, t) and (x', y', z', t') of an event as measured in two frames of reference S and S' which are moving relative to each other. The underlying concepts are best exemplified by considering the simplest example, namely boost in the X-direction.

We shall use the term *boost* to mean relative motion between two frames of reference. Figure 3.1 illustrates a *boost* $c\beta$ of S' relative to S along the X-axis, which we shall write compactly as "boost: $S(c\beta, 0, 0)S'$". Note that

$$\text{boost:}S(c\beta, 0, 0)S' = \text{boost:}S'(-c\beta, 0, 0)S. \tag{3.1}$$

That is, the above arrangement also means a boost $-c\beta$ of S relative to S' along the X'-axis. A more general boost will be

$$\text{boost:}S(c\beta_x, c\beta_y, c\beta_z)S' = \text{boost:}S'(-c\beta_x, -c\beta_y, -c\beta_z)S. \tag{3.2}$$

For all such boosts, we have the following assumptions:

#1 The coordinate axes XYZ of S are parallel to their counterparts $X'Y'Z'$ of S'.

#2 The time origins of both frames are chosen to be the instant (i.e. the clocks in both frames are set to zero hour at the instant) when the origins O and O' of the space frames cross each other.

#3 For the special case of boost:$S(c\beta, 0, 0)S'$, we further assume that S' is moving in the direction of the X-axis with velocity $v = c\beta$ relative to S, and that the X'-axis lies along the X-axis.

We shall call this configuration the *standard configuration.*

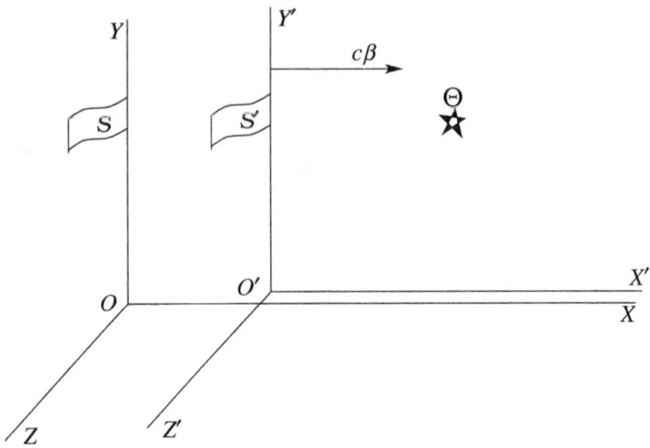

Fig. 3.1. Standard configuration of frames S and S'.

An event "Θ" viewed from these two frames of reference is also shown in Fig. 3.1. The transformation equation that we shall obtain for this particular case is a simplest special case of a general class of relativistic transformation equations, called *Lorentz transformation*, which we shall often abbreviate as LT.

An anomaly exists among the coordinates x, y, z, t in the sense that the last coordinate has a different dimension than the first three. The usual convention in relativity is, therefore, to adopt ct for the time coordinate. This sounds reasonable because of the frame independence of the speed of light. From now on, the coordinates of every event will be written as (ct, x, y, z), with the *time coordinate preceding the space ones*, and each one of the coordinates having the dimension of length. In conformity with this practice, we shall measure all *time intervals in the unit of* ct. We shall, for example, say that a certain event has occurred at the instant ct, or that the time interval between two events "α" and "β" is $c\,\delta t$.

Let us therefore think of an event "Θ" whose coordinates are (ct, x, y, z) in S and (ct', x', y', z') in S'. We shall find a relationship between the two sets of coordinates. For convenience of picturization imagine, the event is a *lightning* which strikes along the X- and X'-axes. We think of these axes as infinitely long straight rods sliding longitudinally along each other (Fig. 3.2). The lightning "Θ" leaves *permanent* marks Q' on the X'-rod and Q on the X-rod. Since y and z coordinates of this event are identically zero in both S and S', we shall ignore them for the time being and write

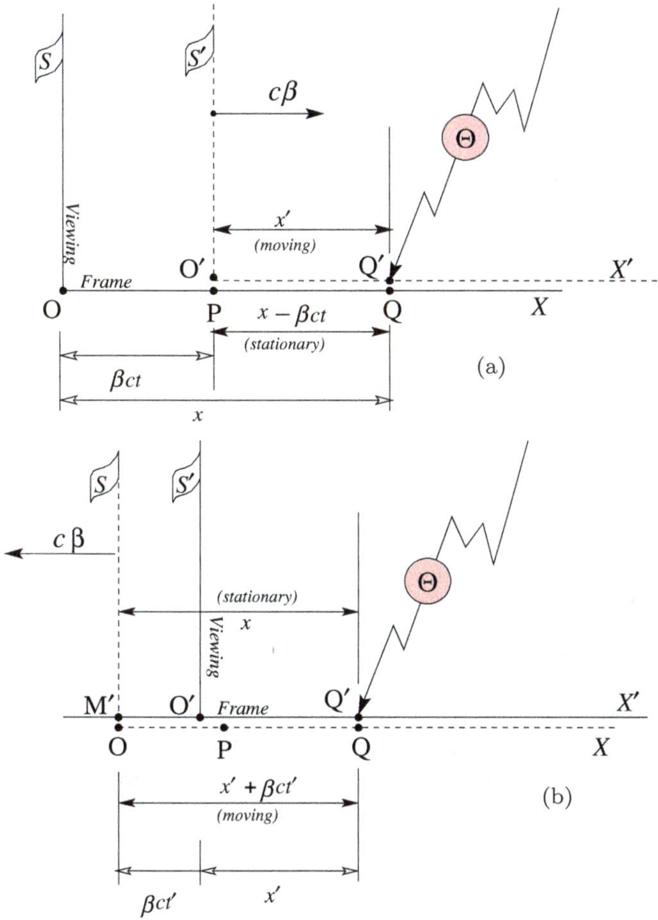

Fig. 3.2. Lightning Θ as seen from S and S'.

the event "Θ" as

$$\text{“}\Theta\text{”} \equiv \text{“}Q' \text{ passes } Q\text{”} = \begin{cases} (ct, x) & \text{in } S, \\ (ct', x') & \text{in } S', \end{cases} \tag{3.3}$$

where $x = \overline{\mathrm{OQ}}$ = proper length of the intercept OQ, and $x' = \overline{\mathrm{O'Q'}}$ = proper length of the intercept $O'Q'$.

Figure 3.2(a) shows a view of the event *as seen from the frame S*. We have represented the *viewing frame* with continuous lines and the *viewed frame* with broken lines. The origin O' of S' is seen to coincide with the mark P on the X-rod at the instant when "Θ" occurs. By this we mean that the event "Ψ" $\equiv$ "O' passes P" is simultaneous with "Θ" in the frame S, i.e. they have the same time coordinate ct in S. From assumption # 3, $\overline{\mathrm{OP}} = \beta ct$. According to (3.3), the "moving segment" $\overline{\mathrm{O'Q'}}$ (moving to the right) and the "stationary segment" $\overline{\mathrm{OQ}}$ have *proper* lengths x' and x, respectively. The proper length of $\overline{\mathrm{OP}} = \beta ct$. Therefore, the proper length of the segment is given by $\overline{\mathrm{PQ}} = \overline{\mathrm{OQ}} - \overline{\mathrm{OP}} = x - \beta ct$. Also, seen from S, the "moving segment" $\overline{\mathrm{O'Q'}}$ coincides with the "stationary segment" $\overline{\mathrm{PQ}}$. Hence, from Eq. (2.27),

$$\begin{aligned} x - \beta ct &= \frac{x'}{\gamma}. \\ \text{Hence,} \quad x' &= \gamma(x - \beta ct). \end{aligned} \tag{3.4}$$

Let us now view the same lightning, i.e. the event "Θ" *from the frame* S', represented in Fig. 3.2(b). The origin O of S, while moving to the left, is seen to coincide with some mark M' on the negative X'-axis at the instant when "Θ" occurs. By this we mean that the event "Φ" $\equiv$ "O passes M'" is simultaneous with "Θ" in the frame S', i.e. they have the same time coordinate ct' in S'. By assumption #3, $\overline{\mathrm{O'M'}} = \beta ct'$. According to (3.3), the "stationary segment" $\overline{\mathrm{O'Q'}}$ and the "moving segment" $\overline{\mathrm{OQ}}$ (moving to the left) have *proper* lengths x' and x, respectively. The "moving segment" $\overline{\mathrm{OQ}}$ and the "stationary segment" $\overline{\mathrm{M'Q'}}$ have, therefore, proper lengths x and $\beta ct' + x'$, respectively, and they coincide when seen from S. Therefore, by using Eq. (2.27) again,

$$\begin{aligned} \beta ct' + x' &= \frac{x}{\gamma}. \\ \text{Hence} \quad x &= \gamma(\beta ct' + x'). \end{aligned} \tag{3.5}$$

Eliminating x' in Eq. (3.5) with the help of Eq. (3.4) we get

$$\begin{aligned} x &= \gamma^2(x - \beta ct) + \gamma\beta ct'. \\ \text{Or} \quad ct' &= \gamma ct - \frac{\gamma^2 - 1}{\gamma\beta}x = \gamma(ct - \beta x), \end{aligned} \tag{3.6}$$

where we have made use of Eq. (2.22).

Alternatively, one can eliminate x in Eq. (3.4) with the help of Eq. (3.5), leading to

$$x' = \gamma^2(\beta ct' + x') - \gamma\beta ct.$$
$$\text{Or} \quad ct = \gamma ct' + \frac{\gamma^2 - 1}{\gamma\beta}x' = \gamma(ct' + \beta x'). \tag{3.7}$$

Equations (3.6) and (3.4) represent the equations of transformation from (ct, x) to (ct', x'). Equations (3.7) and (3.5) constitute the inverse transformation, i.e. from (ct', x') to (ct, x).

The above equations are supplemented by the equations of transformation for the y and z coordinates corresponding to a more general event "Φ" having arbitrary non-zero values for all the four coordinates. Therefore, let a lightning "Φ" strike an arbitrary point T, which for the convenience is identified as the top of a tower. We assign to this event the coordinates (ct, x, y, z) in S and (ct', x', y', z') in S'.

We have shown the event, as viewed from S, in Fig. 3.3. (Note that neither of the S and S' frames is a rest frame of the tower.) Assuming that the base of the tower is on the XZ-plane, the height of the tower should be y in S and y' in S'. According to Rule 4, $y = y'$. By a similar argument, $z = z'$. Let the tower top T be projected to the mark Q

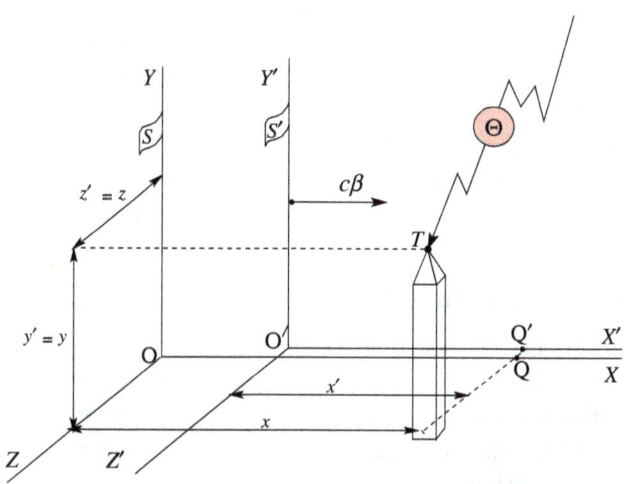

Fig. 3.3. Lightning striking a tower.

on the X-rod and Q' on the X'-rod, as shown in Fig. 3.3. Then the event "Θ" ≡ "Q passes Q'" is the same event as represented in Eq. (3.3) whose coordinate transformation we have just worked out. Summarizing all that we have so far accomplished, we can write the transformation equations of the coordinates for any arbitrary event "Φ" as follows:

(i) Transformation equation from S to S':

$$\boxed{\begin{array}{ll} \text{(a)} & ct' = \gamma(ct - \beta x), \\ \text{(b)} & x' = \gamma(x - \beta ct), \\ \text{(c)} & y' = y, \\ \text{(d)} & z' = z. \end{array}} \tag{3.8}$$

(ii) Transformation equation from S' to S (i.e. the *inverse* of "S to S'"):

$$\boxed{\begin{array}{ll} \text{(a)} & ct = \gamma(ct' + \beta x'), \\ \text{(b)} & x = \gamma(x' + \beta ct'), \\ \text{(c)} & y = y', \\ \text{(d)} & z = z'. \end{array}} \tag{3.9}$$

The transformation equations (3.8) and (3.9) are the simplest examples of Lorentz transformation. We shall refer to them as the *standard Lorentz transformation* corresponding to the standard configuration shown in Fig. 3.2.

Two *inertial frames* of reference S and S' that are connected to each other by a Lorentz transformation — as in (3.8), (3.9), or more generally as in (3.18), (3.19) — have an alternative name: *Lorentz frames.*

3.2. Lorentz Transformation II: General Case

As we found out in Sec. 3.4, one reason why GT is not acceptable within the scheme of special relativity is that it yields different speeds of light in different inertial frames, whereas Corollary 2A in Chapter 2 requires frame independence of the speed of light. It is now incumbent on any relativistic transformation equation to meet this requirement of Corollary 2A.

Consider an event "E", namely "emission of a flash of light", which takes place at the origin of S at $ct = 0$. Consider another frame S' obtained from S by a more general *boost*, as defined at the beginning of Sec. 3.7.

By assumption (b), the event takes place also at the origin of S' and at the time $ct' = 0$ (as measured by S'.) Therefore, the coordinates of "E" are:

$$\text{“}E\text{”}: \left.\begin{array}{ll}(0,0,0,0) & \text{in } S\\ (0,0,0,0) & \text{in } S'\end{array}\right\}.$$

Let "R" represent a subsequent event namely "reception of the flash of light by an observer". Let its coordinates be

$$\text{“}R\text{”}: \left.\begin{array}{ll}(ct,x,y,z) & \text{in } S\\ (ct',x',y',z') & \text{in } S'\end{array}\right\}.$$

Therefore, Corollary 2A in Chapter 2 requires that the following two equations be both satisfied [refer to Eqs. (2.14) and (2.16)]:

$$\begin{aligned} c^2t^2 - (x^2 + y^2 + z^2) &= 0,\\ c^2t'^2 - (x'^2 + y'^2 + z'^2) &= 0. \end{aligned} \tag{3.10}$$

Thus, we *define* Lorentz transformation to be a *linear transformation* between the Cartesian coordinates (ct', x', y', z') and (ct, x, y, z) of any event "P", such that

$$c^2t^2 - (x^2 + y^2 + z^2) = c^2t'^2 - (x'^2 + y'^2 + z'^2). \tag{3.11}$$

(Properly speaking, what we have defined is a homogeneous LT which corresponds to the fact that the clock in either frame is set to zero when the origins O and O′ pass each other.) The adjective "linear" used in the above definition implies that each one of the coordinates (ct', x', y', z') is a linear function of the coordinates (ct, x, y, z) and vice versa. This is the same thing as saying that there exists a matrix $\hat{\Omega}$, with constant elements Ω_{ij} such that

$$\begin{pmatrix} ct' \\ x' \\ y' \\ z' \end{pmatrix} = \begin{pmatrix} \Omega_{00} & \Omega_{01} & \Omega_{02} & \Omega_{03} \\ \Omega_{10} & \Omega_{11} & \Omega_{12} & \Omega_{13} \\ \Omega_{20} & \Omega_{21} & \Omega_{22} & \Omega_{23} \\ \Omega_{30} & \Omega_{31} & \Omega_{32} & \Omega_{33} \end{pmatrix} \begin{pmatrix} ct \\ x \\ y \\ z \end{pmatrix}. \tag{3.12}$$

We shall call this $\hat{\Omega}$ matrix the *Lorentz transformation matrix*. Here note that the rows and columns of the above matrix have been indexed by the numbers $0, 1, 2, 3$. This is in conformity with the indexing convention we shall adopt in Chapter 7. By the way of illustration, we shall retrieve the

LT matrix associated with the boost $c\beta$ in the X-direction, out of Eq. (3.8), which can be rewritten in the following matrix form:

$$\begin{pmatrix} ct' \\ x' \\ y' \\ z' \end{pmatrix} = \begin{pmatrix} \gamma & -\gamma\beta & 0 & 0 \\ -\gamma\beta & \gamma & 0 & 0 \\ 0 & 0 & 1 & 0 \\ 0 & 0 & 0 & 1 \end{pmatrix} \begin{pmatrix} ct \\ x \\ y \\ z \end{pmatrix}. \tag{3.13}$$

Hence, the Lorentz transformation matrix associated with a boost $c\beta$ in the X-direction is as follows:

$$\hat{\Omega} = \begin{pmatrix} \gamma & -\gamma\beta & 0 & 0 \\ -\gamma\beta & \gamma & 0 & 0 \\ 0 & 0 & 1 & 0 \\ 0 & 0 & 0 & 1 \end{pmatrix}. \tag{3.14}$$

We shall now prove that the Lorentz transformation represented by Eq. (3.8), which is also equivalent to Eq. (3.13), satisfies the requirement of Eq. (3.11).

Proof of Eq. (3.11).

$$\begin{aligned} c^2t'^2 - (x'^2 + y'^2 + z'^2) &= \{\gamma(ct - \beta x)\}^2 - \{\gamma(x - \beta ct)\}^2 - y^2 - z^2 \\ &= \gamma^2(1-\beta^2)c^2t^2 - \gamma^2(1-\beta^2)x^2 - y^2 - z^2 \\ &= c^2t^2 - (x^2 + y^2 + z^2). \qquad \text{(QED)} \end{aligned}$$

Let us now take up the case of a general boost $\{S(\boldsymbol{\beta})S'\}$, as illustrated in Fig. 3.4. In this case, the boost velocity is $\boldsymbol{\beta}$, in any arbitrary direction $\mathbf{n}$. That is, the velocity of the origin O' of the frame S' is $\boldsymbol{\beta} = \beta\mathbf{n}$ with respect to the origin O of the frame S. Here $\mathbf{n}$ is a unit vector in the direction of $\boldsymbol{\beta}$.

We have set up a new set of coordinate axes (ξ, ζ, η) to replace (x, y, z) in S, and (ξ', ζ', η') to replace (x', y', z') in S', such that (i) the axes ξ of S and ξ' of S' coincide and are oriented in the direction of $\boldsymbol{\beta}$, (ii) the origins O and O' coincide at $t = t' = 0$. We shall adjust the simple LT given in (3.8) to the general LT pertaining to this general case.

Let $\mathbf{r}$ and $\mathbf{r}'$ represent the radius vectors from O and O', respectively, to the location of the event Θ, as measured in S and S', respectively. We shall

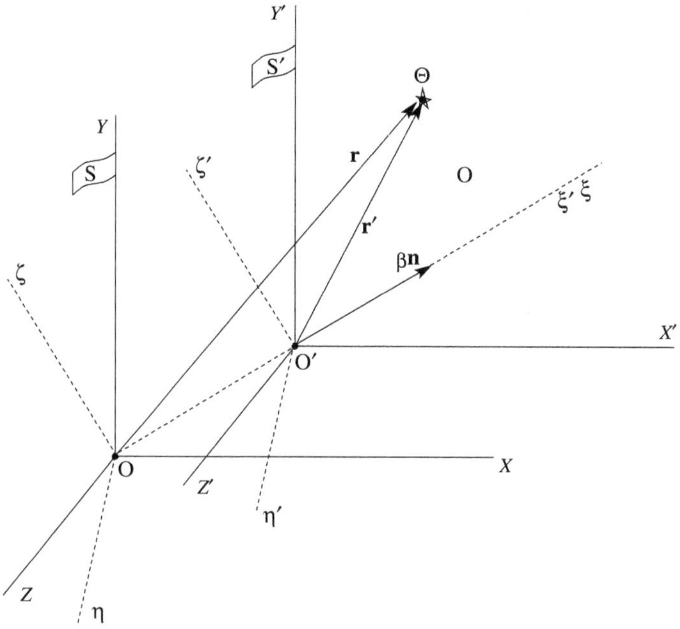

Fig. 3.4. A general boost.

resolve these vectors parallel to and perpendicular to the direction **n**:

$$\mathbf{r} = \xi \mathbf{e}_\xi + \zeta \mathbf{e}_\zeta + \eta \mathbf{e}_\eta = \xi \mathbf{n} + \mathbf{r}_\perp = \mathbf{r}_\parallel + \mathbf{r}_\perp, \tag{3.15}$$

where

$\mathbf{r}_\parallel = \xi \mathbf{n} = \xi \mathbf{e}_\xi$, the component of $\mathbf{r}$ parallel to $\boldsymbol{\beta}$,

$\mathbf{r}_\perp = \zeta \mathbf{e}_\zeta + \eta \mathbf{e}_\eta$, the component of $\mathbf{r}$ perpendicular to $\boldsymbol{\beta}$.

Now, we shall obtain the analogues of Eq. (3.8):

$$\begin{aligned}
(3.8\text{a}) &\Rightarrow ct' = \gamma(ct - \beta\xi) \Rightarrow ct' = \gamma(ct - \boldsymbol{\beta}\cdot\mathbf{r}), \\
(3.8\text{b}) &\Rightarrow \ \xi' = \gamma(\xi - \beta ct) \Rightarrow \mathbf{r}'_\parallel = \xi'\mathbf{n} = \gamma(\xi - \beta ct)\mathbf{n} \\
&\qquad\qquad\qquad\qquad\qquad\quad = \gamma(\mathbf{r}_\parallel - \boldsymbol{\beta} ct), \\
(3.8\text{c}) \text{ and } (3.8\text{d}) &\Rightarrow \ (\zeta', \eta') = (\zeta, \eta) \Rightarrow \mathbf{r}'_\perp = \mathbf{r}_\perp.
\end{aligned} \tag{3.16}$$

From the last two lines,

$$
\begin{aligned}
\mathbf{r}' = \mathbf{r}'_{\parallel} + \mathbf{r}'_{\perp} &= (\gamma \mathbf{r}_{\parallel} + \mathbf{r}_{\perp}) - \gamma \boldsymbol{\beta} ct \\
&= (\mathbf{r}_{\parallel} + \mathbf{r}_{\perp}) + (\gamma - 1)\mathbf{r}_{\parallel} - \gamma \boldsymbol{\beta} ct \\
&= \mathbf{r} + (\gamma - 1)\mathbf{r}_{\parallel} - \gamma \boldsymbol{\beta} ct \\
&= \mathbf{r} + (\gamma - 1)(\mathbf{r} \cdot \mathbf{n})\mathbf{n} - \gamma \boldsymbol{\beta} ct. \qquad (3.17)
\end{aligned}
$$

Noting that $\mathbf{n} = \frac{\boldsymbol{\beta}}{\beta}$, we can summarize as follows. The Lorentz transformation corresponding to the general boost $\{S(\boldsymbol{\beta})S'\}$ has the following form:

$$
\begin{aligned}
ct' &= \gamma(ct - \boldsymbol{\beta}.\mathbf{r}) && (3.18a) \\
&= ct + [(\gamma - 1)ct - \gamma \boldsymbol{\beta} \cdot \mathbf{r}]; && (3.18b) \\
\mathbf{r}' &= (\gamma \mathbf{r}_{\parallel} + \mathbf{r}_{\perp}) - \gamma \boldsymbol{\beta} ct && (3.18c) \\
&= \mathbf{r} + \left[\frac{\gamma - 1}{\beta^2}(\boldsymbol{\beta} \cdot \mathbf{r})\boldsymbol{\beta} - \gamma \boldsymbol{\beta} ct\right] && (3.18d) \\
&= \mathbf{r} + [(\gamma - 1)(\mathbf{r} \cdot \mathbf{n}) - \gamma \beta ct]\mathbf{n}. && (3.18e)
\end{aligned}
$$

We have written each transformation in two or three different forms, so that the reader will have several choices for different applications.

The inverse of the above transformation will correspond to the boost $\{S'(-\boldsymbol{\beta})S\}$, and can be obtained from (3.18) by exchanging $(ct, \mathbf{r})$ with $(ct', \mathbf{r}')$ and replacing $\boldsymbol{\beta}$ with $-\boldsymbol{\beta}$:

$$
\begin{aligned}
ct &= ct' + [(\gamma - 1)ct' + \gamma \boldsymbol{\beta} \cdot \mathbf{r}'], && (3.19a) \\
\mathbf{r} &= \mathbf{r}' + \left[\frac{\gamma - 1}{\beta^2}(\boldsymbol{\beta} \cdot \mathbf{r})\boldsymbol{\beta} + \gamma \boldsymbol{\beta} ct\right]. && (3.19b)
\end{aligned}
$$

We have used the coordinates (ξ, ζ, η) and (ξ', ζ', η') as a tool for obtaining the relation (3.18). Now, we discard them, and interpret $\mathbf{r}$ to mean (x, y, z), $\mathbf{r}'$ to mean (x', y', z'). Moreover, $\boldsymbol{\beta}$ is the boost velocity with components $(\beta_x, \beta_y, \beta z)$, and $\boldsymbol{\beta} \cdot \mathbf{r}$ stands for $x\beta_x + y\beta_y + z\beta_z$.

The Lorentz transformation matrix corresponding to Eq. (3.18) is as follows:

$$\hat{\Omega} = \begin{pmatrix} \gamma & -\gamma\beta_x & -\gamma\beta_y & -\gamma\beta_z \\ -\gamma\beta_x & 1+\dfrac{\gamma-1}{\beta^2}\beta_x^2 & \dfrac{\gamma-1}{\beta^2}\beta_y\beta_x & \dfrac{\gamma-1}{\beta^2}\beta_z\beta_x \\ -\gamma\beta_y & \dfrac{\gamma-1}{\beta^2}\beta_x\beta_y & 1+\dfrac{\gamma-1}{\beta^2}\beta_y^2 & \dfrac{\gamma-1}{\beta^2}\beta_z\beta_y \\ -\gamma\beta_z & \dfrac{\gamma-1}{\beta^2}\beta_x\beta_z & \dfrac{\gamma-1}{\beta^2}\beta_y\beta_z & 1+\dfrac{\gamma-1}{\beta^2}\beta_z^2 \end{pmatrix}. \tag{3.20}$$

To see that "the boost in the X-direction" is a special case of Eqs. (3.18a)–(3.18e), one needs to set $\beta = (\beta, 0, 0)$ in these equations to get back equations in (3.8).

All relativistic formulas must reduce to the corresponding non-relativistic forms in the limits of small velocities. This is a general requirement which we shall use from time to time as one of the checks whenever any new result will be derived. In the present case, the non-relativistic limit implies $\gamma \simeq 1$. Using this approximation in Eq. (3.18), it is easy to see that

$$\begin{aligned} ct' &= ct, \\ \mathbf{r}' &= \mathbf{r} - c\boldsymbol{\beta}t. \end{aligned} \tag{3.21}$$

These are identical to the homogeneous form (2.2), i.e. corresponding to $\mathbf{r}_0 = 0$ if we recognize that $\mathbf{u} = c\boldsymbol{\beta}$.

It will be useful to note at this point that the requirement of Eq. (3.11) is satisfied also by a *pure rotation*, i.e. the transformation in which the (x', y', z') coordinates are obtained from (x, y, z) (or vice versa) by a rotation of the axes XYZ to $X'Y'Z'$, thereby leaving $ct' = ct$. Such a transformation of $(x, y, z) \rightarrow (x', y', z')$ falls under the general class of three-dimensional *orthogonal* transformation (see also Sec. 7.5). In fact, the definition (3.11) is also satisfied by a pure *space reflection*, i.e., $x' = -x, y' = -y, z' = -z, ct' = ct$ and a pure *time reversal*, i.e. $x' = x, y' = y, z' = z', ct' = -ct$. What we shall call Lorentz transformation in this book will include (a) pure boost, (b) pure rotation and (c) a combination of boost and rotation, but will exclude space reflection and time reversal. Such a transformation is normally called *proper* Lorentz transformation. It will be a useful exercise for the reader to show that the effect of two successive boosts is in general equal to one boost and one rotation, so that boost and rotation are in fact inseparable.[a]

3.3. Simple Applications of Lorentz Transformation

In order to illustrate the meaning of LT, we shall use Eqs. (3.8) and (3.9) to retrieve the time dilation and length contraction formulas derived in Secs. 2.6 and 2.7.

Consider two events "A" and "B" having coordinates

$$\text{“}A\text{”} = \begin{cases} (ct, x, y, z) & \text{in } S, \\ (ct', x', y', z') & \text{in } S', \end{cases}$$

$$\text{“}B\text{”} = \begin{cases} (ct + c\Delta t, x + \Delta x, y + \Delta y, z + \Delta z) & \text{in } S, \\ (ct' + c\Delta t', x' + \Delta x', y' + \Delta y', z' + \Delta z') & \text{in } S'. \end{cases}$$

We can obtain the transformations of the coordinates of "B" and "A" and take the difference to get the transformation of the coordinate differences:

$$\begin{cases} c\Delta t' = \gamma(c\Delta t - \beta \Delta x), \\ \Delta x' = \gamma(\Delta x - \beta c\Delta t), \\ \Delta y' = \Delta y, \\ \Delta z' = \Delta z, \end{cases} \tag{3.22a}$$

$$\begin{cases} c\Delta t = \gamma(c\Delta t' + \beta \Delta x'), \\ \Delta x = \gamma(\Delta x' + \beta c\Delta t'), \\ \Delta y = \Delta y', \\ \Delta z = \Delta z'. \end{cases} \tag{3.22b}$$

These equations mean that the *coordinate differences follow the same Lorentz transformation as the coordinates themselves.*

Now consider two events "A" and "B" which occur at the same spatial location (x', y', z'), but at two different times ct' and $ct' + c\Delta t'$, as seen from the frame S'. In that case $\Delta x' = 0$, so that $\Delta t'$ is the proper time between these events. The corresponding time interval Δt measured in the frame S is the "improper" time, and is obtained from the first one of transformation

[a] See Ref. [34, Sec. 2.8].

equations given in (3.22b),

$$\Delta t = \gamma \Delta t' = \gamma \Delta \tau. \tag{3.23}$$

We recall Rulc 2 given in Eq. (2.23). Note that we have used the symbol $\Delta\tau$ to mean proper time between the two events.

Let us elucidate with a commonplace example. A train arrives at Bhopal at 10 hours, and at Nagpur at 18 hours. These two arrivals are examples of the events "A" and "B" considered here. They occur at the *same* spatial location (x', y', z') with reference to the frame S' of the train but at different spatial locations with reference to the frame S of a station master which is fixed on the ground. The time interval $\Delta t' = \Delta\tau = 6$ hours as noted by a passenger of the train is the proper time between reaching Bhopal and reaching Nagpur. The corresponding time Δt that is recorded by the station master is in this case "improper time" and will be more than 6 hours.

Let us now consider the length measurement of a moving rod AB using shadow-graph in a laboratory S, as illustrated in Fig. 2.8. Two flashes of light emitted simultaneously by the flash guns G_1 and G_2 graze past the endpoints A and B of the speeding rod and etches marks P_1, P_2 on the experiment table, as illustrated in Fig. 3.5.

Two events are involved in this measurement: "A" = "A coincides with P_1" and "B" = "B coincides with P_2". The flash guns are riveted to the experiment table in the Lab frame S and the points P_1, P_2 are fixed points on this table. The distance between them is $\Delta x = L$ in S. Also they take simultaneously in S (but not in S'), so that $\Delta t = 0$.

The rest frame S' of the rod is moving with velocity βc with respect to S. The endpoints A and B of the rod are the locations of the events "A" and "B" as seen from S'. The distance $\Delta x' = L_0$ between these events

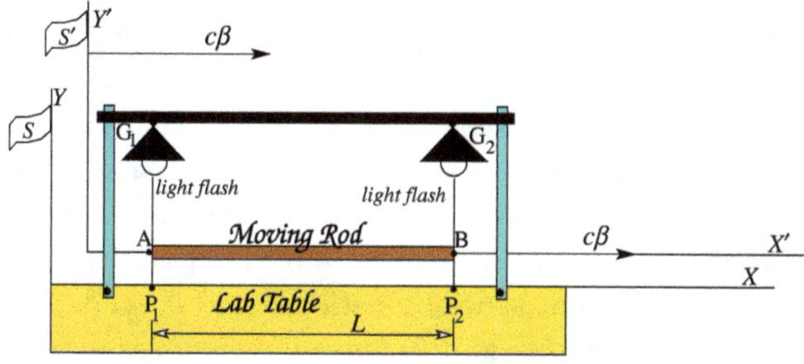

Fig. 3.5. Length measurement using LT.

is the "proper length" of the rod, which is now determined from the second one of transformation equations given in (3.22a).

$$\Delta x' = \gamma \Delta x, \quad \text{or} \quad L_0 = \gamma L. \tag{3.24}$$

We recall back the length contraction formula (2.27).

3.4. Time-Like, Light-Like, Space-Like Intervals

The coordinate difference between two events satisfies the Lorentz transformation according to Eq. (3.22). Hence, the property (3.11) should also be shared by the coordinate difference, i.e.,

$$c^2\Delta t^2 - (\Delta x^2 + \Delta y^2 + \Delta z^2) = c^2\Delta t'^2 - (\Delta x'^2 + \Delta y'^2 + \Delta z'^2) \stackrel{\text{def}}{=} \Delta s^2. \tag{3.25}$$

We have denoted this *invariant* quantity by the symbol Δs^2. The quantity Δs as written above will be called the *square of the interval*, or (in order to be brief) just the *interval*, between the two events. This interval will be called

- *time-like*, if $\Delta s^2 > 0$;
- *space-like*, if $\Delta s^2 < 0$;
- *light-like*, if $\Delta s^2 = 0$.

The significance of these strange names can be explained as follows.

- If the interval is *time-like*, then we can find a frame of reference S', moving with a boost velocity $c\beta$; $\beta < 1$, with respect to S, such that the events occur at the *same* spatial location, say the origin, but at different instants of time, say, t_A and $t_B = t_A + \Delta\tau$, where $\Delta\tau$ is the *proper time* between the events. In this case, $\Delta x' = \Delta y' = \Delta z' = 0$ and $c\Delta t' = c\Delta\tau$. In other words, the *coordinate difference has only time component, but no space component.* Hence, the name time-like.

  Note that, for a time-like separation between two events, $\Delta\tau = \frac{\Delta s}{c}$.
- If the interval is *space-like*, then we can find a frame of reference S', moving with a boost velocity $c\beta$; $\beta < 1$, with respect to S, such that the events occur at the *same time*, say at $t' = 0$. That is, they are simultaneous, but occur at different spatial locations, say, x'_A and $x'_B = x'_A + \Delta l$. In other words, the *coordinate difference has only space component, but no time component.* Hence, the name space-like.

  Note that, for a space-like separation between two events, *there is no proper time*, because Δs is imaginary.

- If the interval is *light-like*, then there is no frame of reference in which the events can occur at the same spatial location. In other words, the two events cannot be connected by any frame of reference. Only a light signal, e.g. "θ_1" = a radio message is sent; "θ_2" = the same message is received, can connect the events. In this case, the proper time $\Delta\tau = 0$.

3.5. Relativistic Doppler Formula

As another interesting application of LT, we shall obtain a relativistic formula for the Doppler effect. Figure 3.6 shows a transmission tower of height h moving in the X-direction with velocity $c\beta$. At the top of the tower is the transmitter K. Fixed on the ground is the receiving station O. Moreover, S and S' are, respectively, the rest frames of the receiving station and the transmitter. Consider the following events:

- "Θ_K" = "the transmitter transmits a sharp beep signal";
- "Φ_O" = "the receiver receives that signal".

We write event coordinates as (ct, x, y), leaving out the z coordinate. Noting that $x = \beta ct$, we write

$$\text{“}\Theta_K\text{”} = \begin{cases} (ct', 0, h) & \text{in } S', \\ (ct, \beta ct, y) & \text{in } S. \end{cases}$$

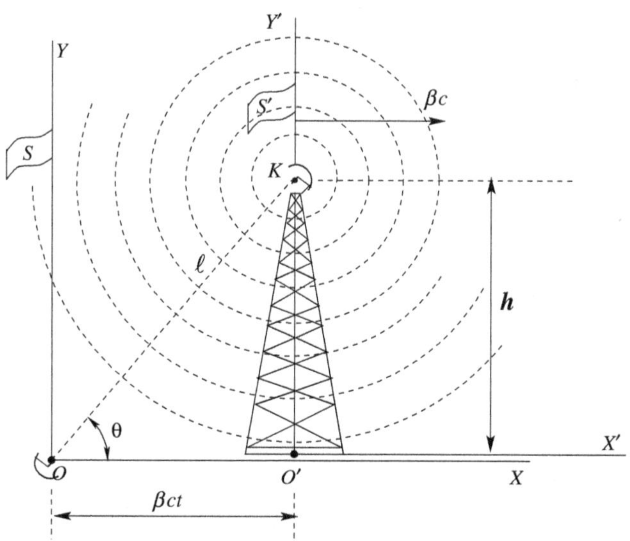

Fig. 3.6. Explaining Doppler effect.

Applying the Lorentz transformation equation (3.9) to the time coordinate and to the y coordinate, we get

$$ct = \gamma ct'; \quad y = h. \tag{3.26}$$

Since the reception takes place at the origin of S, we can write its coordinates as "Φ_O" $= (cT, 0, 0)$ in S, where T is the time when the signal is received. The signal has propagated from K to O, a distance of $\sqrt{x^2 + h^2}$ in time $(T - t)$. Therefore

$$(\beta ct)^2 + h^2 = c^2(T - t)^2. \tag{3.27}$$

Using Eq. (3.26), we get

$$\gamma^2\beta^2c^2t'^2 + h^2 = c^2(T - \gamma t')^2.$$

Simplifying,

$$T^2 + t'^2 - 2\gamma Tt' = \frac{h^2}{c^2}, \tag{3.28}$$

which is the relation between the transmission time t' and the reception time T, each measured at its respective station.

Now, let there be a continuous beam of a sinusoidal radio wave transmitted from the moving tower, whose frequency is measured to be f_O at the transmission station and f at the receiving station. The wave consists of a series of crests separated by a time interval which equals $dt' = \frac{1}{f_O}$ and $dT = \frac{1}{f}$ as measured at the transmitting and at the receiving station, respectively. Therefore, by differentiating (3.28) we get the required relation between f_O and f:

$$\frac{dT}{dt'} = \frac{\gamma T - t'}{T - \gamma t'}. \tag{3.29}$$

It follows from Eqs. (3.26) and (3.27) that $T - \gamma t' = T - t = \sqrt{\frac{x^2+y^2}{c^2}} = \frac{l}{c}$, where l= length of the hypotenuse $\overline{\text{OK}}$. Also

$$\begin{aligned}\gamma T - t' &= \gamma T - \frac{t}{\gamma} = \frac{1}{\gamma}(\gamma^2 T - t)\\ &= \frac{1}{\gamma}\left(\frac{T}{1-\beta^2} - t\right) = \gamma[(T - t) + \beta^2 t] = \frac{\gamma}{c}(l + \beta x).\end{aligned}$$

Therefore, from (3.29),

$$\frac{dT}{dt'} = \gamma(1 + \beta\cos\theta), \tag{3.30}$$

where θ is the angle of elevation of the transmitter when viewed from O. In other words

$$f = \frac{f_0}{\gamma(1 + \beta\cos\theta)}. \tag{3.31}$$

Equation (3.31) represents the relativistic Doppler formula. The angle θ is to be interpreted as the angle between the direction of motion of the source of light and the ray line. The reader must have realized that this formula is valid only for light transmitted by a moving source.

We shall write $\tilde{\Gamma} \equiv \frac{f}{f_0}$ and specialize the formula for three special cases:

$$\begin{aligned}
&\theta = 0 \text{ longitudinal; source receding} && \tilde{\Gamma} = \sqrt{\frac{1-\beta}{1+\beta}},\\
&\theta = \frac{\pi}{2} \text{ transverse; source vertically up} && \tilde{\Gamma} = \frac{1}{\gamma},\\
&\theta = \pi \text{ longitudinal; source approaching} && \tilde{\Gamma} = \sqrt{\frac{1+\beta}{1-\beta}}.
\end{aligned} \tag{3.32}$$

For non-relativistic Doppler effect, we set $\gamma = 1$, and get

$$f = \frac{f_0}{(1 + \beta\cos\theta)}, \tag{3.33}$$

which is valid when the transmission tower (or the source of the wave) is moving with speed $v \ll c$. This formula is valid for sound as well if we substitute for c the speed of sound [11].

The simple look of the Doppler formula (3.31) is misleading. In actual application, it can present conceptual and mathematical challenges. We have presented two examples for a better appreciation of the formula. One of them is a detailed workout in Sec. 8.9.5. The other one is Exercise R3 for the reader, in Sec. 8.10.

3.6. Worked Out Problems I

Problem 3.1. Let R and R' be two straight rods sliding past each other with relative speed $c\beta$ as shown in Fig. 3.7. Prove the following theorem as a corollary of Theorem 2.1.

Theorem 3.1. *Let the segment* $\overline{\mathrm{AB}}$ *of* R *coincide with the segment* $\overline{\mathrm{A'B'}}$ *of* R' *when viewed from* R *(Fig. 3.7(a)), and with the segment* $\overline{\mathrm{A'B''}}$ *of* R'

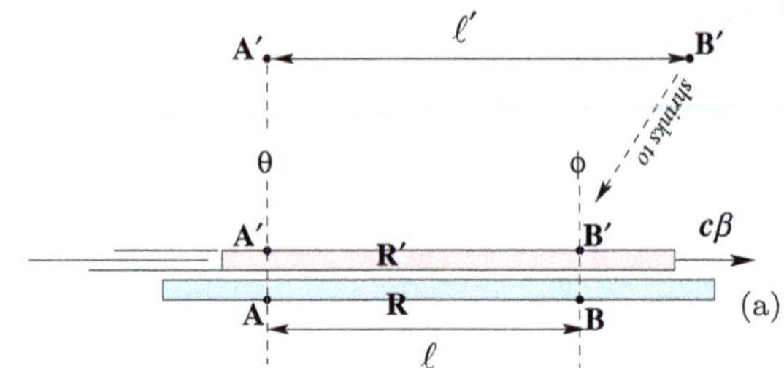

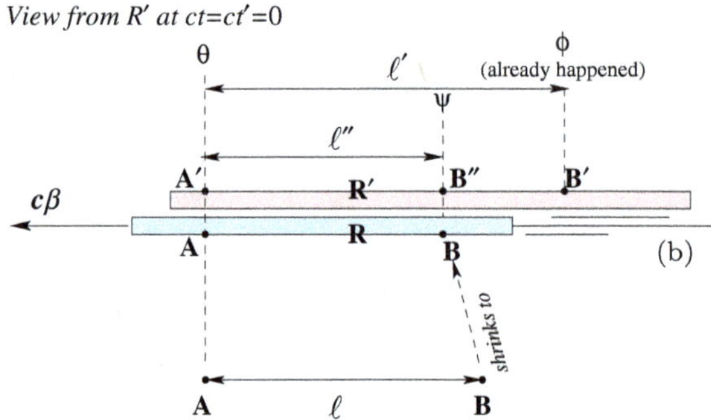

Fig. 3.7. Problem 3.1.

when viewed from R' (Fig. 3.7(b)). If ℓ' and ℓ'' be the proper lengths of $\overline{\mathrm{A'B'}}$ and $\overline{\mathrm{A'B''}}$, respectively, then

$$\ell'' = \frac{1}{\gamma^2}\ell'. \tag{3.34}$$

Solution to Problem 3.1

The proper length $\overline{\mathrm{A'B'}}$ shrinks to proper length $\overline{\mathrm{AB}}$ when seen from R. The proper length $\overline{\mathrm{AB}}$ shrinks to proper length $\overline{\mathrm{A'B''}}$ when seen from R'. Let ℓ be the proper length of the segment $\overline{\mathrm{AB}}$.

Hence, by Theorem 2.1,

$$\overline{\mathrm{AB}} = \frac{\overline{\mathrm{A'B'}}}{\gamma}, \quad \text{or,} \quad \ell = \frac{\ell'}{\gamma},$$

$$\overline{\mathrm{A'B''}} = \frac{\overline{\mathrm{AB}}}{\gamma}, \quad \text{or,} \quad \ell'' = \frac{\ell}{\gamma},$$

Therefore,

$$\ell'' = \frac{1}{\gamma}\left(\frac{\ell'}{\gamma}\right) = \frac{1}{\gamma^2}\ell'.$$

Problem 3.2. Prove the following theorem as a corollary of Theorem 3.1.

Theorem 3.2. *Consider two events "θ" and "ϕ" occurring at the locations (A, B) on R and (A', B') on R'. Let these events be* simultaneous *when viewed from R (so that $ct_\theta = ct_\phi$). Then, the time interval between these events, when viewed from R', is given by*

$$c(t'_\theta - t'_\phi) = \beta\ell', \tag{3.35}$$

where ℓ' is the proper length of the segment $\overline{\mathrm{A'B'}}$.

Solution to Problem 3.2

(a) *Solution using length contraction formula*
In the following, we shall write (A : A′) to mean the event: A passes A′ = A′ passes A.

Consider the following three events $\theta = (\mathrm{A : A'})$, $\phi = (\mathrm{B : B'})$, $\psi = (\mathrm{B : B''})$.

Seen from R, θ and ϕ are simultaneous, so that $ct_\theta = ct_\phi$.

Seen from R', θ and ψ are simultaneous, so that $ct'_\theta = ct'_\psi$.

Seen from R', as the rod R moves to the left, the event ϕ precedes the event ψ.

According to Theorem 3.1, and Fig. 3.7, the proper length of $\overline{\mathrm{B''B'}}$ equals $\ell' - \ell'' = (1 - \frac{1}{\gamma^2})\ell' = \beta^2\ell'$.

The mark B on the rod R, moving to the left with velocity $c\beta$, covers up the proper distance $\overline{\mathrm{B''B'}}$ in time $\Delta t = \frac{\beta^2\ell'}{c\beta} = \frac{\beta\ell'}{c}$, and the event ψ occurs after the event ϕ, so that $t'_\psi > t'_\phi$.

Therefore with respect to the frame R'

$$c(t'_\theta - t'_\phi) = c(t'_\psi - t'_\phi) = \beta l'.$$

(b) *Solution using Lorentz transformation*
We have pictured the standard configuration in Fig. 3.8. The "moving" rod R' is stationary in the frame S', and the rod R is stationary in the frame S. The events θ and ϕ are simultaneous in S. Their coordinates (ct, x) in

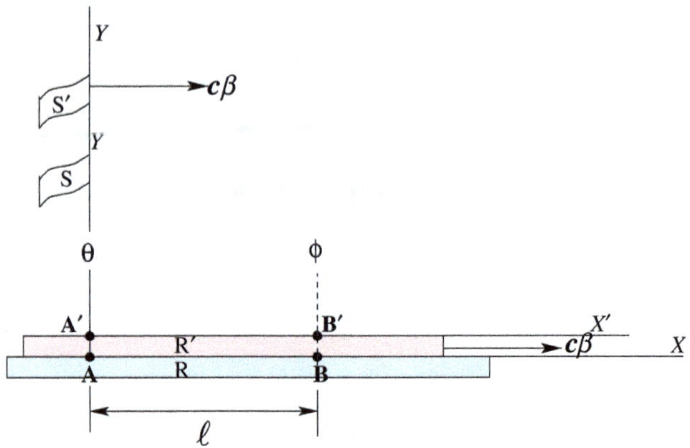

Fig. 3.8. Problem 2.2.

S and S' are written as follows:

$$\theta = (ct_\theta, x_\theta) = (0,0), \quad \phi = (ct_\phi, x_\phi) = (0,\ell) \quad \text{in } S \tag{3.36a}$$

$$\theta = (ct'_\theta, x'_\theta) = (0,0), \quad \phi = (ct'_\phi, x'_\phi) \quad \text{in } S' \tag{3.36b}$$

$$\text{Apply LT:} \quad ct'_\phi = \gamma(ct_\phi - \beta x_\phi) = \gamma(0 - \beta\ell) = -\gamma\beta\ell. \tag{3.36c}$$

$$x'_\phi = \gamma(x_\phi - \beta ct_\phi) = \gamma(\ell - \beta \times 0) = \gamma\ell. \tag{3.36d}$$

Note that $x'_\phi = \ell'$, i.e. the proper length between A' and B'. Therefore,

$$\text{from Eq. (3.36d): } \ell' = \gamma\ell;$$

$$\text{from Eqs. (3.36b) and (3.36c): } c(t'_\theta - t'_\phi) = 0 - (-\gamma\beta\ell) = \gamma\beta\ell = \beta\ell'. \tag{3.37}$$

Problem 3.3. Prove the following theorem.

Theorem 3.3. *Let "θ", "ϕ" and "ψ" be three events, such that "θ" and "ϕ" occur at the same spatial location in S, "θ" and "ψ" at the same spatial location in S'. If the time intervals, as measured in S', are T'_1 between "θ" and "ϕ", and T'_2 between "θ" and "ψ", and if "ϕ" and "ψ" are* simultaneous *in S, then*

$$T'_2 = \frac{T'_1}{\gamma^2}. \tag{3.38}$$

Compare Eq. (3.38) with Eq. (3.34). (See Example 3.5 for illustration of this theorem.)

Solution to Problem 3.3

(a) *Solution using time-dilation formula*

In Fig. 3.9, we have explained the configurations of the frames S and S' at the three events (θ, ϕ, ψ), which we identify with three lightning strokes striking at different times as shown in the figure. We take the timing of the event θ to be $T_0 = T'_0 = 0$, that of the event ϕ to be T_1, T'_1 and of the event

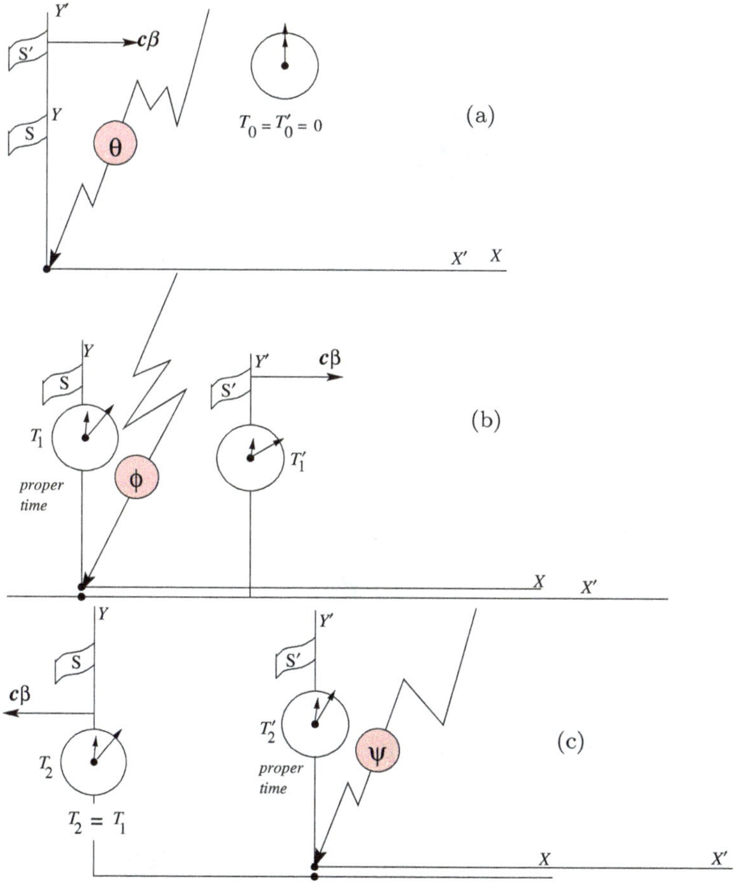

Fig. 3.9. Problem 2.3.

ψ to be T_2, T_2', with respect to S and S', respectively. Then

$$\begin{aligned}
&T_1 = \text{proper time between } \theta \text{ and } \phi, \text{ so that } T_1 = T_1'/\gamma \\
&T_2' = \text{proper time between } \theta \text{ and } \psi, \text{ so that } T_2' = T_2/\gamma \\
\text{Also,}\quad &T_1 = T_2 \text{ by assumption} \qquad (3.39)\\
\text{Hence,}\quad &T_1'/\gamma = \gamma T_2' \\
\text{Or,}\quad &T_2' = T'^1/\gamma^2.
\end{aligned}$$

(b) *Solution using LT*

$$\begin{aligned}
&\theta = (0.0);\ \phi = (cT_1, 0);\ \psi = (cT_2, x_\phi) \quad \text{in } S\\
&\theta = (0.0);\ \phi = (cT'_1, x'_\phi);\ \psi = (cT_2', 0) \quad \text{in } S'\\
\text{Apply LT:}\quad &cT_1' = \gamma(cT_1 - \beta \times 0) = \gamma c T_1 \qquad (3.40)\\
\text{Apply inverse LT:}\quad &cT_2 = \gamma(cT_2' + \beta \times 0) = \gamma c T_2'\\
\text{By assumption:}\quad &T_2 = T_1. \quad \text{Hence, } cT_1' = \gamma c T_2 = \gamma^2 c T_2'.
\end{aligned}$$

Problem 3.4. (a) Rewrite Eq. (3.12) in the form:

$$X' = \hat{\Omega} X, \tag{3.41}$$

where

$$X' = \begin{pmatrix} ct' \\ x' \\ y' \\ z' \end{pmatrix}, \quad X = \begin{pmatrix} ct \\ x \\ y \\ z \end{pmatrix}$$

are column matrices. Let $\tilde{X}' = (ct', x', y', z'), \tilde{X} = (ct, x, y, z)$ be transposes of X' and X, respectively, and let

$$\hat{G} = \begin{pmatrix} 1 & 0 & 0 & 0 \\ 0 & -1 & 0 & 0 \\ 0 & 0 & -1 & 0 \\ 0 & 0 & 0 & -1 \end{pmatrix}. \tag{3.42}$$

(This $\hat{G}$ is identical with the metric tensor to be discussed in Sec. 7.9.3.) Show that Eq. (3.11) can be expressed compactly as follows:

$$\tilde{X}'\hat{G}X' = \tilde{X}\hat{G}X. \tag{3.43}$$

(b) Hence, establish the following important property of any LT matrix:

$$\hat{\Omega}^T \hat{G} \hat{\Omega} = \hat{G}, \tag{3.44}$$

where $\hat{\Omega}^T$ represents the transpose of $\hat{\Omega}$.

Solution to Problem 3.4

(a)

$$\text{l.h.s.} = (ct', x', y', z') \begin{pmatrix} 1 & 0 & 0 & 0 \\ 0 & -1 & 0 & 0 \\ 0 & 0 & -1 & 0 \\ 0 & 0 & 0 & -1 \end{pmatrix} \begin{pmatrix} ct' \\ x' \\ y' \\ z' \end{pmatrix}$$

$$= c^2t'^2 - x'^2 - y'^2 - z'^2.$$

Similarly, the r.h.s $= c^2t^2 - x^2 - y^2 - z^2$.

By assumption l.h.s = r.h.s. Hence the identity.

(b)

$$\tilde{X'} = (ct', x', y', z') = (\hat{\Omega}X)^{\mathrm{T}} = \tilde{X}\hat{\Omega}^T,$$

$$\tilde{X}'\hat{G}X' = (\tilde{X}\hat{\Omega}^T)\hat{G}(\hat{\Omega}X) = \tilde{X}(\hat{\Omega}^T\hat{G}\hat{\Omega})X = \tilde{X}\hat{G}X. \tag{3.45}$$

Hence, $\hat{\Omega}^T\hat{G}\hat{\Omega} = \hat{G}$.

Problem 3.5. Let "θ_1", "θ_2" be simultaneous events in S occurring along the X-axis. Show that they remain simultaneous events in S' under the boost: $S(0, c\beta, 0)S'$, or more generally under the boost: $S(0, c\beta_y, c\beta_z)S'$.

Solution to Problem 3.5

Refer to the Lorentz transformation (3.18) for the general case. Note how time transforms:

$$ct' = \gamma(ct - \boldsymbol{\beta} \cdot \mathbf{r}). \tag{3.46}$$

Suppose the first event occurs at the common origin of the frames S and S' when the time is set to zero in both frames. For the second event $\beta \cdot \mathbf{r} = 0$ according to our assumption. Hence, $ct' = 0$.

Problem 3.6. As shown in Eq. (3.22) the coordinate differences between the events "θ_1", "θ_2" undergo the same LT as the coordinates. We shall rewrite the same equation more clearly as follows.

$$c(t'_2 - t'_1) = \gamma[c(t_2 - t_1) - \beta(x_2 - x_1)], \tag{3.47a}$$

$$(x'_2 - x'_1) = \gamma[(x_2 - x_1) - \beta c(t_2 - t_1)], \tag{3.47b}$$

$$(y'_2 - y'_1) = (y_2 - y_1), \tag{3.47c}$$

$$(z'_2 - z'_1) = (z_2 - z_1). \tag{3.47d}$$

Using the above transformation equations, prove the following statements:

(a) There exist frames A and B having *inverse* time relations for the events "θ_1" and "θ_2" (i.e. if "θ_1" happens *before* "θ_2" in A, then "θ_1" happens *after* "θ_2" in B), if and only if there exists a frame S in which these events are *simultaneous.* [In fact there exists an infinity of such frames. See Example 3.11.]

(b) The temporal sequence of the events "θ_1" and "θ_2" is the same in all frames (i.e. if "θ_1" occurs before "θ_2" in some frame A, then the same must be true in all frames) if and only if there exists a frame S in which these events occur at the same spatial location, in which case the least time interval between the events is the (proper) time measured in S.

Solution to Problem 3.6

(a) Set $c(t_2 - t_1) = 0$ in Eq. (3.50a). Get

$$\begin{gathered} c(t'_2 - t'_1) = -\gamma\beta(x_2 - x_1). \\ \text{Therefore, } t'_2 - t'_1 \lesseqgtr 0 \quad \text{for } \beta \text{ positive/negative.} \end{gathered} \tag{3.48}$$

(b) Set $(x_2 - x_1) = 0$ in Eq. (3.50a). Get

$$\begin{gathered} c(t'_2 - t'_1) = \gamma c(t_2 - t_1). \\ \text{Therefore, } t'_2 - t'_1 > 0 \text{ for any value of } \gamma, \text{ since } \gamma \geq 1. \end{gathered} \tag{3.49}$$

3.7. Illustrative Numerical Examples I

Take $c = 3 \times 10^8$ m/sec for numerical exercises in this chapter. We have used "light-second" (abbreviated as lt-sec) as an alternative unit of length. 1 lt-sec $= 3 \times 10^8$ m. While working out the exercises, the reader may find it convenient to convert "lt-sec" unit to "meter" unit by multiplying with c. For example, 5.2 lt-sec $= 5.2c$ m.

Examples 3.1–3.6 allude to an imaginary superfast (and superlong) space-train called Cosmos Express shown in Fig. 3.10. It is coasting with a uniform speed $\beta = \frac{4}{5}$. Its (proper) length is $\ell = 6 \times 10^8$ m (i.e. 2 lt-sec). Along its straight line route lie two space stations named Andromeda and

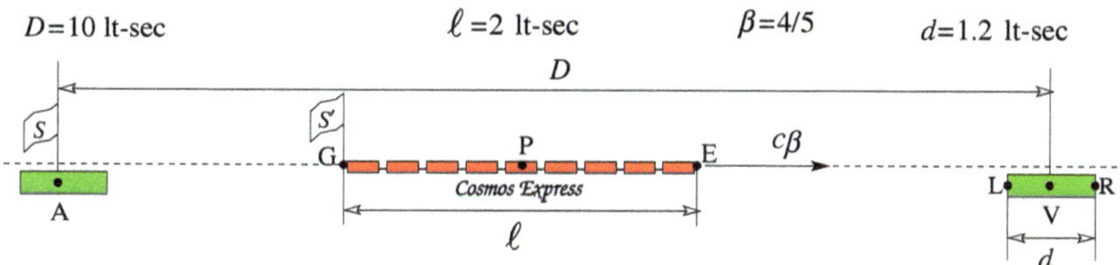

Fig. 3.10. Cosmos Express running between S and V.

Vega (no relative velocity between them) manned by the respective station masters Amar and Vivek, to be referred as Mr A and Mr V. Their offices (shown by the letters A and V) are $D = 3 \times 10^9\,\text{m}$ (i.e. 10 lt-sec) apart. Peter, the passenger (to be referred to as Mr P), is travelling in the centre coach. The intervening terrain between A and V being treacherous (occasional attacks by space pirates), Peter's mother prays for his safe passage, as she sits quietly at A. Clocks in the space station frame S and the train's frame S' are set to zero, i.e. $t = t' = 0$, when "P passes A". At the same instant "mother starts praying".

In the following questions, all primed quantities (e.g. T', D') refer to measurements in S', and the unprimed quantities to those in S. All messages between individuals or stations are radio messages (so that they are transmitted with the speed of light). Answer the questions asked in Examples 3.1–3.5 by applying Rules 1, 2 and 3 (i.e. without applying LT).

Example 3.1. "Mother finishes praying" at $t = T_1$, when "P passes V" (i.e. when Peter has reached Vega), according to clocks in S. At the same instant "she sends her good wishes" to Peter. "Peter receives her message" at $t = T_3$, $t' = T_3'$. Obtain T_1 and T_3.

<u>*Solution.*</u> Let us first identify the events involved, and their coordinates (x, ct) in S and (x', ct') in S'. We use the symbol † to mean that we are not interested in this particular coordinate:

$$\Theta_0 = \text{"P passes A"} \; (x = 0, ct = 0); \; (x' = 0, ct' = 0);$$

$$\Theta_1 = \text{"Mother finishes prayer"} \; (x = 0, ct = cT_1); \; (x' = \dagger, ct' = cT_1');$$

$$\Theta_2 = \text{"P passes V"} \; (x = D, ct = cT_2); \; (x' = 0, ct' = cT_2');$$

$$\Theta_3 = \text{"P receives message"} \; (x = \dagger, ct = cT_3); \; (x' = \dagger, ct' = cT_3').$$

Note that "Mother finishes praying" and "Mother sends message" are the same event Θ_1. In time T_1, the train has moved a distance D with speed βc. Also, the events Θ_1 and Θ_2 are simultaneous events in S. Hence,

$$T_1 = T_2 = \frac{D}{c\beta} = \frac{10c\,\text{m}}{\frac{4}{5}c\,\text{m/s}} = 12.5\,\text{s}.$$

The message is travelling at speed c between Θ_3 and Θ_1. It is sent when P has already moved a distance D. Therefore, the message travels an extra distance D over the train which is travelling at speed βc. Hence,

$$c(T_3 - T_1) = D + \beta c(T_3 - T_1), \Rightarrow T_3 - T_1 = \frac{D}{(1-\beta)c} = \frac{10c\,\text{m}}{\frac{1}{5}c\,\text{m/s}} = 50\,\text{s}.$$

Hence, $T_3 = T_1 + 50 = 62.5\,\text{s}$.

Note: None of the rules of relativity has been used in finding the above three answers.

Example 3.2. Determine T_3', using Rule 2.

<u>*Solution.*</u> Note that T_3' is the proper time between the events Θ_3 and Θ_0 (because Mr P is present at both events). We shall use Eq. (2.23) to relate T_3' and the "improper time" T_3 between the same events which has been already found out in the last question. For this, we need the Lorentz factor γ:

$$\gamma = \frac{1}{\sqrt{1-\beta^2}} = \frac{1}{\sqrt{1-\frac{16}{25}}} = \frac{5}{3} \Rightarrow T_3' = \frac{T_3}{\gamma} = \frac{62.5}{\frac{5}{3}} = 37.5\,\text{s}.$$

Example 3.3. From the time T_3', Peter determines the time $t' = T_1'$, when his "mother finished praying and sent the message". Find out T_1'. [Hint: Peter does not know relativity, but knows that his mother has been receding from him at the speed $\beta = \frac{4}{5}$.]

<u>*Solution.*</u> In the time T_1', which is the time interval between Θ_0 and Θ_1 as measured in S', the point A has moved with velocity $-\beta c$ (i.e. in the negative X'-direction) a certain distance a, i.e. from $x' = 0$ to $x' = -a$. Therefore,

$$a = \beta c T_1' = \frac{4}{5} c T_1'.$$

In the subsequent time interval $T_3' - T_1'$ (measured in S'), P remains stationary at $x' = 0$ in his rest frame S', while the message travels from $x' = -a$

to $x' = 0$ with speed c. Therefore,

$$c(T_3' - T_1') = a = \beta c T_1'. \Rightarrow T_1' = \frac{T_3'}{1+\beta} = \frac{37.5}{\frac{9}{5}} = 20.83\,\text{s}.$$

Example 3.4. Find the time $t' = T_2'$ when "P passes V". [Hint: The line segment $\widehat{AV}$ has a contracted length.]

<u>*Solution.*</u> At time $t' = T_2'$ the point A has moved in the negative X' direction at a speed of βc to cover a distance D' which is the contracted length of the proper length D. Therefore,

$$\beta c T_2' = D' = \frac{D}{\gamma} \Rightarrow T_2' = \frac{D}{\beta c \gamma} = \frac{10c\,\text{m}}{\frac{4}{5}c \times \frac{5}{3}\,\text{m/s}} = 7.5\,\text{s}.$$

Example 3.5. Summarize your answers obtained so far in the following tabular form:

Duration of mother's prayer

$$\begin{cases} \text{according to } S \text{ frame, } T_1 = \ldots \\ \text{according to } S' \text{ frame, } T_1' = \ldots \end{cases}$$

Duration of Peter's journey from S to V

$$\begin{cases} \text{according to } S \text{ frame, } T_2 = \ldots \\ \text{according to } S' \text{ frame, } T_2' = \ldots \end{cases}$$

Verify that Rule 2 is illustrated in both sets of answers, namely $T_1 = \frac{1}{\gamma}T_1'; T_2' = \frac{1}{\gamma}T_2$.

<u>*Solution.*</u> The proper and "improper" times between Θ_0 and Θ_1 are T_1 and T_1', respectively. So we expect that $\frac{T_1}{T_1'} = \frac{1}{\gamma}$. Similarly, the proper and "improper" times between Θ_0 and Θ_2 are T_2' and T_2, respectively. So we expect that $\frac{T_2'}{T_2} = \frac{1}{\gamma}$.

$$T_1 = 12.5\,\text{s}; \quad T_1' = 20.83\,\text{s}; \quad \frac{T_1}{T_1'} = \frac{3}{5} = \frac{1}{\gamma}.$$

$$T_2 = 12.5\,\text{s}; \quad T_2' = 7.5\,\text{s}; \quad \frac{T_2'}{T_2} = \frac{3}{5} = \frac{1}{\gamma}.$$

Hence, our expectations are confirmed.

Example 3.6. If the length of the space station V is 1.2 lt-sec, how long does the length of Cosmos Express take to traverse the length of V (a) according to S? (b) according to S'? Answer by applying Rule 3.

Solution. (a) Let t be the required time in S. In this time, Cosmos Express travelling with speed βc moves a distance equal to the length of the platform which is the proper length d plus the length of the train which is the "improper" length ℓ'. Hence

$$\ell' = \frac{\ell}{\gamma} = \frac{2c\,\mathrm{m}}{\frac{5}{3}} = 1.2c\,\mathrm{m}, \tag{3.50a}$$

$$\beta ct = \ell' + d = 1.2c + 1.2c = 2.4c\,\mathrm{m}, \tag{3.50b}$$

$$t = \frac{\ell' + d}{\beta c} = \frac{2.4}{\beta} = \frac{2.4}{\frac{4}{5}} = 3\,\mathrm{s}. \tag{3.50c}$$

(b) Let t' be the require time in S. In this time, platform V travelling with velocity $-\beta c$ moves a distance equal to the length of the train which is the proper length ℓ plus the length of the platform which is the "improper" length d'. Hence

$$d' = \frac{d}{\gamma} = \frac{1.2c\,\mathrm{m}}{\frac{5}{3}} = 0.72\,c\,\mathrm{m}, \tag{3.51a}$$

$$\beta ct' = \ell + d' = 2c + 0.72c = 2.72c\,\mathrm{m}, \tag{3.51b}$$

$$t' = \frac{\ell + d'}{\beta c} = \frac{2.72}{\beta} = \frac{2.72}{\frac{4}{5}} = 3.4\,\mathrm{s}. \tag{3.51c}$$

Example 3.7. Obtain the answer in (b) of the last question by applying LT.

Solution. We have shown the two boundary points of the station V, the left boundary L and the right boundary R. The office V is located at midpoint between L and R. In order to apply LT, we have to identify the two events involved, namely, Θ_L="E passes L" and Θ_R = "G passes R". The space and time intervals between these two events, as measured in S are: $\Delta x = d$ and $c\Delta t = \frac{\ell + d'}{\beta}$ which is same as the time t obtained in part (a) of the last problem. We shall obtain the values of these coordinates in S' applying the transformation of coordinate differences as given in Eq. (3.22)

$$c\Delta t' = \gamma(c\Delta t - \beta\Delta x) = \gamma\left[\frac{\ell' + d}{\beta} - \beta d\right] = \frac{\gamma\ell'}{\beta} + \gamma\left(\frac{1}{\beta} - \beta\right)d$$

$$= \frac{1}{\beta}\left[\gamma\ell' + \frac{d}{\gamma}\right] = \frac{\ell + d'}{\beta}.$$

We have used Eqs. (2.22), (3.50a), (3.51a). We get the same answer as in (3.51c).

Example 3.8. Muons (denoted by the symbol μ^-) are charged particles like electrons, but about 207 times heavier. They are present abundantly in the earth's upper atmosphere, being caused due to bombardment of nuclei by cosmic rays. A muon decays spontaneously into an electron and a pair of neutrinos, the decay mean-life being $\tau_0 = 2.3 \times 10^{-6}$ sec in muon's rest frame. However, they live longer as they travel through the atmosphere with relativistic speed very close to c. Their prolonged life is one confirmation of the time dilation formula.

A balloon is sent up to a height of 2 km in the atmosphere to measure the flux density of muons of average speed $0.995c$. If a detector carried on the balloon counts 10 muons per minute, how many muons per minute, on the average, will the same apparatus count at the sea level? Assume that the muons are travelling vertically downward and that no muon is stopped by the atmosphere. [Hint: The number of muons fall off exponentially as $e^{-\frac{t}{t_0}}$, where t_0 is the mean-life and t is the time of flight of the muons, both measured in the experimenter's frame.]

<u>*Solution.*</u> Let N_0 be the number of muons to be detected by the balloon at the reference point A which is located 2 km above sea level. Let N_s be the same number at the sea level. The number of muons decreases with travel time t from the point A downward, and is given as $N(t) = N_0 e^{-\frac{t}{t_0}}$, where t_0 is the mean life in *Earth frame.* The mean life τ_0 given in the problem is the "proper" lifetime of an average muon between the events "birth" and "death". The two mean lives t_0 and τ_0 are related by the time dilation formula.

$$\gamma = \frac{1}{\sqrt{1-(0.995)^2}} = 10,$$

$$t_0 = \gamma\tau_0 = 10 \times 2.3 \times 10^{-6} = 2.3 \times 10^{-5}\,\text{s},$$

$$t = \text{travel time of } \mu^- = \frac{2 \times 10^3}{0.995 \times 3 \times 10^8} = 0.67 \times 10^{-5}\,\text{s},$$

$$N_s = N_0 e^{-\frac{t}{t_0}} = 10 \times e^{-\frac{0.67}{2.3}} = 7.47 \text{ per minute.}$$

Example 3.9. Two events "A" and "B", when viewed from S, are separated by a distance $d = 2$ lt-sec and a time interval t as given below:

(a) $t = 6\,\text{sec}$;
(b) $t = 2\,\text{sec}$;
(c) $t = 1\,\text{sec}$.

(i) Is there a frame of reference S', for each one of the above three cases, in which the events occur simultaneously? If so, determine the boost velocity $c\beta$ of S' with respect to S.
(ii) Is there a frame of reference S', for each one of the above three cases, in which the events occur at the same spatial location? If so, determine the boost velocity $c\beta$ of S' with respect to S and the proper time between the events.

Solution. For this and subsequent solution to problems in this chapter, we shall write $\Delta\overrightarrow{\mathbf{r}}$ to mean coordinate difference between two points B and A. Actually, this quantity is a *4-vector*, stretching from the event A to the event B. The meaning and significance will be explained in Sec. 7.8.

(a) Let the coordinate difference between the two events be $\Delta\overrightarrow{\mathbf{r}} = \mathrm{B} - \mathrm{A}$. Then

$$\begin{aligned} & \quad (ct, x) \\ \Delta\overrightarrow{\mathbf{r}} &= (6c, 2c) \\ \Delta s^2 &= (6c)^2 - (2c)^2 = (4\sqrt{2}\,c)^2 \\ \Delta\tau &= \text{proper time} = 4\sqrt{2}\,\text{sec}. \end{aligned}$$

The interval is *time-like*. Hence, we have the following conditions:

(i) The events *cannot occur simultaneously* in any frame.
(ii) Yes, there exists a frame S' in which the events occur at the *same spatial location*. Let βc be the boost velocity. Then,

$$\begin{aligned} \Delta x' &= 0 = \gamma(\Delta x - \beta\Delta\, ct) \\ \Rightarrow \beta &= \frac{\Delta x}{\Delta\, ct} = \frac{2c}{6c} = \frac{1}{3}; \\ \gamma &= \frac{1}{\sqrt{1 - \dfrac{1}{9}}} = \sqrt{\frac{9}{8}} = \frac{3}{2\sqrt{2}}. \end{aligned}$$

There is a *proper time* $\Delta\tau$ between the events. Let us find it.

$$c\,\Delta\tau = \gamma(\Delta\, ct - \beta\,\Delta x) = \frac{3}{2\sqrt{2}}\left(6 - \frac{1}{3} \times 2\right) c = 4\sqrt{2}\,c \text{ lt-sec.}$$

Hence, $\Delta\tau = 4\sqrt{2}\,\text{sec}$.

$\dfrac{\Delta\tau}{\Delta t} = \dfrac{4\sqrt{2}}{6} = \dfrac{2\sqrt{2}}{3} = \dfrac{1}{\gamma}$, consistent with the time-dilation formula (2.23).

(b) $\Delta\vec{\mathbf{r}} = (2c, 2c)$; $\Delta s^2 = 0$. The interval is *light-like.*

(i) No. (ii) No.

(c) $\Delta\vec{\mathbf{r}} = (1c, 2c)$; $\Delta s^2 = -3c^2$. The interval is *space-like.*

(i) Yes, there exists a frame S' in which the events occur *simultaneously.* Let us find the boost velocity for the frame S'.

$$c\,\Delta t' = 0 = \gamma(\Delta\, ct - \beta\,\Delta x)$$
$$\Rightarrow \beta = \frac{\Delta\, ct}{\Delta x} = \frac{c}{2c} = \frac{1}{2};$$
$$\gamma = \frac{1}{\sqrt{1-\frac{1}{4}}} = \frac{2}{\sqrt{3}}.$$
$$\Delta x' = \frac{2}{\sqrt{3}}(2 - \tfrac{1}{2} \times 1)c = \sqrt{3}\,c = \text{distance between A and B in } S'.$$

This is consistent with the length contraction formula (2.26):

$$\frac{\Delta x}{\Delta x'} = \frac{2}{\sqrt{3}} = \gamma.$$

(ii) No.

Example 3.10. Two lightnings strike, leaving marks A and B on a straight railway track *simultaneously* according to the Station Master (Mr. S). A traveller (Mr T) travelling in a long Einsteinian railway train moving with velocity $0.6c$ records the timings of the events using his instruments. Let the distance between A and B be $L = 2.4 \times 10^3$ km, as measured by Mr S.

(a) Did the events occur simultaneously according to the measurements of Mr T? If not, which occurred earlier?
(b) What is the time interval between these events according to Mr T?
(c) What is the length of the track segment $\overline{\mathrm{AB}}$ according to Mr T?
(d) The lightnings also left marks on the wheels A′, B′ of the train. What is the length L' of the segment $\overline{\mathrm{A'B'}}$ (i) according to Mr T? (ii) according to Mr S?

<u>*Solution.*</u>

(a) No.
(b) $L = 2.4 \times 10^3\,\text{km} = 0.8 \times 10^{-2}$ lt-sec. $= \frac{4}{5}\,c \times 10^{-2}\,\text{m}$.
$\beta = 0.6 = 3/5$; $\gamma = \frac{1}{\sqrt{1-\frac{9}{25}}} = 5/4$.

$$\text{In S:}\begin{cases} & (ct, x) \\ A = & (0,0) \\ B = & \left(0, \frac{4}{5}c \times 10^{-2}\right) \end{cases}$$

$$\text{In T:}\begin{cases} ct'_A = 0 \\ ct'_B = \frac{5}{4}\left(0 - \frac{3}{5} \times \frac{4}{5}c \times 10^{-2}\right) = -\frac{3}{5}c \times 10^{-2} \\ t'_B - t'_A = -\frac{3}{5} \times 10^{-2}\,\text{sec.} \end{cases} \tag{3.52}$$

The time interval $t'_B - t'_A$ is *negative*. Therefore, B occurs *before* A.

(c) The ground segment $\overline{\text{AB}}$ is *at rest* with respect to Mr S. He measures the *proper* length L for this segment. In contrast, Mr T sees the ground segment $\overline{\text{AB}}$ to be *moving* to the left. Therefore, he measures the *improper* length L' for the same segment.

By Eq. (2.26): $L = \gamma L'$. Hence $L' = L/\gamma = 2.4 \times 10^3 \times \frac{4}{5} = 1.92 \times 10^3$ km.

(d) The train segment $\overline{\text{A}'\text{B}'}$ is *at rest* with respect to Mr. T. He measures the *proper* length ℓ for this segment. In contrast, Mr. S sees the train segment $\overline{\text{A}'\text{B}'}$ to be *moving* to the right. Therefore he measures the *improper* length L for the same segment.

By Eq. (2.26): $\ell = \gamma L = \frac{5}{4} \times 2.4 \times 10^3 \times \frac{4}{5} = 3 \times 10^3$ km.

Note that $\frac{\ell}{L'} = \gamma^2$, as per Theorem 3.1.

Example 3.11. An event "A" $= (0,0,0,0)$ is followed by a second event "B" $= (1,2,0,0)$, 1 sec later, the coordinates being measured with respect to a frame S in lt-sec. Is there a frame of reference S' in which

(a) "A" and "B" are simultaneous?
(b) the event "B" is followed by the event "A", 1 sec later, i.e. $t'_A - t'_B = 1$?
(c) the events "A" and "B" occur at the same spatial location?

Find the boost velocity $c\beta$ of S' relative to S in each of the above cases (if such a frame S' exists).

<u>*Solution.*</u> $\Delta\vec{\mathbf{r}} = (1,2)$ in S. Hence, $\Delta s^2 = 1 - 4 = -3$. The interval is *space-like*.

(a) Yes.

In the second frame $\Delta t' = 0$. Hence, $0 = \gamma(1 - \beta \times 2) \Rightarrow \beta = \frac{1}{2}$.

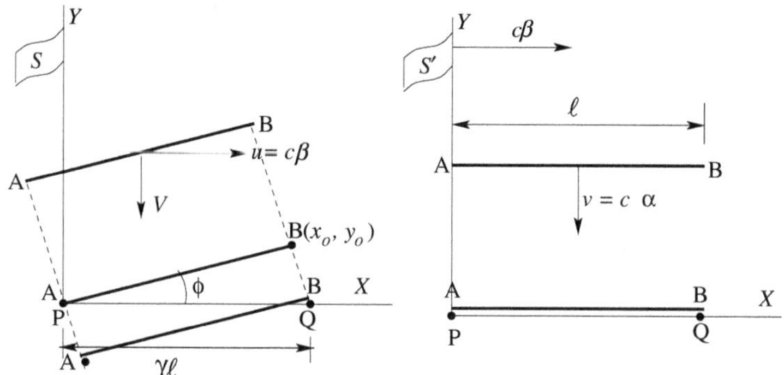

Fig. 3.11. Rod falling vertically.

(b) Yes. To find this frame, note that $c\Delta t' = -1$. Hence,

$$-1 = \gamma(1 - \beta \times 2)$$

$$\text{Or,} \quad -\frac{1}{\gamma} = 1 - 2\beta$$

$$\text{Squaring both sides:} \quad 1 - \beta^2 = 1 + 4\beta^2 - 4\beta$$

$$\text{Or,} \quad 5\beta^2 - 4\beta = 0$$

$$\text{Or,} \quad \beta = 0, \frac{4}{5}$$

$$\beta = 0, \text{ gives } \Delta t = 1. \text{ Same order}$$

$$\beta = \frac{4}{5} \text{ gives } \Delta t = -1. \text{ Reverse order.}$$

(b) No, because the interval is space-like.

Example 3.12. Consider a straight horizontal rod moving vertically downward with uniform speed V as seen from frame S' (see Fig. 3.11). Its ends A and B will therefore strike the horizontal floor (represented by the X-axis) *simultaneously* according to S'. However, when seen from S, the ends strike the floor at different times, and, therefore, the bar is not horizontal. Find the angle that the bar makes with the horizontal (measure positive angle anticlockwise from the X-axis.)

<u>*Solution.*</u> To find the angle ϕ, we need to know the coordinates of the ends A and B at the same instant of time (i.e. simultaneously) in S.

We know that the two events θ_{AP} = "A touches P" and θ_{BQ} = "B touches Q" are simultaneous in S'. Their coordinates in S' and S are as follows:

$$\theta_{AP} = \begin{cases} (ct, x), \\ (0,0) \text{ in } S', \\ (0.0) \text{ in } S, \end{cases} \quad \theta_{BQ} = \begin{cases} (ct, x), \\ (0,\ell) \text{ in } S', \\ (ct_{BQ}, x_{BQ}) \text{ in } S. \end{cases}$$

Using inverse Lorentz transformation equation (3.9), we get the values of (ct_{BQ}, x_{BQ}) in S:

$$ct_{BQ} = \gamma(0 + \beta\ell) = \gamma\beta\ell,$$

$$x_{BQ} = \gamma(\ell + 0) = \gamma\ell.$$

Let us look at the rod from the frame S. At $ct = 0$, the left end A of the rod is at the origin, i.e. at the *spatial coordinates* $(0,0)$. Let the right end B of the rod at the same instant be at the *spatial coordinates* (x_0, y_0). At ct_{BQ}, the right end of the rod is touching the floor. Therefore, in time $\{ct_{BQ} = \gamma\beta\ell\}$, the end B has undergone a spatial displacement from (x_0, y_0) to $(\gamma\ell, 0)$.

Seen from S, all points of the rod have a velocity v. The horizontal and the vertical components of this velocity can be found from the velocity addition formula (4.23): $v_x = c\beta; v_y = \frac{-V}{\gamma} = -c\frac{\nu}{\gamma}$. We have set ν as the dimensionless velocity, like β. $V = c\nu$.

Therefore,

$$x_0 + \left(\frac{\gamma\beta\ell}{c}\right)(\beta c) = \gamma\ell \Rightarrow x_0 = \gamma\ell(1 - \beta^2) = \frac{\ell}{\gamma},$$

$$y_0 + \left(\frac{\gamma\beta\ell}{c}\right)\left(-c\frac{\nu}{\gamma}\right) = 0 \Rightarrow y_0 = \beta\ell\nu.$$

Therefore, $\tan\phi = \dfrac{y_0}{x_0} = \beta\gamma\nu.$

Example 3.13. Distant galaxies are known to be receding from ours at very large speeds, as evidenced by the shift in the frequencies of the spectral lines from atoms towards lower values, a phenomenon known as *red shift.* A distant galaxy is recognized to be receding at a speed of $0.5c$ along the line of sight, from the measurement of the red shift of the sodium D_2 line of wavelength 5890 Å. Find the measured value of the wavelength.

Solution. Use the Doppler shift formula (3.31). Note that $\frac{\lambda}{\lambda_o} = \frac{\nu_o}{\nu}$, and $\beta = \frac{1}{2}; \gamma = \frac{2}{\sqrt{3}}; \theta = 0°$. Therefore,

$$\lambda = \gamma(1 + \beta\cos 0^o)\lambda_o = \sqrt{3} \times 5890 = 10,201 \text{ Å}.$$

Chapter 4

Relativistic Mechanics

The previous chapter elucidated the relativity principle by first enunciating the basic postulates and then showing their immediate consequences. One of these consequences, namely the Lorentz transformation, has played a crucial role in reshaping classical mechanics by redefining the fundamental quantities, like energy and momentum, and then modifying Newton's equations of motion. We propose to sketch in this chapter some of these revolutionary developments and structure relativistic mechanics as an outgrowth of Lorentz transformation. Central to this development is the velocity addition formula, which forms the topic of the next section.

We shall adopt the following convention. The boost velocity, i.e. the velocity of S' with respect to S will be denoted by $\mathbf{u} = c\boldsymbol{\beta}$ and the velocity of a moving particle will be denoted by $\mathbf{v} = c\boldsymbol{\nu}$ in S and $\mathbf{v}' = c\boldsymbol{\nu}'$ in S'.

4.1. Relativistic Form of Velocity Transformation

The relativistic velocity transformation formula, generally known by its popular name *Velocity Addition Formula*, is a relativistic generalization of the Galilean velocity transformation formula shown in Eq. (2.8c). We shall first obtain the simplest special case by considering the boost:$S(c\beta, 0, 0)S'$, corresponding to the *standard configuration* shown in Fig. 3.1. In Fig. 4.1, we have shown a moving particle, and its radius vectors $\mathbf{r}$ and $\mathbf{r}'$ and its velocities $\mathbf{v}$ and $\mathbf{v}'$, both with respect to the frames S and S'.

Let us consider a particle moving along a certain trajectory which appears as Σ to S and Σ' to S' (Figs. 4.1(b) and 4.1(c)). Consider two points A and B on the trajectory of the particle and lying infinitely close to each other. The radius vectors of these points are $\mathbf{r}$ and $\mathbf{r} + \mathbf{dr}$ in S and

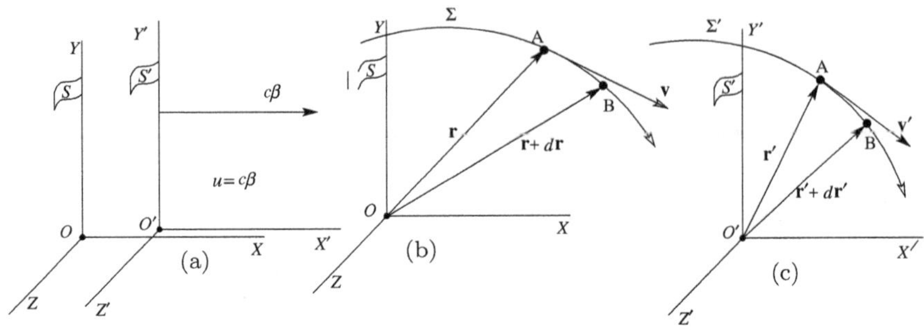

Fig. 4.1. Trajectory of a particle in the frames S and S'.

$\mathbf{r}'$ and $\mathbf{r}' + \mathbf{dr}'$ in S'. These points are reached at times t and $t + dt$ in S and t' and $t' + dt'$ in S'. We can, therefore, say that "the particle arrives at A" and "the particle arrives at B" are two events, which can be designated by the symbols "Θ_A" and "Θ_B", respectively, and whose coordinates are as follows:

$$\text{“}\Theta_A\text{”} = \begin{cases} (ct, \mathbf{r}) = (ct, x, y, z) & \text{in } S, \\ (ct', \mathbf{r}') = (ct', x', y', z') & \text{in } S', \end{cases} \tag{4.1}$$

$$\text{“}\Theta_B\text{”} = \begin{cases} (ct + c\,dt, \mathbf{r} + \mathbf{dr}) & \\ \quad = (ct + c\,dt, x + dx, y + dy, z + dz) & \text{in } S, \\ (ct' + c\,dt', \mathbf{r}' + \mathbf{dr}') & \\ \quad = (ct' + c\,dt', x' + dx', y' + dy', z' + dz') & \text{in } S'. \end{cases} \tag{4.2}$$

The coordinate differentials transform according to Lorentz transformation, as pointed out in Eq. (3.22). Therefore,

$$c\,dt' = \gamma(c\,dt - \beta\,dx), \tag{4.3a}$$

$$dx' = \gamma(dx - \beta c\,dt), \tag{4.3b}$$

$$dy' = dy, \tag{4.3c}$$

$$dz' = dz. \tag{4.3d}$$

Let the velocity of the particle at the event "Θ" be $\mathbf{v} = \frac{d\mathbf{r}}{dt}$ as it appears to S, and $\mathbf{v}' = \frac{d\mathbf{r}'}{dt'}$ as it appears to S'. The Cartesian components of these vectors are as follows:

$$\mathbf{v} = (v_x, v_y, v_z) = \left(\frac{dx}{dt}, \frac{dy}{dt}, \frac{dz}{dt}\right),$$
$$\mathbf{v}' = (v'_x, v'_y, v'_z) = \left(\frac{dx'}{dt'}, \frac{dy'}{dt'}, \frac{dz'}{dt'}\right). \tag{4.4}$$

Note that, in order to determine the velocity components in a given frame of reference, we have used the length and time intervals pertaining to that frame of reference only. We can now utilize Eq. (4.3) to connect the two sets of velocity components as follows:

$$v'_x = \frac{dx'}{dt'} = \frac{c\gamma(dx - \beta c\, dt)}{\gamma(c\, dt - \beta\, dx)}$$
$$= \frac{c\left\{\dfrac{dx}{dt} - \beta c\right\}}{c - \dfrac{\beta dx}{dt}} = \frac{v_x - u}{1 - \frac{v_x u}{c^2}}, \tag{4.5}$$

where $u \equiv c\beta$. In the same manner, we can obtain the transformation equations for the other components. Writing them together, we get the velocity transformation for this special case, namely, boost: $S(v,0,0)S'$ as

$$\boxed{v'_x = \frac{v_x - u}{1 - \frac{v_x u}{c^2}}, \quad v'_y = \frac{v_y}{\gamma\left(1 - \frac{v_x u}{c^2}\right)}, \quad v'_z = \frac{v_z}{\gamma\left(1 - \frac{v_x u}{c^2}\right)}.} \tag{4.6}$$

Alternatively, if we had started from the inverse transformation equations (3.9), we would have obtained the following result, which is the inverse of Eq. (4.6):

$$\boxed{v_x = \frac{v'_x + u}{1 + \frac{v'_x u}{c^2}}, \quad v_y = \frac{v'_y}{\gamma\left(1 + \frac{v'_x u}{c^2}\right)}, \quad v_z = \frac{v'_z}{\gamma\left(1 + \frac{v'_x u}{c^2}\right)}.} \tag{4.7}$$

Equation (4.7) effectively shows the result of adding velocity $\mathbf{u}$ to $\mathbf{v}$, whereas Eq. (4.6) tells us the result of subtracting $\mathbf{u}$ from $\mathbf{v}$. These results are distinctively different from their Galilean counterparts. These equations are often referred to as the *velocity addition formulas* (subtraction of $\mathbf{u}$ is same as addition of $-\mathbf{u}$).

We have now to verify that the transformation equations (4.6) and (4.7) satisfy two essential requirements of all relativistic results, namely that (i) they must reduce to the corresponding non-relativistic result in the limit of small velocities; and that (ii) if a particle is moving with speed c in the frame S', then it should move with the same speed c in the second frame S, and vice versa (so that light propagates with the same speed c in both frames). With regard to requirement (i), let us record here the *non-relativistic limit* criterion

$$\text{When } v \ll c, \quad \beta \to 0, \quad \gamma \to 1 \quad \text{(N.R. limit)}. \tag{4.8}$$

Applying the criterion (4.8) to Eq. (4.6), so that $\frac{v_x \beta}{c} \to 0$, one gets

$$v'_x = v_x - u, \quad v'_y = v_y, \quad v'_z = v_z. \tag{4.9}$$

This is the Galilean result (2.5).

To meet the requirement (ii), we set $\mathbf{v}' = (c, 0, 0)$ in Eq. (4.6) and get

$$v_x = \frac{c+u}{1+\frac{cu}{c^2}} = c, \quad v_y = \frac{0}{1+\frac{cu}{c^2}} = 0, \quad v_z = \frac{0}{1+\frac{cu}{c^2}} = 0, \tag{4.10}$$

so that $\mathbf{v} = (c, 0, 0)$.

We shall now illustrate the velocity addition formula by showing that the speed of light c is the ultimate speed, i.e. no material particle can be accelerated beyond the speed of light. Suppose a particle is moving along the X-axis with speed $0.99c$ relative to the laboratory frame S. Imagine a second frame S' which is also moving along the X-axis with velocity $0.99c$ relative to the Lab frame so that the particle is at rest in S'. The particle is now accelerated (from the present zero velocity in S') to velocity $0.99c$ along the X'-axis in this second frame S'. According to the velocity addition formula (4.7), therefore, the resulting velocity of the particle in the laboratory frame S would be

$$v = \frac{0.99c + 0.99c}{1 + 0.99 \times 0.99} = 0.9999494c, \quad \text{along the } X\text{-axis.}$$

Therefore, our velocity addition formula effectively tells us that if we add velocity $0.99c$ to $0.99c$, we get $0.9999494c$, instead of $1.98c$. However hard we may try to increase the speed of a particle, which has nearly reached the speed of light, it will still be never able to reach, or surpass, the speed of light.

We can make the argument more convincing by analyzing the motion of a particle which is under a "constant acceleration" a in its instantaneous rest frame (to be abbreviated as *IRF*) along the X-axis. The term *instantaneous rest frame* — also called the *comoving Lorentz frame* — means an *inertial frame* in which the particle is, at least momentarily, at rest. We shall explain the concept of IRF with the diagram shown in Fig. 4.2.

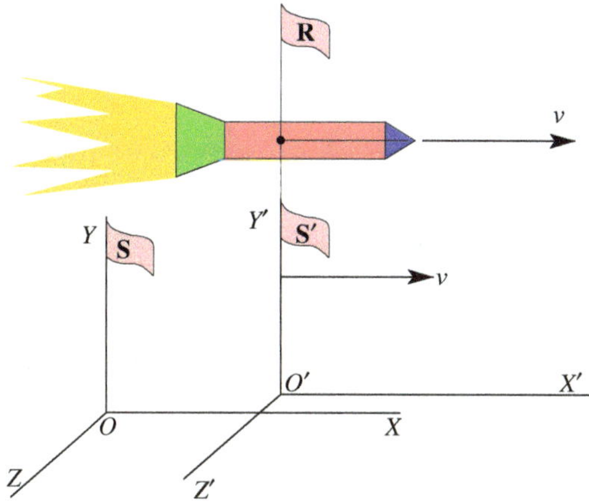

Fig. 4.2. Explaining Instantaneous Rest Frame, shown as S'.

Imagine an accelerating particle, e.g. a rocket. Its own frame of reference (i.e. the frame in which it is always at rest) is an *accelerating* frame. At a certain event, say "Θ: rocket passes a space station A", the rocket releases an iron cage with $X'Y'Z'$-axes welded on it. We label this frame as S'. This frame S' is not in acceleration. Hence it is an IF whose velocity is the same as that of R at the event "Θ". Then S' is an IRF of R at the event "Θ". We can establish a Lorentz connection between S and S' through the boost $\{S(\beta, 0, 0)S'\}$, where $\beta = v/c$. The following paragraphs will amplify the concept of IRF.

In Fig. 4.3(a), we have shown the rectilinear trajectory of a particle accelerating along the X-axis, as seen from the laboratory frame S. The particle passes space stations A and B with speeds v and $v + dv$ at times t and $t + dt$. Call these events "Θ_A" and "Θ_B", respectively.

Figure 4.3(b) shows the IRF S' of the particle at the event "Θ_A". In this frame, the events "Θ_A" and "Θ_B" take place at the times t' and $t'+dt'$, the particle's speed at these events being 0 and dv', respectively.

$$dv' \equiv a\,dt'. \tag{4.11}$$

According to the velocity addition formula (4.7), therefore,

$$v + dv = \frac{dv' + v}{1 + \frac{v\,dv'}{c^2}} \simeq (dv' + v)\left(1 - \frac{v\,dv'}{c^2}\right) \simeq v + \left(1 - \frac{v^2}{c^2}\right) dv',$$

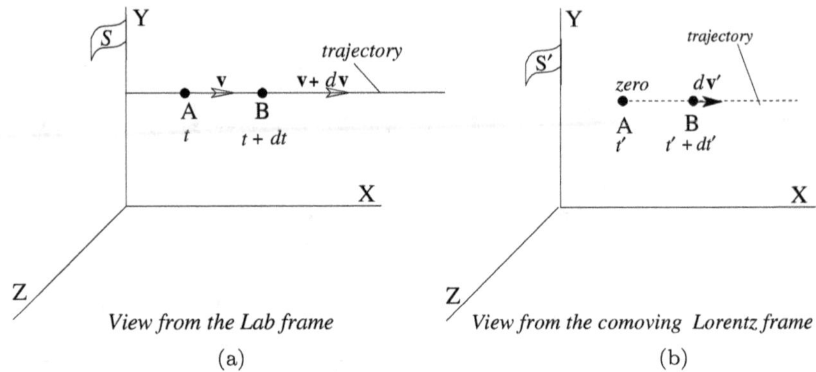

Fig. 4.3. Lab frame and comoving Lorentz frame.

so that

$$dv = \left(1 - \frac{v^2}{c^2}\right) a\,dt'. \tag{4.12}$$

Applying the LT to the coordinate differentials (using the *inverse* of Eqs. (4.3)), we get

$$c\,dt = \frac{1}{\sqrt{1 - \frac{v^2}{c^2}}}\left(c\,dt' + \frac{v}{c}\,dx'\right). \tag{4.13}$$

Now, the particle, which was momentarily at rest at t', has covered the distance dx' under a constant acceleration a in the subsequent time interval dt'. Therefore, $dx' = \frac{1}{2}a(dt')^2 \simeq 0$, so that, from (4.13),

$$dt = \frac{1}{\sqrt{1 - \frac{v^2}{c^2}}}\,dt'. \tag{4.14}$$

We shall introduce a new Lorentz factor, namely one associated with the velocity of a particle, and call it *dynamic Lorentz factor*, to be symbolized by capital gamma:

$$\boxed{\Gamma \equiv \frac{1}{\sqrt{1 - \frac{v^2}{c^2}}}.} \tag{4.15}$$

Note that the time interval dt' is measured in the IRF. Hence it is the same as the proper time $d\tau$ between the events "Θ_A" and "Θ_B". By virtue of (4.15), Eq. (4.14) can be written as

$$\boxed{dt = \Gamma dt' = \Gamma\, d\tau,} \tag{4.16}$$

which is the same as the relation between proper time and improper time given in Eq. (2.23). Equation (4.16) will find many applications in the discussions to follow.

Compare the upper case Γ of (4.15) with the lower case gamma γ defined in Eq. (2.20) to represent the Lorentz factor associated with the *boost velocity*, which we had called *boost Lorentz factor*.

We can now rewrite Eq. (4.12) as

$$dv = \left(1 - \frac{v^2}{c^2}\right) a\, d\tau \tag{4.17a}$$

$$= \left(1 - \frac{v^2}{c^2}\right)^{\frac{3}{2}} a\, dt = \frac{a}{\Gamma^3}\, dt. \tag{4.17b}$$

It is seen from Eq. (4.17b) that the acceleration of the particle, as seen from the laboratory frame, is constantly reducing, and is given as follows:

$$a_{\text{Lab}} = \frac{dv}{dt} = \left(1 - \frac{v^2}{c^2}\right)^{\frac{3}{2}} a = \frac{a}{\Gamma^3}. \tag{4.18}$$

Note that as the particle reaches relativistic speed, finally approaching c, $\Gamma \to \infty$, and $a_{\text{Lab}} \to 0$. Even at a constant acceleration in its IRF, the particle does not accelerate at all in the Lab frame, as it approaches the speed of light. The formula (4.18) therefore safeguards one of the principles of relativity, that is $v \lessapprox c$.

If we rewrite Eq. (4.17) as

$$\frac{dv}{\left(1 - \frac{v^2}{c^2}\right)^{\frac{3}{2}}} = a\, dt \tag{4.19}$$

and integrate, subject to the initial condition: $v = 0$ at $t = 0$, we get the expression for the velocity in the Lab frame S:

$$\frac{v}{\sqrt{1 - \frac{v^2}{c^2}}} = at. \tag{4.20}$$

Squaring both sides and simplifying, we get

$$v = \frac{c}{\left\{1 + \frac{c^2}{(at)^2}\right\}^{\frac{1}{2}}}. \tag{4.21}$$

The result shows that a particle under "constant acceleration in its instantaneous rest frame" will asymptotically approach the speed c in any IF, say S, but will never reach it. See Fig. 4.5 below.

The presence of c^2 in the denominators of Eqs. (4.6) and (4.7) make them appear somewhat clumsy. To make them look neater, we shall rewrite them in terms of dimensionless velocities $\boldsymbol{\beta}, \boldsymbol{\nu}$, defined as

$$\mathbf{u} = c\boldsymbol{\beta} = c\,(\beta, 0, 0), \quad \mathbf{v} = c\boldsymbol{\nu} = c\,(\nu_x, \nu_y, \nu_z). \tag{4.22}$$

With these modifications, Eqs. (4.6) and (4.7) take the following dimensionless forms:

$$\nu'_x = \frac{\nu_x - \beta}{1 - \nu_x\beta}, \quad \nu'_y = \frac{\nu_y}{\gamma(1 - \nu_x\beta)}, \quad \nu'_z = \frac{\nu_z}{\gamma(1 - \nu_x\beta)}, \tag{4.23a}$$

$$\nu_x = \frac{\nu'_x + \beta}{1 + \nu'_x\beta}, \quad \nu_y = \frac{\nu'_y}{\gamma(1 + \nu'_x\beta)}, \quad \nu_z = \frac{\nu'_z}{\gamma(1 + \nu'_x\beta)}. \tag{4.23b}$$

4.2. Relativistic Form of Acceleration Transformation

Let

$$\begin{aligned} \boldsymbol{a} &= \frac{d\mathbf{v}}{dt} = c\frac{d\boldsymbol{\nu}}{dt} \equiv c\dot{\boldsymbol{\nu}} = c\,(\dot{\nu}_x, \dot{\nu}_y, \dot{\nu}_z), \\ \boldsymbol{a}' &= \frac{d\mathbf{v}'}{dt'} = c\frac{d\boldsymbol{\nu}'}{dt'} \equiv c\dot{\boldsymbol{\nu}}' = c\,(\dot{\nu}'_x, \dot{\nu}'_y, \dot{\nu}'_z) \end{aligned} \tag{4.24}$$

represent the acceleration vector and its Cartesian components in S and S', respectively. Note that the "dot" ($\dot{}$) means $\frac{d}{dt}$ in the first line and $\frac{d}{dt'}$ in the second line. Then

$$\dot{\nu}'_i = c\frac{d\nu'_i}{c\,dt'} = c\frac{d\nu'_i/dt}{c\,dt'/dt}, \quad i = x, y, z. \tag{4.25}$$

Differentiating (4.23b) with respect to t

$$\begin{aligned} \frac{d\nu'_x}{dt} &= \frac{\dot{\nu}_x(1 - \beta^2)}{(1 - \nu_x\beta)^2} \\ &= \frac{\dot{\nu}_x}{\gamma^2(1 - \nu_x\beta)^2}, \\ \frac{d\nu'_y}{dt} &= \frac{\dot{\nu}_y(1 - \nu_x\beta) + \nu_y\dot{\nu}_x\beta}{\gamma(1 - \nu_x\beta)^2}. \end{aligned} \tag{4.26}$$

Using (4.3a) $\dfrac{c\,dt'}{dt} = \gamma c(1 - \beta\nu_x)$.

We wrap up (4.25) and (4.26) to obtain the following acceleration transformation formulas:

$$\dot{\nu}'_x = \frac{\dot{\nu}_x}{[\gamma(1 - \nu_x\beta)]^3}, \tag{4.27a}$$

$$\dot{\nu}'_y = \frac{\dot{\nu}_y(1 - \nu_x\beta) + \nu_y\dot{\nu}_x\beta}{\gamma^2(1 - \nu_x\beta)^3}, \tag{4.27b}$$

$$\dot{\nu}'_z = \frac{\dot{\nu}_z(1 - \nu_x\beta) + \nu_z\dot{\nu}_x\beta}{\gamma^2(1 - \nu_x\beta)^3}. \tag{4.27c}$$

Let us specialize (4.27) to rectilinear motion, along the X-axis. In this case, only the line (4.27a) is relevant. We drop the subscript "$_x$", and write:

$$\begin{aligned} \dot{\nu}' &= \frac{\dot{\nu}}{[\gamma(1 - \nu\beta)]^3} \quad \text{for boost: } S(\beta, 0, 0)S', \\ \dot{\nu} &= \frac{\dot{\nu}'}{[\gamma(1 + \nu'\beta)]^3} \quad \text{for boost: } S'(-\beta, 0, 0)S. \end{aligned} \tag{4.28}$$

In the second line, let S' represent IRF. Then $\nu' = 0$. We get

$$\dot{\nu} = \frac{\dot{\nu}'}{\gamma^3} = \frac{\dot{\nu}'}{\Gamma^3}, \tag{4.29}$$

as in (4.18), except that now the acceleration $c\dot{\nu}'$ in the IRF is variable.

4.3. A New Definition of Momentum

One of the consequences of the GT, as we saw in Sec. 2.2.1, is that the acceleration **a** of a particle is the same in all inertial frames. This invariance of acceleration is lost under the LT, as shown in Eq. (4.27). An immediate consequence of the frame-dependence of acceleration is that if we express Newton's second law of motion as $\mathbf{F} = m\mathbf{a}$, and treat the mass m as invariant, in line with Newtonian assumption, then force also becomes frame dependent. A constant force $\mathbf{F}$ in one inertial frame would then appear as a variable force in another inertial frame.

A more serious consequence of the velocity addition formula is that the laws of conservation of energy and momentum, which were seen in Sec. 2.2.2 to be valid in all inertial frames under the GT, now appear to break down under the LT. It turns out that if the defining expressions for energy E and momentum $\mathbf{p}$ are modified to meet the requirement of the ultimate speed c, then the frame independence of the conservation laws can be restored.

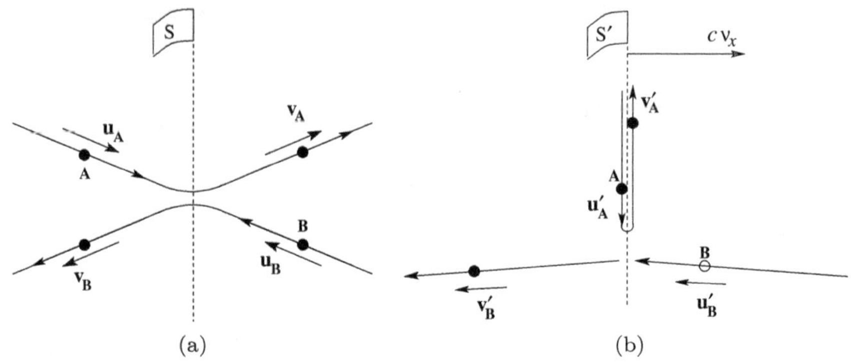

Fig. 4.4. Symmetric collision

In the rest of this section, we shall consider a thought experiment to seek a pointer to the definition of linear momentum. With this in mind, let us consider the following example. An observer S watches an elastic collision experiment with two billiard balls A and B, each of mass m_o, and assumed to be moving with relativistic speeds (Fig. 4.4(a)). For the sake of simplicity, we assume the collision to be taking place on the XY-plane. The velocities of the ball, before and after the collision, are denoted as $\mathbf{u}_A, \mathbf{u}_B$ and $\mathbf{v}_A, \mathbf{v}_B$, respectively. We have assumed that the initial speeds of the two balls are equal and that they make equal angles with the X-axis. A straight forward application of the energy–momentum conservation equations (2.10) and (2.11) would lead to the symmetry relations:

$$u_A = u_B = v_A = v_B. \tag{4.30}$$

Of the above four velocities, only $\mathbf{v}_A$ has both X and Y components positive. We shall write these components as $c\nu_x$ and $c\nu_y$, respectively, so that $\boldsymbol{\nu}_A = (\nu_x, \nu_y)$ is the dimensionless velocity of A after the collision. Referring to Fig. 4.4(a), we can now write the following components for the velocities involved:

$$\begin{aligned} \mathbf{u}_A &= c(\nu_x, -\nu_y), \quad & \mathbf{v}_A &= c(\nu_x, \nu_y), \\ \mathbf{u}_B &= c(-\nu_x, \nu_y), \quad & \mathbf{v}_B &= c(-\nu_x, -\nu_y). \end{aligned} \tag{4.31}$$

Now imagine another observer in a frame S' which is moving relative to S with speed $u_{Ax} = c\nu_x$ in the direction of the X-axis (Fig. 4.4(b)).

Using the velocity addition formulas (4.23), we obtain the following components for the velocity vectors in S':

$$
\begin{aligned}
u'_{Ax} &= \frac{u_{Ax} - u_{Ax}}{1 - u_{Ax}^2/c^2} = \frac{\nu_x - \nu_x}{1 - \nu_x^2}c = 0,\\
u'_{Ay} &= \frac{u_{Ay}}{\gamma(1 - u_{Ax}^2/c^2)} = -\frac{\nu_y}{\gamma(1 - \nu_x^2)}c,\\
v'_{Ax} &= \frac{v_{Ax} - u_{Ax}}{1 - v_{Ax}u_{Ax}/c^2} = \frac{\nu_x - \nu_x}{1 - \nu_x^2}c = 0,\\
v'_{Ay} &= \frac{v_{Ay}}{\gamma(1 - u_{Ax}v_{Ax}/c^2)} = \frac{\nu_y}{\gamma(1 - \nu_x^2)}c,\\
u'_{Bx} &= \frac{u_{Bx} - u_{Ax}}{1 - v_{Bx}u_{Ax}/c^2} = \frac{-2\nu_x}{1 + \nu_x^2}c,\\
u'_{By} &= \frac{u_{By}}{\gamma(1 - u_{Ax}^2/c^2)} = \frac{\nu_y}{\gamma(1 + \nu_x^2)}c,\\
v'_{Bx} &= \frac{v_{Bx} - u_{Ax}}{1 - v_{Ax}u_{Ax}/c^2} = \frac{-2\nu_x}{1 + \nu_x^2}c,\\
v'_{By} &= \frac{v_{By}}{\gamma(1 - u_{Ax}v_{Ax}/c^2)} = -\frac{\nu_y}{\gamma(1 + \nu_x^2)}c,
\end{aligned}
\tag{4.32}
$$

where

$$
\gamma = \frac{1}{\sqrt{1 - \nu_x^2}}. \tag{4.33}
$$

Hence,

$$
\begin{aligned}
\mathbf{u'}_A &= c\left(0, -\frac{\nu_y}{\gamma(1 - \nu_x^2)}\right), \quad \mathbf{v'}_A = c\left(0, \frac{\nu_y}{\gamma(1 - \nu_x^2)}\right),\\
\mathbf{u'}_B &= \frac{c}{1 + \nu_x^2}\left(-2\nu_x, \frac{\nu_y}{\gamma}\right), \quad \mathbf{v'}_B = \frac{c}{1 + \nu_x^2}\left(-2\nu_x, -\frac{\nu_y}{\gamma}\right).
\end{aligned}
\tag{4.34}
$$

We shall now put the *non-relativistic* momentum expression:

$$
\mathbf{p} = m_o\mathbf{v} \quad (\textit{Not Valid in Relativity}) \tag{4.35}
$$

to the momentum conservation test in the frame S' (conservation of momentum and energy in S had been assumed for obtaining Eqs. (4.30) and (4.31)). From Eqs. (4.34), the total momentum of the system in the frame S', before and after the collision, are seen to be as follows:

$$\begin{aligned}
\mathbf{P}'^{(\text{initial})} &= m_\circ(\mathbf{u}'_A + \mathbf{u}'_B) \\
&= m_\circ c\left(0 - \frac{2\nu_x}{1+\nu_x^2}, -\frac{\nu_y}{\gamma(1-\nu_x^2)} + \frac{\nu_y}{\gamma(1+\nu_x^2)}\right) \\
&= m_\circ c\left(-\frac{2\nu_x}{1+\nu_x^2}, -\frac{2\nu_x^2\nu_y}{\gamma(1-\nu_x^4)}\right). \\
\mathbf{P}'^{(\text{final})} &= m_\circ(\mathbf{v}'_A + \mathbf{v}'_B) \\
&= m_\circ c\left(0 - \frac{2\nu_x}{1+\nu_x^2}, \frac{\nu_y}{\gamma(1-\nu_x^2)} - \frac{\nu_y}{\gamma(1+\nu_x^2)}\right) \\
&= m_\circ c\left\{-\frac{2\nu_x}{1+\nu_x^2}, \frac{2\nu_x^2\nu_y}{\gamma(1-\nu_x^4)}\right\}.
\end{aligned}$$

Thus, $\mathbf{P}'^{(\text{final})} \neq \mathbf{P}'^{(\text{initial})}$, suggesting that the momentum conservation law breaks down if we adopt the definition (4.35). This calls for a revised definition for linear momentum.

One may get a clue along the following lines. The transformation $\mathbf{u} = \left(\frac{dx}{dt}, \frac{dy}{dt}, \frac{dz}{dt}\right) \to \mathbf{u}' = \left(\frac{dx'}{dt'}, \frac{dy'}{dt'}, \frac{dz'}{dt'}\right)$ due to change of reference frame is a double transformation in the sense that the LT converts both the numerator dx and the denominator dt according to Eq. (4.3). It might have been better to adopt instead of the coordinate time dt, some other measure of the time interval which would remain invariant under the LT. The natural choice would be the proper time interval $d\tau$ of the particle's own motion.

Referring to the trajectory of the particle shown in Fig. 4.1, we can think of a clock attached to the particle and being carried with it. The timing read on this clock is the *proper time of the particle's own motion* (proper time, because the particle's clock is present all along its own trajectory). This clock reads the timings at the points A and B (i.e. at the events Θ_A and Θ_B) to be τ and $\tau + d\tau$, respectively. Therefore, the proper time interval between Θ_A and Θ_B is $d\tau$. Dividing the spatial intervals between these events by this proper time, we get a new measure of the velocity, which we shall call *proper velocity* and write as $\mathbf{v}_{\text{prop}}$. The components of this velocity in the frames S and S' will be

$$\mathbf{v}_{\text{prop}} = \left(\frac{dx}{d\tau}, \frac{dy}{d\tau}, \frac{dz}{d\tau}\right), \quad \mathbf{v}'_{\text{prop}} = \left(\frac{dx'}{d\tau}, \frac{dy'}{d\tau}, \frac{dz'}{d\tau}\right), \tag{4.36}$$

respectively.

Since a comoving inertial frame of the particle at the event Θ_A is moving with velocity $\mathbf{v}$ with respect to S, and $\mathbf{v}'$ with respect to S', it follows from Eq. (2.23) that

$$d\tau = \frac{dt}{\Gamma} = \frac{dt'}{\Gamma'}, \tag{4.37}$$

where

$$\Gamma = \frac{1}{\sqrt{1 - \frac{v^2}{c^2}}}, \quad \Gamma' = \frac{1}{\sqrt{1 - \frac{v'^2}{c^2}}} \tag{4.38}$$

are the dynamic Lorentz factors in the frames S and S', respectively, as defined in Eq. (4.15). Therefore, from Eqs. (4.36) and (4.37),

$$\mathbf{v}_{\text{prop}} = \Gamma \mathbf{v}, \quad \mathbf{v}'_{\text{prop}} = \Gamma' \mathbf{v}' \tag{4.39}$$

Hence we try the following definition for momentum:

$$\mathbf{p} = m_o \mathbf{v}_{\text{prop}} = \Gamma m_o \mathbf{v}. \tag{4.40}$$

For non-relativistic speeds, $\Gamma \to 1$, so that the definition (4.40) converges to (4.35).

We shall now show that this new definition (4.40) will ensure momentum conservation in the example illustrated through Fig. 4.4. For this we shall first compute all the Γ-factors, in the S' frame, corresponding to the velocities written in Eq. (4.34). It is easily seen that the Γ-factors corresponding to $\mathbf{u}'_A$ and $\mathbf{v}'_A$ are equal and therefore can be represented by a common symbol Γ'_A. Similarly, the Γ-factors corresponding to $\mathbf{u}'_B$ and $\mathbf{v}'_B$ are equal and can be represented by another symbol Γ'_B. These factors can be computed with the help of Eqs. (4.34) and (4.33) as follows:

$$\begin{aligned}
\frac{1}{\Gamma'^2_A(v'_A)} &= 1 - \frac{v'^2_A}{c^2} = 1 - \frac{\nu_y^2}{\gamma^2(1-\nu_x^2)^2} = 1 - \frac{\nu_y^2}{1-\nu_x^2} = \frac{1-\nu_x^2-\nu_y^2}{1-\nu_x^2}, \\
\frac{1}{\Gamma'^2_B(v'_B)} &= 1 - \frac{v'^2_B}{c^2} = 1 - \frac{1}{(1+\nu_x^2)^2}\left(4\nu_x^2 + \frac{\nu_y^2}{\gamma^2}\right) \\
&= \frac{1}{(1+\nu_x^2)^2}\left[(1+\nu_x^4+2\nu_x^2) - \{4\nu_x^2 + (1-\nu_x^2)\nu_y^2\}\right] \\
&= \frac{(1-\nu_x^2)(1-\nu_x^2-\nu_y^2)}{(1+\nu_x^2)^2}.
\end{aligned}$$

Therefore

$$\Gamma'_A = \sqrt{\frac{1-\nu_x^2}{1-\nu_x^2-\nu_y^2}} = \frac{\gamma(1-\nu_x^2)}{\sqrt{1-\nu_x^2-\nu_y^2}},$$
$$\Gamma'_B = \frac{(1+\nu_x^2)}{\sqrt{(1-\nu_x^2)(1-\nu_x^2-\nu_y^2)}} = \frac{\gamma(1+\nu_x^2)}{\sqrt{1-\nu_x^2-\nu_y^2}}. \tag{4.41}$$

The new definition (4.40) yields the following momenta for A and B before the collision.

$$\mathbf{p}_A'^{(\text{initial})} = \Gamma'_A \mathbf{u}'_A\, m_\circ = \frac{\gamma(1-\nu_x^2)}{\sqrt{1-\nu_x^2-\nu_y^2}}\left(0, -\frac{\nu_y}{\gamma(1-\nu_x^2)}\right) m_\circ c$$
$$= \left(0, -\frac{\nu_y}{\sqrt{1-\nu_x^2-\nu_y^2}}\right) m_\circ c,$$
$$\mathbf{p}_B'^{(\text{initial})} = \Gamma'_B \mathbf{u}'_B\, m_\circ = \frac{\gamma(1+\nu_x^2)}{\sqrt{1-\nu_x^2-\nu_y^2}}\left(-2\nu_x, \frac{\nu_y}{\gamma}\right) \frac{m_\circ c}{1+\nu_x^2}$$
$$= \left(-\frac{2\gamma\nu_x}{\sqrt{1-\nu_x^2-\nu_y^2}}, \frac{\nu_y}{\sqrt{1-\nu_x^2-\nu_y^2}}\right) m_\circ c.$$

Hence, the initial momentum of the system is

$$\mathbf{P}'^{(\text{initial})} = \left\{-\frac{2\gamma\nu_x}{\sqrt{1-\nu_x^2-\nu_y^2}}, 0\right\} m_\circ c. \tag{4.42}$$

Similarly, the momenta of the particles after the collision are as follows:

$$\mathbf{p}_A'^{(\text{final})} = \Gamma'_A \mathbf{v}'_A\, m_\circ c = \frac{\gamma(1-\nu_x^2)}{\sqrt{1-\nu_x^2-\nu_y^2}}\left(0, \frac{\nu_y}{\gamma(1-\nu_x^2)}\right) m_\circ c$$
$$= \left(0, \frac{\nu_y}{\sqrt{1-\nu_x^2-\nu_y^2}}\right) m_\circ c,$$
$$\mathbf{p}_B'^{(\text{final})} = \Gamma'_B \mathbf{v}'_B\, m_\circ c = \frac{\gamma(1+\nu_x^2)}{\sqrt{1-\nu_x^2-\nu_y^2}}\left(-2\nu_x, -\frac{\nu_y}{\gamma}\right) \frac{m_\circ c}{1+\nu_x^2}$$
$$= \left(-\frac{2\gamma\nu_x}{\sqrt{1-\nu_x^2-\nu_y^2}}, -\frac{\nu_y}{\sqrt{1-\nu_x^2-\nu_y^2}}\right) m_\circ c.$$

Hence, the final momentum of the system is

$$\mathbf{P}'^{(\text{final})} = \left(-\frac{2\gamma\nu_x}{\sqrt{1-\nu_x^2-\nu_y^2}}, 0 \right) m_o c. \tag{4.43}$$

From Eqs. (4.42) and (4.43), we find that the new definition (4.40) of linear momentum will conserve linear momentum in both S and S'. In Sec. 4.6, we shall prove this statement for a more general collision, and in Sec. 8.4, we shall arrive at the same definition of momentum in a more natural way while discussing 4-momentum. For the time being we shall confirm this new definition through the following equation:

$$\mathbf{p} = \Gamma m_o \mathbf{v} = \Gamma m_o c \boldsymbol{\beta}, \tag{4.44}$$

where $\mathbf{v}$ is the velocity of the particle. We have used a subscript 0 under m to underscore the fact that m_o is the *intrinsic mass*, often also called the *rest mass* of the particle, for reasons that will be explained.

Frequently, the Γm_o is called the *relativistic mass*. We shall reserve the symbol m (i.e. m without subscript) to mean relativistic mass:

$$m \equiv \Gamma m_o = \frac{m_o}{\sqrt{1-\frac{v^2}{c^2}}}. \tag{4.45}$$

In contrast with m, the rest mass m_o measures the inertia of the particle when the particle is momentarily at rest, i.e. the inertia of the particle when the particle is either at rest, or is being accelerated from its (momentary) rest position. We can alternatively express momentum in the old style:

$$\mathbf{p} = m\mathbf{v}, \tag{4.46}$$

where the mass m appearing now is, however, relativistic mass. It is a *variable mass* and its relation to the *invariant rest mass* m_o is given by the expression (4.45).

Later in this chapter (see Eq. (4.79)) we shall show that the "total energy" E (sum of rest energy and the kinetic energy) of a particle is proportional to the relativistic mass m. In fact $E = mc^2$. Therefore, m is also a measure of the particle's energy. Hence a plot of m vs. β is also a plot of E vs. β

Variation of momentum p and energy E with velocity has been plotted in Fig. 4.6, below. In this figure, we have indicated the relativistic momentum p using the relativistic formula (4.44), and the non-relativistic momentum p_{NR}, obtained by using the N.R. formula (4.35). It can be noted that the two plots are almost identical in the region $\beta \lessapprox 0.3$, suggesting that we can use N.R. formulas for momentum and energy all the way upto $\approx 0.3\,c$, without causing appreciable error.

4.4. Force

Newton's second law of motion is a law as well as a quantitative definition of force. Following this law, the force **F** acting on a particle is measured as the rate of change of its momentum **p**:

$$\boxed{\mathbf{F} = \frac{d\mathbf{p}}{dt} = \frac{d(\Gamma m_o \mathbf{v})}{dt} = \frac{d(m\mathbf{v})}{dt},} \tag{4.47}$$

where m is the relativistic mass as defined in Eq. (4.45).

With the definition of force as adopted in Eq. (4.47), the proportionality between force **F** and acceleration $\boldsymbol{a}$ is lost. In particular, **F** and $\boldsymbol{a}$ are no longer parallel. We shall establish a relation between the two, as shown in Eq. (4.82) below.

We shall now illustrate the second law by working out the effect of a *constant* force **F** on a particle of rest mass m_o which is at rest at $t = 0$. Since the resulting motion is one dimensional, we shall drop the vector symbol. From Eq. (4.47),

$$\frac{dp}{dt} = F, \quad \text{a constant.} \tag{4.48}$$

Integrating, and using the initial condition: $p = 0$ when $t = 0$, we get

$$p = Ft. \tag{4.49}$$

However, from Eq. (4.44),

$$p = \Gamma m_o v = \Gamma \beta m_o c, \tag{4.50}$$

where β and Γ are defined in Eq. (4.15). From (4.49) and(4.50),

$$\Gamma\beta = \frac{F}{m_o c} t. \tag{4.51}$$

Also, analogous to Eq. (2.22), we have

$$\frac{\beta^2}{1-\beta^2} = \Gamma^2\beta^2. \tag{4.52}$$

Therefore, from (4.51) and (4.52),

$$\frac{1}{\beta^2} = 1 + \left\{\frac{m_o c}{Ft}\right\}^2. \tag{4.53}$$

Simplifying, we get

$$v = c\beta = \frac{c}{\sqrt{1 + \left(\frac{m_o c}{Ft}\right)^2}}. \tag{4.54}$$

Note from Eq. (4.54) that as $t \to \infty, \beta \to 1$. That is, the velocity of the particle approaches c asymptotically with time. Also note that the result shown in Eq. (4.54) is identical with that in Eq. (4.21), if we set $a = \frac{F}{m_o}$. The convergence of the two results suggests that a *relativistic particle under a constant force is undergoing a constant acceleration in its instantaneous rest frame.*

We can integrate Eq. (4.54) to obtain an expression for the distance x covered by the particle in time t:

$$x = \int_0^t v\,dt = c\int_0^t \beta\,dt = \int_0^t \frac{c\,dt}{\sqrt{1 + \left(\frac{m_o c}{Ft}\right)^2}} \tag{4.55}$$

or

$$x = \frac{m_o c^2}{F}\left[\sqrt{1 + \left(\frac{Ft}{m_o c}\right)^2} - 1\right]. \tag{4.56}$$

We can make the formulas (4.54) and (4.56) look less formidable by defining a characteristic time

$$\tau_o \equiv \frac{m_o c}{F}. \tag{4.57}$$

Formulas (4.54) and (4.56) will now take the following easier looking forms:

$$v = c\beta = \frac{c}{\sqrt{1 + \left(\frac{\tau_o}{t}\right)^2}}, \tag{4.58}$$

$$x = c\tau_o\left[\sqrt{1 + \left(\frac{t}{\tau_o}\right)^2} - 1\right]. \tag{4.59}$$

All relativistic formulas should satisfy one important test, namely they should yield the corresponding old familiar results in the *non-relativistic limit*:

$$\text{N.R. limit} \Rightarrow \ v \ll c, \quad \beta \to 0, \quad \Gamma \to 1. \tag{4.60}$$

On the other extreme we have the *ultra-relativistic limit.*

$$\text{U.R. limit} \Rightarrow \ v \simeq c, \quad \beta \to 1, \quad \Gamma \to \infty. \tag{4.61}$$

In the present case, the criterion (4.60) is realized during the early times $t \ll \tau$, when the speed gained by the particle is still small compared to the speed of light, and the momentum Ft imparted to it by the force F is small compared to $m_o c$. Applying this criterion to Eq. (4.56), it is easily seen that

$$x \simeq \frac{m_o c^2}{F}\left[\left\{1+\frac{1}{2}\left(\frac{Ft}{m_o c}\right)^2\right\}-1\right]=\frac{1}{2}\left(\frac{F}{m_o}\right)t^2. \tag{4.62}$$

Here F/m_o is the acceleration a, so that we get back our familiar non-relativistic kinematic result: $x = \frac{1}{2}at^2$.

If we apply the approximation (4.60) in Eq. (4.54), we get

$$v = c\beta = \frac{c}{\left(\frac{m_o c}{Ft}\right)\sqrt{1+\left(\frac{Ft}{m_o c}\right)^2}} \simeq \frac{F}{m_o}t\left[1-\frac{1}{2}\left\{\frac{Ft}{m_o c}\right\}^2\right] \simeq \frac{F}{m_o}t, \tag{4.63}$$

which is again the old familiar non-relativistic result.

Before leaving this context it can be useful to carry a "numerical feel" of the relativistic formulas just derived, by considering two extreme examples of commonly encountered forces. In the first example, we shall consider a particle subjected to a constant force equal to the force of gravity (as experienced near the surface of the earth). The force on the particle is then $F = m_o g$, where $g = 9.81\,\mathrm{m/s^2}$. Therefore, the characteristic time is

$$\begin{aligned}\tau_o &= \frac{m_o c}{F} = \frac{c}{g} = \frac{3\times 10^8\,\mathrm{m/s}}{9.81\,\mathrm{m/s^2}}\\ &= 3.06\times 10^7\,\mathrm{sec} = 354\,\mathrm{days} \simeq 1\,\mathrm{year}.\end{aligned}$$

If we measure time in years, then Eqs. (4.58) and (4.63) will give:

$$\beta = \frac{v}{c} = \frac{t}{\sqrt{1+t^2}} \quad \text{(General result)}, \tag{4.64}$$

$$\beta = t \quad \text{(N.R. result)}. \tag{4.65}$$

The pre-relativistic result (4.65) tells us that the speed of the particle should increase linearly and would equal the speed of light in 1 year, twice that speed in 2 years, and so on, and will keep on increasing without limit. The relativistic formula (4.64), on the other hand, predicts a linear increase in speed in the beginning, followed by an asymptotic approach to the ultimate speed c at the end. It predicts a speed of $0.71c$ at the end of 1 year, $0.89c$ after 2 years, $0.95c$ after 3 years, and so on, so that the rate of

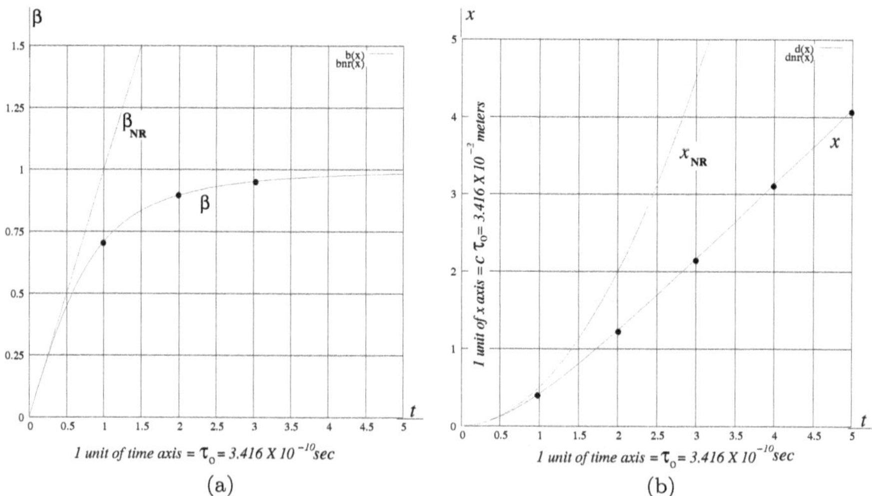

Fig. 4.5. Velocity β and displacement x as functions of time t under a constant force.

increase slows down as time progresses. The $v-t$ relationship is illustrated in Fig. 4.5.

We have just seen that it takes 1 year to reach 71% of the speed of light under a force equal to the terrestrial force of gravity. Therefore, if a rocket engine keeps firing continuously to generate a thrust equal to the weight of the space vehicle for one full year, it will come somewhat close to the speed of light. The above example therefore serves to illustrate how impossible it is for a space traveller to attain a speed anywhere near that of light.

If, however, the above example suggests that special relativity has nothing to do with terrestrial experiments and down-to-earth applications in physics, then the following example may correct such a notion.

Linear accelerators of earlier designs like the Van de Graff accelerator and Cockcroft Walton accelerator were used mostly to accelerate protons, deuterons, alpha-particles and other ions. Each of these machines can be designed to generate a potential difference of several million volts across the length of a tube called the accelerating tube. The charged particles required to be accelerated are allowed to fall through the potential difference between the ends of the tube.

Let us consider a Van de Graff accelerator having a length of 1 m and a potential difference of 5 millions volts. We shall use this machine to accelerate an electron. We shall, therefore, need to calculate the velocity attained by the electron after traversing the length of the tube and the time required to cover this distance. It is given that the rest mass of an electron is $m_o = 9.11 \times 10^{-31}$ kg. Charge of the electron is $e = 1.6 \times 10^{-19}$ coul. Therefore, the force on the electron is

$$F = eE = 1.6 \times 10^{-19} \times 5 \times 10^6 = 8 \times 10^{-13} \text{ N}.$$

The characteristic time is

$$\tau_o = \frac{9.11 \times 10^{-31} \times 3 \times 10^8}{8 \times 10^{-13}} = 3.416 \times 10^{-10} \text{ sec.}$$

From Eq. (4.58), we can obtain the speed of the electron at different times as follows:

$$v = 0.707c \quad \text{at } t = \tau_o = 3.416 \times 10^{-10} \text{ sec},$$
$$v = 0.89c \quad \text{at } t = 2\tau_o = 6.832 \times 10^{-10} \text{ sec},$$
$$v = 0.95c \quad \text{at } t = 3\tau_o = 10.248 \times 10^{-10} \text{ sec}.$$

We have plotted β vs. t and x vs. t in Fig. 4.5. The time axis has been calibrated in the unit of τ_o. In Fig. 4.5(a), we have indicated the velocity achieved after time intervals of $\tau_o, 2\tau_o, 3\tau_o$. In Fig. 4.5(b), the vertical axis represents displacement x, and has been calibrated in the unit of $c\tau_o$. We have indicated the displacement x at times $\tau_o, 2\tau_o, 3\tau_o, 4\tau_o, 5\tau_o$. It can be seen that after time $t = 2\tau_o$, the $x - t$ graph is almost a straight line.

The time required to cover the distance of 1 m can be calculated from Eq. (4.59) using the value of τ_o. It works out to be $t = 10.71\tau_o = 36.59 \times 10^{-10}$ sec. The velocity of the electron after accelerating through this distance is now obtained from (4.58) to be $0.99565c$.

Note that the electron gains a speed equal to 99.56% of the speed of light by falling through a potential difference of 5 million volts. The kinetic energy gained by the electron in this process is 5 million electron volts, or 5 MeV. We shall take up the concept of relativistic energy in the next section.

4.5. Energy

Let us review how an expression for kinetic energy can be derived in non-relativistic physics. For simplicity, we first consider a particle of mass m in one-dimensional motion under a constant force F. The acceleration is $a = F/m$ and is a constant. If the particle starts from rest from the origin and moves along the X-axis, then we have the familiar kinematic equation for the velocity v reached after traversing a distance x: $v^2 = 2ax$, so that $Fx = max = \frac{1}{2}mv^2$.

The quantity Fx is the work done by the force F in displacing the particle through a distance x. Using the conservation of energy principle, one may argue that this quantity Fx of potential energy must have been lost by some agent (which may a gravitational or electric field, a machine, human muscle, or something else) and has been transferred to the particle in the form of its kinetic energy. Therefore, the kinetic energy K of the particle is as follows:

$$K = \text{work done in raising the velocity of the particle from 0 to } v$$

$$= Fx = \frac{1}{2}mv^2. \quad \text{(N.R. result)} \tag{4.66}$$

For obtaining the same result from a more general force $\mathbf{F}$ which has arbitrarily variable magnitude and direction, we proceed as follows:

$$\begin{aligned} K &= \text{work done} = \int \mathbf{F}\cdot d\mathbf{r} = m\int \frac{d\mathbf{v}}{dt}\cdot \mathbf{v}dt \\ &= m\int \mathbf{v}\cdot d\mathbf{v} = \frac{1}{2}m\int d(\mathbf{v}\cdot\mathbf{v}) = \int d\left(\frac{1}{2}mv^2\right). \end{aligned} \tag{4.67}$$

In the above equation, the integration is carried out over the path along which particle has moved. Since the velocity of the particle is assumed to be zero at the beginning of the path, the above integration gives the same non-relativistic result, namely

$$K = \frac{1}{2}mv^2 \quad \text{(N.R. result)}. \tag{4.68}$$

We shall now apply a similar argument to a relativistic particle. To start with, we first consider the simplest example, namely a constant force F applied on a particle along the X-axis. The work done in moving the particle over a distance x is, as before, Fx. In this process, the velocity of the particle increases from 0 to v. It is seen from Eqs. (4.53) and (4.52) that

$$\begin{aligned} &\left[\frac{m_o c}{Ft}\right]^2 = \frac{1}{\beta^2} - 1 = \frac{1}{\Gamma^2\beta^2}, \\ &\Gamma^2\beta^2 + 1 = \Gamma^2. \end{aligned} \tag{4.69}$$

Combining the above two relations, we get

$$\Gamma^2 = 1 + \left[\frac{Ft}{m_o c}\right]^2. \tag{4.70}$$

It now follows with the help of Eq. (4.56) that

$$K = Fx = (\Gamma - 1)m_o c^2. \tag{4.71}$$

One can derive the above relation for a more general case. Consider the applied force $\mathbf{F}$ to have arbitrary magnitude and direction so that the particle will now move along a more general curvilinear trajectory. Using the defining equations (4.47) and (4.44) for force and momentum, we get

$$W = \text{work done} = \int \mathbf{F} \cdot d\mathbf{r} = m_o \int \frac{d(\Gamma \mathbf{v})}{dt} \cdot \mathbf{v} dt. \tag{4.72}$$

The right-hand side of (4.72) will readily transform to the form shown in (4.71) with the help of the following identity:

$$\frac{d(\Gamma \mathbf{v})}{dt} \cdot \mathbf{v} = c^2 \frac{d\Gamma}{dt}. \tag{4.73}$$

To prove Eq. (4.73), we first note that

$$\frac{1}{\Gamma^2} = 1 - \frac{v^2}{c^2}. \tag{4.74}$$

Differentiating either side, we get

$$\frac{2c^2}{\Gamma^3} \frac{d\Gamma}{dt} = \frac{d(v^2)}{dt}. \tag{4.75}$$

Therefore,

$$\begin{aligned} \frac{d(\Gamma \mathbf{v})}{dt} \cdot \mathbf{v} &= \frac{d\Gamma}{dt} \mathbf{v} \cdot \mathbf{v} + \Gamma \frac{d\mathbf{v}}{dt} \cdot \mathbf{v} = v^2 \frac{d\Gamma}{dt} + \frac{\Gamma}{2} \frac{d(v^2)}{dt} \\ &= \left[v^2 + \frac{c^2}{\Gamma^2} \right] \frac{d\Gamma}{dt}. \end{aligned}$$

Also note from (4.74) that $v^2 + \frac{c^2}{\Gamma^2} = c^2$. Hence the identity (4.73) is proved. (QED)

From Eqs. (4.72) and (4.73), we have

$$W = m_o c^2 \int_1^\Gamma d\Gamma = (\Gamma - 1)m_o c^2. \tag{4.76}$$

Here we have used the fact that at $t = 0, v = 0$ so that $\Gamma = 1$. The kinetic energy is therefore given by the earlier expression (4.71). We rewrite this as follows:

$$K = W = (\Gamma - 1)m_o c^2. \tag{4.77}$$

We shall show that the above expression for kinetic energy indeed reduces to the non-relativistic expression (4.68) under the condition (4.60)

$$\Gamma - 1 = \frac{1}{\sqrt{1-\beta^2}} - 1 \simeq 1 + \frac{1}{2}\beta^2 - 1 = \frac{1}{2}\beta^2. \tag{4.78a}$$

Hence,

$$K = (\Gamma - 1)m_o c^2 \simeq \frac{1}{2}m_o\beta^2 c^2 = \frac{1}{2}m_o v^2. \quad \text{(N.R. result)} \tag{4.78b}$$

The quantity $E = K + m_o c^2$ is called the *total energy* E of a "free particle" having velocity $\mathbf{v}$:

$$\boxed{\begin{aligned} &E = K + m_o c^2 = \Gamma m_o c^2, \\ \text{or}\quad &E = mc^2, \end{aligned}} \tag{4.79}$$

where m is the relativistic mass, as defined in Eq. (4.45). In contrast with the total energy mc^2, the quantity $m_o c^2$ is called the *rest energy*, or the *rest mass energy* of the particle. When $v \ll c$, the above equation reduces to

$$E = K + m_o c^2 \simeq \frac{1}{2}m_o v^2 + m_o c^2 \quad \text{(N.R. formula).} \tag{4.80}$$

Note that E = kinetic energy + mass energy, *always*, even in the N.R. limit, as we shall find in many applications of the energy formula in the sequel.

In Sec. 4.6, we shall present a satisfactory explanation why mc^2 is looked upon as the total energy of a relativistic particle.

It is seen from (4.79) that the Γ-factor represents *the ratio between the total energy and the rest energy of a particle.* Note from Eqs. (4.60) and (4.61) that for a non-relativistic particle $\Gamma \simeq 1$, and the total energy is almost exclusively its rest energy. For an ultra-relativistic particle (see the examples in the following paragraphs), $\Gamma \gg 1$. Thus, the Γ-factor is sometimes used as an index of the relativistic "magnitude" of high energy particles.

In Fig. 4.6, we have plotted variation of momentum p and the energy E with velocity.

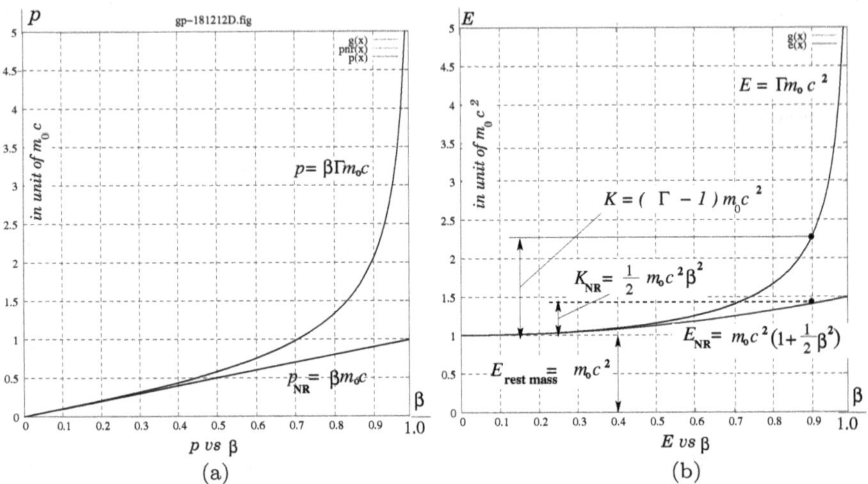

Fig. 4.6. Momentum and energy as functions of velocity.

The rate of change of Γ is proportional to the power ϖ delivered by the force $\mathbf{F}$, as we can see using Eqs. (4.77) and (4.72). (You can pronounce the symbol ϖ as "varpi" — a variant of π.)

$$\varpi = \mathbf{F} \cdot \mathbf{v} = \frac{dW}{dt} = \frac{dK}{dt} = \frac{dE}{dt} = m_o c^2 \frac{d\Gamma}{dt}. \tag{4.81}$$

The relation $\mathbf{F} \cdot \mathbf{v} = \frac{dK}{dt}$ is valid when the rest mass m_0 of the particle moving under the force $\mathbf{F}$ remains constant. We shall in future encounter a situation when this assumption will not be valid. We shall then avoid Eq. (4.81).

As we had remarked following Eq. (4.47) that $\mathbf{F}$ and $\boldsymbol{a}$ are not parallel. We shall obtain a relation between the two, using the relation (4.81), and the definition of force and momentum as given in (4.47) and (4.46):

$$\mathbf{F} = \frac{d\mathbf{p}}{dt} = \frac{d}{dt}(\Gamma m_o \mathbf{v}) = m_o \left[\Gamma \frac{d\mathbf{v}}{dt} + \mathbf{v}\frac{d\Gamma}{dt}\right] = m\boldsymbol{a} + \frac{\varpi \mathbf{v}}{c^2}. \tag{4.82}$$

The following relationship between the total energy E and the momentum $\mathbf{p}$ is extremely useful:

$$\boxed{E^2 = c^2 p^2 + m_o^2 c^4.} \tag{4.83}$$

In the N.R. limit, the above formula approximates to the formula (4.80), as the reader can easily verify.

To prove (4.83), we shall use Eq. (4.79), the identity (4.69), and Eq. (4.44).

$$E^2 = \Gamma^2 m_o^2 c^4 = (1+\Gamma^2\beta^2) m_o^2 c^4 = m_o^2 c^4 + p^2c^2. \qquad \text{(QED)}$$

To specialize Eq.(4.83) for a massless particle, like photon, we set $m_o = 0$ and get the corresponding energy–momentum relation

$$\boxed{E = cp, \quad \text{for a photon.}} \tag{4.84}$$

As a simple application of Eq. (4.83), we shall calculate the momenta and the velocities of the radioactive particles emitted during the beta-decay of a typical gamma-emitter (commonly used in college physics laboratories), namely ^{60}Co. ^{60}Co emits a beta-particle of kinetic energy 0.31 MeV, which is followed by two gamma-rays of energy 1.33 MeV and 1.17 MeV, respectively. The rest energy of the beta-particle (i.e. electron) is known to be $m_o c^2 = 0.51$ MeV. Therefore, from (4.79), $E = 0.31 + 0.51 = 0.82$ MeV

Hence, $pc = \sqrt{E^2 - m_o^2c^4} = \sqrt{(0.82)^2 - (0.51)^2} = 0.64$ MeV.

If one uses MeV/c as a unit of momentum, then $p = 0.64$ MeV/c.

To compute the velocity of the emitted electron, we shall first compute the Γ-factor, and use the following relation (which follows from Eq. (4.74):

$$\beta = \frac{\sqrt{\Gamma^2 - 1}}{\Gamma}. \tag{4.85}$$

The following estimates are obtained using (4.79) and (4.85): $\Gamma = \frac{E}{m_o c^2} = \frac{0.82}{0.51} = 1.61$; $\beta = 0.78$.

The above example shows that the velocity of the beta-particles emitted by ^{60}Co and other radioactive isotopes are fairly relativistic. In the present case, the velocity of the emitted beta-particle is $v = 0.78c$, that is, about 78% of the speed of light. It also follows from Eq. (4.83) that the momenta of the two photons emitted in the radioactive decay of ^{60}Co are 1.33 MeV/c and 1.17 MeV/c, respectively (because the rest mass of a photon is zero).

As a second example, we shall consider the beta decay of ^{34}Cl which decays by the emission of a positron of energy 4.5 MeV. As before, $E = 4.50 + 0.51 = 5.01$ MeV.

Hence, $pc = \sqrt{(5.01)^2 - (0.51)^2} = 4.98$ MeV, so that $p = 4.98$ MeV/c.

Also, $\Gamma = \frac{5.01}{0.51} = 9.8$, implying highly relativistic nature of the emitted positron. Using Eq. (4.15), one now gets $\beta = 0.995$. That is, the velocity of the emitted particle is 99.5% of the speed of light.

The above examples serve to illustrate that relativity is not just a Utopian idea, far removed from real life encounters. Even undergraduate students routinely experiment with electrons and positrons travelling with almost the speed of light, in their college laboratories. At the same time it should also be remembered that it is not so easy to obtain heavier particles, like protons, alpha-particles and heavy ions, at relativistic speeds. The following example will illustrate this point.

Consider an alpha-particle emitted from the radioactive decay of the isotope ^{212}Po. It has a kinetic energy of 8.78 MeV. This is about the highest energy with which any particle is emitted from natural radioactivity. The rest mass energy of the alpha-particle is $m_o c^2 = 3727.23\,\text{MeV}$.

Therefore, $\Gamma = \frac{T+m_o c^2}{m_o c^2} = \frac{8.78+3727.23}{3727.23} = 1.0024$.

Thus, $\Gamma \simeq 1$, so that the emitted alpha is non-relativistic. Since $\beta \ll 1$, we can use the approximation (4.78), and obtain the velocity of the alpha-particle approximately as follows:

$$\beta \simeq \sqrt{2(\Gamma - 1)} = 0.0693.$$

The speed of an alpha-particle of energy 8.78 MeV is now seen to be about 7% of the speed of light.

Consider protons that have been accelerated to an energy of 1 TeV, which is equal to 10^{12} ev. Such particles are ultra-relativistic ($\Gamma = 1067$). We shall establish the relativistic nature of a more modest energy proton, say, one of energy 10 GeV, i.e. 10^4 MeV. The rest energy of a proton is $m_o c^2 = 938.247\,\text{MeV}$. Therefore, $E = 10938\,\text{MeV}$. One can calculate the momentum and velocity using the relations (4.83) and (4.85). The final answer is as follows: $p = 10.9\,\text{GeV}/c$; $\Gamma = 11.66$; $\beta = 0.996$.

4.6. Energy–Momentum Conservation Law

In Sec. 4.3, we proposed a new definition for momentum, through Eq. (4.44), with a promise to provide a general proof later that such a definition would ensure Lorentz invariance of the momentum conservation law. (The term Lorentz invariance would mean preservation of a certain relationship, law or equation after the corresponding quantities have been transformed by Lorentz transformation, following a change of the frame of reference.) At the time of defining total energy of a particle through Eq. (4.79), we also promised to justify the nomenclature "total energy" in a subsequent section. We shall now redeem these pledges. Perhaps, the reader will be surprised to discover that, unlike in non-relativistic physics, the conservation of momentum and energy are not two disjointed concepts, but that one implies the other. If we demand that momentum be conserved in *all* frames of reference, that itself will guarantee conservation of total energy, and vice versa.

Before establishing the Lorentz invariance of the conservation laws, it will be necessary to establish the Lorentz transformation of energy and momentum. We shall, for simplicity, consider boost: $S(c\beta, 0, 0)S'$ as represented by the configuration shown in Fig. 4.1(a). Let $\mathbf{v} = \frac{d\mathbf{r}}{dt}$ and $\mathbf{v}' = \frac{d\mathbf{r}'}{dt'}$ represent the velocity of a particle of rest mass m_o, as measured from the Lorentz frames S and S', respectively. Then, according to the defining Eqs. (4.44) and (4.79), its momentum and energy in these two frames will be

$$\begin{cases} \mathbf{p} = m_o\Gamma\mathbf{v}, & E = m_o\Gamma c^2, \quad \text{in } S. \\ \mathbf{p}' = m_o\Gamma'\mathbf{v}', & E' = m_o\Gamma' c^2, \quad \text{in } S', \end{cases} \tag{4.86}$$

where Γ and Γ' represent the Lorentz-factors associated with the velocities of the particle in S and S', respectively. It follows from Eq. (4.37) that

$$\mathbf{v} = \frac{d\mathbf{r}}{dt} = \frac{1}{\Gamma}\frac{d\mathbf{r}}{d\tau}; \Rightarrow \Gamma\mathbf{v} = \frac{d\mathbf{r}}{d\tau}. \tag{4.87}$$

Hence

$$\begin{aligned} \mathbf{p} &= m_o\frac{d\mathbf{r}}{d\tau} = m_o\left(\frac{dx}{d\tau}, \frac{dy}{d\tau}, \frac{dz}{d\tau}\right), & \frac{E}{c} &= m_o\frac{c\,dt}{d\tau}, \\ \mathbf{p}' &= m_o\frac{d\mathbf{r}'}{d\tau} = m_o\left(\frac{dx'}{d\tau}, \frac{dy'}{d\tau}, \frac{dz'}{d\tau}\right), & \frac{E'}{c} &= m_o\frac{c\,dt'}{d\tau}. \end{aligned} \tag{4.88}$$

Using the Lorentz transformation equations (4.3) for the coordinate differentials, the momentum and the total energy in the frame S' are now expressed as a linear combination of both of these quantities in S (for convenience we shall use E/c, instead of E. Note that E/c has the same dimension as momentum p).

$$\begin{cases} p'_x = m_o\dfrac{dx'}{d\tau} = m_o\gamma\left(\dfrac{dx}{d\tau} - \beta\dfrac{c\,dt}{d\tau}\right) = \gamma\left(p_x - \beta\dfrac{E}{c}\right), \\ p'_y = p_y, \quad p'_z = p_z, \\ \dfrac{E'}{c} = m_o\dfrac{c\,dt'}{d\tau} = m_o\gamma\left(\dfrac{c\,dt}{d\tau} - \beta\dfrac{dx}{d\tau}\right) = \gamma\left(\dfrac{E}{c} - \beta p_x\right). \end{cases} \tag{4.89}$$

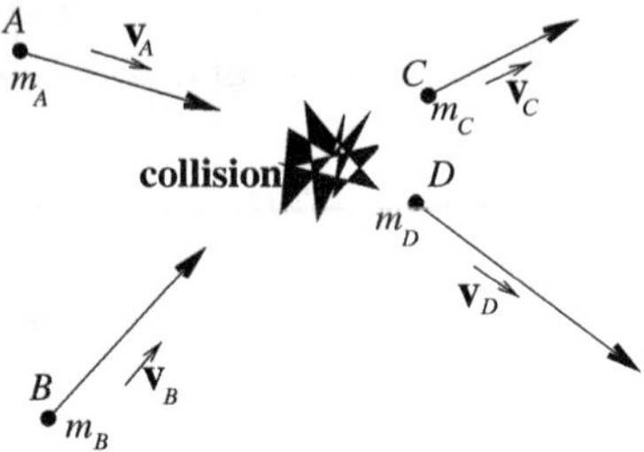

Fig. 4.7. Collision of two particles.

The inverse of the above transformation follows without much ado:

$$\left.\begin{aligned} p_x &= \gamma\left(p'_x + \beta\frac{E'}{c}\right), \\ p_y &= p'_y, \\ p_z &= p'_z, \\ \frac{E}{c} &= \gamma\left(\frac{E'}{c} + \beta p'_x\right). \end{aligned}\right\} \tag{4.90}$$

A comparison with Eq. (3.8) shows that the set $(E/c, p_x, p_y, p_z)$ transforms, under a change of the frame of reference, in the same way as the set (ct, x, y, z).

With the above preparation we now reconsider the collision phenomenon, earlier discussed in Sec. 2.2.2, once again. Figure 4.7 (which is a reproduction of Fig. 2.2) describes the setting. Two particles A and B come and collide and result in the emergence of two other particles C and D from the scene of the collision:

$$A + B \to C + D.$$

This kind of process occurs frequently in the sub-atomic world, as illustrated in the famous example

$$\pi^- + p \to \Lambda + K^0,$$

in which a pi-meson collides with a proton to create a K-meson and a lambda particle.

As in Sec 2.3, let us assume that the total momentum of the system, as measured in the frame S, is the same before the collision as after it. Since momentum is a vector, the above statement implies conservation of its x, y and z components separately. Using the definition (4.44) of momentum, we get the following three equations:

$$\begin{aligned} p_{Ax} + p_{Bx} &= p_{Cx} + p_{Dx}, \\ p_{Ay} + p_{By} &= p_{Cy} + p_{Dy}, \\ p_{Az} + p_{Bz} &= p_{Cz} + p_{Dz}. \end{aligned} \tag{4.91}$$

In the above, p_{Ax} stands for the x-component of the momentum of the particle A, B, etc. Using the transformation equation (4.90), we can now write the above momentum components in terms of quantities measured in the frame S'. This results in the following equation:

$$\gamma\left(p'_{Ax} + \beta\frac{E'_A}{c}\right) + \gamma\left(p'_{Bx} + \beta\frac{E'_B}{c}\right) = \gamma\left(p'_{Cx} + \beta\frac{E'_C}{c}\right) + \gamma\left(p'_{Dx} + \beta\frac{E'_D}{c}\right),$$

or

$$(p'_{Ax} + p'_{Bx} - p'_{Cx} - p'_{Dx}) + \beta\left(\frac{E'_A}{c} + \frac{E'_B}{c} - \frac{E'_C}{c} - \frac{E'_D}{c}\right) = 0.$$

Since β is arbitrary, the above equation separates into the following two equations.

$$\begin{cases} E'_A + E'_B = E'_C + E'_D, \\ p'_{Ax} + p'_{Bx} = p'_{Cx} + p'_{Dx}. \end{cases} \tag{4.92}$$

We note from Eqs. (4.91) and (4.92) that conservation of the x component of linear momentum in the frame S implies conservation of both x momentum and total energy in the frame S'. We would have come to the same conclusion for the other components of momentum had we started this exercise from a general boost. We therefore summarize as follows. Conservation of momentum in any one frame implies a wider conservation law in all other frames, which brings within its fold not only momentum conservation, but also energy conservation, that is, conservation of "total energy" as defined by Eq. (4.79). Momentum and energy become inseparable from each other in relativity, a fact we shall see more transparently in Sec. 12.3.

We shall rewrite this very important concept in the form of more general equations for a fuller comprehension of its significance. Suppose there are n particles, which are marked for identification as $1, 2, \ldots, n$. They are allowed to interact. Following the interaction, they may either retain their original identities as particles numbered $1, 2, \ldots, n$, or may change into N other particles, which may now be marked by numbers $n+1, n+2, \ldots, n+N$. Considering the more general case of N particles resulting from the interaction ($N = n$, and the particle marked i is identical with the one marked $i + n$ in the special case where the particles do not change in the interaction process), we can now use the definitions (4.44) and (4.79) to rewrite the conservation laws, as implied by Eq. (4.92), in any general frame of reference S, as follows:

$$\sum_{i=1}^{n} m_{0i}\Gamma_i \mathbf{v}_i = \sum_{i=n+1}^{n+N} m_{0i}\Gamma_i \mathbf{v}_i, \tag{4.93a}$$

$$\sum_{i=1}^{n} m_{0i}\Gamma_i c^2 = \sum_{i=n+1}^{n+N} m_{0i}\Gamma_i c^2. \tag{4.93b}$$

In the above $\Gamma_i \equiv \frac{1}{\sqrt{1-\frac{v_i}{c^2}}}$. Equations (4.93a) and (4.93b) together represent conservation of four quantities namely p_x, p_y, p_z and E. One lesson from Eq. (4.93 b) is that it is Γ times mass, and *not* mass alone, which is a conserved quantity in relativity. (The word "mass" will always mean rest mass m_o, unless relativistic mass m is explicitly implied.) Conservation of mass, a guiding principle of chemistry, wilts away under the onslaught of relativity:

$$\sum_{i=1}^{n} m_{0i} \neq \sum_{i=n+1}^{n+N} m_{0i} \quad \text{(in general)}.$$

$$\text{Mass before reaction} \neq \text{Mass after reaction} \quad \text{(in general)}.$$

The reader should now see the reason for calling $E = mc^2 = \Gamma m_o c^2$ the *total energy* in the definition (4.79). It is this $\Gamma m_o c^2$ which is conserved in all natural processes. Since $(\Gamma - 1)m_o c^2$ has already been shown to be the kinetic energy, the balance $m_o c^2$ is supposed to be the potential energy of the particle manifested in the form of its rest mass. This rest mass potential energy can be converted into kinetic energy, or some other form of energy. Nature provides an abundance of such examples.

Most of the practical examples of rest mass potential energy being converted into kinetic energy lie in the domain of nuclear physics. An awesome example is the *fission* of uranium. The ^{235}U nucleus captures a slow neutron and spontaneously breaks up into a number of smaller fragments, releasing in this process an enormous amount of energy. This fission reaction is utilized not only in the detonation of the atom bomb, but also in the extraction of useful energy in most of the nuclear power reactors. Even though the products of this reaction are not the same in different fissions of the same ^{235}U isotope, they generally consist of two medium weight nuclei (one of them with a mass number between 85 and 104, and the other with a mass number between 130 and 149), along with about two to three neutrons and about six to seven electrons. One typical such reaction is the following:

$$^{235}\text{U} + n \;\rightarrow\; ^{95}\text{Mo} + ^{139}\text{La} + 7e^- + 2n.$$

The right-hand side of the above reaction has *less* mass than the left-hand side, so that mass has not been conserved in the reaction. The energy equivalence of the difference between the rest mass before the reaction and the rest mass after the reaction is the energy released in a single fission. It goes into the kinetic energy of the fragments. When there are some 10^{23} nuclear fissions taking place in a fraction of a second, the resulting fission fragments will be dashing about with their own shares of the released energy. The net effect is then a violent liberation of heat.

In the table below, we have listed the parents as well as the products of the above fission reaction, their rest masses in atomic mass unit, and converted the mass difference into energy, so as to provide a numerical illustration of how mass is converted into kinetic energy.[a] We have used the conversion factor: $1\,\text{a.m.u.} \times c^2 = 931.16\,\text{MeV}$.

	Particle	Mass (in a.m.u.)	$m_o c^2$ (in MeV)
Before	U-235	235.0439	
	n	1.0087	
	Total	236.0526	219,802.73
After	Mo-95	94.9058	
	La-139	138.9061	
	2n	2×1.0087	
	7e	7×0.00055	
	Total	235.8331	219,598.39
	Difference	0.2195	204.34 (0.09% of the original mass energy)

As another example of mass–energy conversion, we shall consider a *fusion* reaction, in which four hydrogen nuclei, i.e. protons, fuse to form a helium nucleus:

$$4p \;\rightarrow\; ^4\text{He} + 2e^+,$$

[a]A good description of the fission and the fusion processes have been discussed by Samuel Gladstone [12]. The mass values have been taken from the same book.

where e^+ represents positron, having the same rest mass as an electron. We have the following table.

	Particle	Mass (in a.m.u.)	$m_o c^2$ (in MeV)
Before	4p	4×1.0078 $= 4.0312$	 3753.692
After	He-4 2e	4.0030 2×0.00055	
	Total	4.0031	3727.526
	Difference	0.0281	26.166 (0.7% of the original mass energy)

We have provided more examples in the suggested problems at the end of this chapter to give the reader a thorougher familiarity with the energy-momentum conservation principle.

If, the examples cited so far have given the reader a wrong notion that these principles are applicable only in physics and not in chemistry, then we shall consider the example of the hydrogen atom to clear this misunderstanding. Actually the mass of a hydrogen atom is less than the mass of its constituents, namely, a proton and an electron. However, this difference, converted into mass energy, is only 13.6 eV. Compared to the rest mass energy of the hydrogen atom, which is about 938.36 MeV, this 13.6 eV is very negligible, only about $1.45 \times 10^{-6}\%$ of hydrogen's rest energy. This mass is so negligibly gained, or lost, in every chemical reaction, that a chemist does not need to pay any attention to it. However, it will be useful to remember that when a candle burns, or ice melts, these processes are accompanied by a very small change in the masses. The end products are, lighter in the first example and heavier in the second one, although the differences are so tiny that it will be very difficult to measure them.

4.7. The Centre of Mass of a System of Particles: The Zero Momentum Frame

The centre of mass (C.M.) of a system of particles plays an important role in non-relativistic classical mechanics,[b] especially in a many-body system,

[b]Somnath Datta [8, Chapter 12]. The CM has been defined on p. 470, the rigid body motion, especially the precession of a spinning top with diagrams on pp. 513–521, the two-body system with several applications on pp. 521–533.

like the motion of a rigid body (which is made up of a very large number of atomic particles), and in two-body systems. like the Earth–Moon pair moving in the gravitational field of the Sun.

The CM frame of such a system is the inertial frame in which the CM is stationary, even though the individual particles may be moving with arbitrary velocities. Also the number of the constituent particles, and their velocities, may differ following a reaction or a collision, and yet the CM of the system will move on with a constant, unchanged velocity.

The relativistic counterpart of the classical CM is defined along parallel lines, except that the invariant mass of each constituent particle is now replaced by the relativistic mass of such particles.[c]

Let us consider system of N particles, having rest masses, radius vectors and velocities $\{m_{0i}, \mathbf{r}_i, \mathbf{v}_i;\ i = 1, 2, 3, \ldots, N\}$, with respect to an inertial frame S. We define the location $\mathbf{r}_{\text{cm}}$ and the velocity $\mathbf{v}_{\text{cm}}$ of the CM as follows:

$$\mathbf{r}_{\text{cm}} = \frac{\sum_{i=0}^{N} m_i \mathbf{r}_i}{\sum_{i=0}^{N} m_i}, \tag{4.94a}$$

$$\mathbf{v}_{\text{cm}} = \frac{d\mathbf{r}_{\text{cm}}}{dt} = \frac{\sum_{i=0}^{N} m_i \mathbf{v}_i}{\sum_{i=0}^{N} m_i} = \frac{\mathbf{P}}{M}, \tag{4.94b}$$

$$\text{where,}\quad m_i = \Gamma_i m_{0i} = \text{the relativistic mass of the particle } i, \tag{4.94c}$$

$$\mathbf{P} = \sum_{i=0}^{N} m_i \mathbf{v}_i = \text{Total momentum of the system}, \tag{4.94d}$$

$$M = \sum_{i=0}^{N} m_i = \text{Total relativistic mass of the system}. \tag{4.94e}$$

For zero mass particles, like photons of light frequency ν, $m_i = h\nu/c^2$.

Since $\mathbf{P}$ and E are each conserved for an isolated system, these two values should remain unchanged following a chemical or nuclear reaction. Therefore, it follows from Eq. (4.94b) that the velocity $\mathbf{v}_{\text{cm}}$ should remain unchanged in such a reaction. In order to clarify the concept of CM, and its uniform motion in the relativistic case, we have worked out a few examples in Problem 4.7.

---

[c]Rindler [4, Example 6.5, p. 126].

The zero momentum (ZM) frame is the relativistic analogue of the CM frame of N.R. mechanics. It is the inertial frame in which $\mathbf{P} = \mathbf{0}$.

To make a transition from the Lab frame to the ZM frame one has to make a Lorentz transformation of the energy–momentum 4-vector. We have taken up this topic seriously in Sec. 8.8.1.

4.8. The Twin Paradox

An example which is often cited for illustrating the relative aspect of time is the Twin Paradox. It is a riddle around the ageing rates between the brother who stays home and his counterpart who undertakes a space voyage. But then, why should there be any ageing difference, given that all inertial frames are equal? The clue lies in recognizing that one of the brothers is necessarily in an accelerating frame.

There are several versions of this paradox (see References) most of which allude to a very brief but violent acceleration at the beginning, at the turn-around point, and at the end of the journey. This tends to take away attention from the accelerating aspect which is the crux of the riddle. Therefore, we shall choose a version of the paradox (see J D Jackson Problem 11.4) in which the acceleration is moderate but steady and lasts through the entire duration of the voyage.

Ram and Sam are twins who have been living in an inertial space lab S (shown as inertial frame S in Fig. 4.8) since childhood. Ram being the more adventurous of the two, decided to take a space odyssey when they were 20 years old. He boards a rocket R (shown as frame R) which *accelerates* at a *constant* rate a *with respect to a comoving Lorentz frame* R. After a time span of n years, as recorded on his watch, Ram, wanting to come back, starts *decelerating* at the same rate a for another n years to reach zero speed with respect to S. Then he *turns around*, accelerates homeward at the same rate a for n years, then decelerates for another n years and safely lands at their old Space Lab to reunite with his brother Sam. Ram is now $20 + 4n$ years old. How old is Sam?

Let us first set up the basic differential equation required for finding the answer.

Let P and Q be two adjacent points along the path of the rocket at distances x and $x + dx$ from the starting point O of Ram's journey. Let v and $v + dv$ represent the velocities of Ram with respect to S while passing P ("event P") and Q ("event Q"), respectively. Let $R_0(\Theta_P)$ represent the co-moving Lorentz Frame at the event "Θ_P". We shall assign the following coordinates to the above pair of events with respect to the three frames of reference.

$$\text{“}\Theta_P\text{”} = \left\{ \begin{array}{ll} (ct, x) & \text{in } S \\ (ct', 0) & \text{in } R_0(\Theta_P) \\ (c\tau, 0) & \text{in } R \end{array} \right\}, \tag{4.95}$$

$$\text{“}\Theta_Q\text{”} = \left\{ \begin{array}{ll} (ct + c\,dt, x + dx) & \text{in } S \\ (ct' + c\,dt', dx') & \text{in } R_0(\Theta_P) \\ (c\tau + c\,d\tau, 0) & \text{in } R \end{array} \right\}. \tag{4.96}$$

Noting that the frame R is moving with infinitesimal speed at the event "Θ_Q" relative to the frame $R_0(\Theta_P)$ it follows that $d\tau = dt'$. Therefore, from Eq. (4.17a)

$$\frac{dv}{1 - \frac{v^2}{c^2}} = a\,d\tau, \tag{4.97}$$

which is the required differential equation for this problem.

In Fig. 4.9 we have shown the path followed by Ram. We have demarcated his route into four segments namely, outward acceleration along OA,

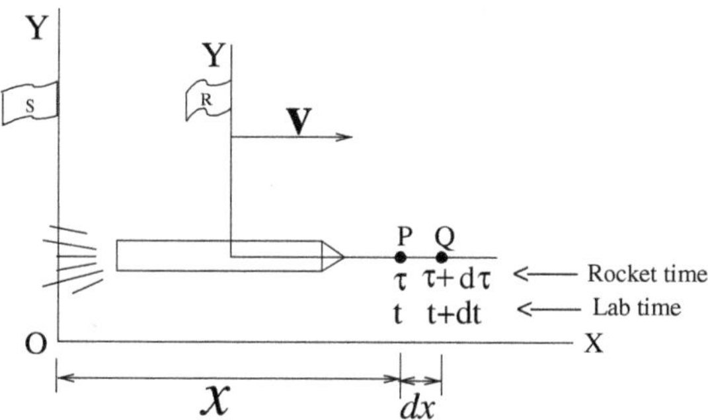

Fig. 4.8. Sam's frame S and Ram's frame R.

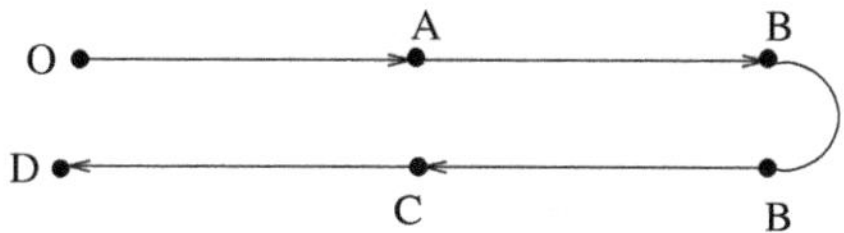

Fig. 4.9. Round trip route of Ram.

deceleration along AB, turning around at B, homeward acceleration along BC, followed by deceleration along CD, and finally return home at D.

Let us consider the first segment of his journey, namely to O to A. Let t and τ be the times measured in the frames S and R, respectively, corresponding to the event that "Ram has travelled a distance x from O" on this segment of his tour. The velocity of Ram at this point is obtained by integrating Eq. (4.97).

$$\int_0^v \frac{dv}{1-\frac{v^2}{c^2}} = a\int_0^\tau d\tau, \tag{4.98}$$

$$\text{or} \quad \tanh\frac{a\tau}{c} = \frac{v}{c}. \tag{4.99}$$

Noting that

$$dt = \gamma\, d\tau, \tag{4.100}$$

where

$$\gamma = \frac{1}{\sqrt{1-\frac{v^2}{c^2}}} = \frac{1}{\sqrt{1-\tanh^2\frac{a\tau}{c}}} = \cosh\frac{a\tau}{c}, \tag{4.101}$$

$$\text{so that} \quad v = \frac{dx}{dt} = \frac{1}{\gamma}\frac{dx}{d\tau}, \tag{4.102}$$

$$\text{we get} \quad \frac{dx}{d\tau} = c\tanh\frac{a\tau}{c}\cosh\frac{a\tau}{c} = c\sinh\frac{a\tau}{c}. \tag{4.103}$$

Integrating again,

$$x = c\int_0^\tau \sinh\frac{a\tau}{c}d\tau = \frac{c^2}{a}\left[\cosh\frac{a\tau}{c} - 1\right]. \tag{4.104}$$

To find t we integrate Eq. (4.100) and get

$$t = \int_0^t \gamma\, d\tau = \int_0^t \cosh\frac{a\tau}{c}d\tau = \frac{c}{a}\sinh\frac{a\tau}{c}. \tag{4.105}$$

Let the time recorded on the clock of Ram at the instant when he has reached A be τ_0. Then the time and space coordinates (ct_A, x_A) of the event "Ram has reached A", with respect to the frame S, are obtained by replacing τ by τ_0 in Eqs. (4.104) and (4.105).

$$x_A = \frac{c^2}{a}\left[\cosh\frac{a\tau_0}{c} - 1\right], \tag{4.106a}$$

$$t_A = \frac{c}{a}\sinh\frac{a\tau_0}{c}. \tag{4.106b}$$

The velocity of Ram at this point follows from Eqs. (4.99):

$$v_A = c\tanh\frac{a\tau_0}{c}. \tag{4.107}$$

The equation of motion for the second leg, i.e. AB, is obtained by replacing a by $-a$ in (4.97). Integrating we get

$$\int_{v_A}^{v}\frac{dv}{1-\frac{v^2}{c^2}} = -a\int_{\tau_A}^{\tau} d\tau. \tag{4.108}$$

Now making use of Eq. (4.107), we get

$$v = c\tanh\left[\frac{a(2\tau_0-\tau)}{c}\right], \quad \tau_0 \le \tau \le 2\tau_0, \tag{4.109}$$

$$\gamma = \cosh\left[\frac{a(2\tau_0-\tau)}{c}\right], \quad \tau_0 \le \tau \le 2\tau_0. \tag{4.110}$$

To find the corresponding time measured in S, we integrate as in Eq. (4.105)

$$\int_{t_A}^{t} dt = \int_{\tau_A}^{\tau}\gamma d\tau \tag{4.111}$$

and get

$$t = 2t_A - \frac{c}{a}\sinh\left[\frac{a(2\tau_0-\tau)}{c}\right], \quad \tau_0 \le \tau \le 2\tau_0. \tag{4.112}$$

To obtain x, we proceed similarly and get:

$$\int_{x_A}^{x} dx = \int_{\tau_0}^{\tau}\gamma v\, d\tau,$$

$$\text{or} \quad x = \frac{c^2}{a}\left[2\cosh\left(\frac{a\tau_0}{c}\right) - \cosh\left(\frac{a(2\tau_0-\tau)}{c}\right) - 1\right],$$

$$\tau_0 \le \tau \le 2\tau_0. \tag{4.113}$$

Setting $\tau = \tau_0$ in Eqs. (4.109), (4.110) and (4.111), we get the coordinates and the velocity of Ram at B:

$$t_B = 2t_A, \quad x_B = 2x_A, \quad v_B = 0. \tag{4.114}$$

The return journey is a mirror reflection of the outward journey. Hence, the duration and the furthest distance of Ram's space odyssey, as measured by Sam are as follows:

$$\begin{cases} T = \dfrac{4c}{a}\sinh\left(\dfrac{a\tau_0}{c}\right), \\ D = \dfrac{2c^2}{a}\left[\cosh\left(\dfrac{a\tau_0}{c}\right) - 1\right]. \end{cases} \tag{4.115}$$

In contrast, the travel time, as measured by Ram himself, is

$$T_0 = 4\tau_0. \tag{4.116}$$

For a numerical feel of the problem, the reader may consider a to be equal to the acceleration due to gravity near the surface of the earth, and take τ_0 to be equal to 5 years, so that Ram returns home at the age of 40. This will then lead to the following estimates:

$$D = 16 \times 10^{14}\,\text{km !!} \quad T = 335\,\text{years !!!!}$$

4.9. Compton Scattering

As an illustration of the energy–momentum conservation laws, we shall cite one example of historical significance in atomic physics. It is the scattering of X-rays by atomic electrons resulting in a shift in the wavelength of the scattered X-rays — an effect known as Compton Scattering. Since the outer electrons (with binding energy of a few electron volts) of the atom are chiefly responsible for Compton scattering and since the X-ray photons have comparatively higher energies (K_α X-ray photons from Tungsten have energy of 69.83 keV), these electrons are assumed to be at rest (i.e. zero energy) before collision — an assumption that simplifies calculations greatly. Figure 4.10 shows the scattering mechanism schematically. The incoming photon is scattered at an angle $-\theta$ with the direction of incidence (which is taken to be along the X-axis), after hitting an electron, which is knocked off at an angle of ϕ.

The energy of a photon of frequency ν is given as

$$E(\nu) = h\nu, \tag{4.117}$$

where h is Planck's constant. The magnitude of the momentum of the same photon (whose rest mass is zero) follows from Eq. (4.84) to be

$$p(\nu) = \frac{h\nu}{c}. \tag{4.118}$$

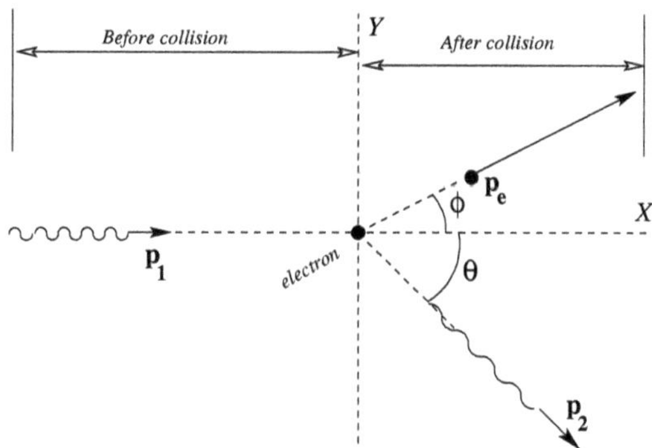

Fig. 4.10. Compton scattering. Schematic diagram.

Let us first write down the momentum conservation equation. The momentum of the photon, before and after the collision, is given by the vectors $\mathbf{p}_1$ and $\mathbf{p}_2$ respectively, whose magnitudes are, according to Eq. (4.118), $p_1 = \frac{h\nu_1}{c}$, $p_2 = \frac{h\nu_2}{c}$, where ν_1, ν_2 are the frequencies of the X-ray photon before and after the scattering. If $\mathbf{p}_e$ is the momentum of the recoiling electron, then the momentum conservation statement is that

$$\begin{aligned} \mathbf{p}_1 &= \mathbf{p}_2 + \mathbf{p}_e, \\ \text{or} \quad \mathbf{p}_e &= \mathbf{p}_1 - \mathbf{p}_2. \end{aligned} \tag{4.119}$$

Squaring either side,

$$p_e^2 = p_1^2 + p_2^2 - 2p_1p_2\cos\theta. \tag{4.120}$$

Now we shall obtain the energy conservation equation. Letting the rest mass of the electron to be $m_\circ$, and using Eq. (4.78), we get the initial and the final energy of the system as

$$\begin{aligned} E_{\text{initial}} &= h\nu_1 + E_{\text{e,rest}} = p_1c + m_\circ c^2, \\ E_{\text{final}} &= h\nu_2 + E_{\text{e,moving}} = p_2c + \sqrt{m_\circ^2c^4 + p_e^2c^2}. \end{aligned}$$

The energy conservation therefore means

$$p_1c + m_\circ c^2 = p_2c + \sqrt{(m_\circ c^2)^2 + p_e^2c^2}.$$

Simplifying we get

$$p_e^2 = (p_1 - p_2)^2 + 2(p_1 - p_2)m_\circ c. \tag{4.121}$$

From (4.120) and (4.121):

$$(p_1 - p_2)^2 + 2\,(p_1 - p_2)\, m_o c = p_1^2 + p_2^2 - 2p_1 p_2 \cos\theta, \tag{4.122a}$$

or $$(p_1 - p_2)\, m_o c = p_1 p_2 (1 - \cos\theta), \tag{4.122b}$$

or $$h(\nu_1 - \nu_2) m_o = \left(\frac{h}{c}\right)^2 \nu_1 \nu_2 (1 - \cos\theta), \tag{4.122c}$$

or $$c\left(\frac{1}{\nu_2} - \frac{1}{\nu_1}\right) = \frac{h}{m_o c}(1 - \cos\theta). \tag{4.122d}$$

From Eq. (4.122c), we get the change in the gamma ray energy:

$$h\nu_1 - h\nu_2 = \frac{h\nu_1\, h\nu_2}{m_o c^2}(1 - \cos\theta). \tag{4.123}$$

From Eq. (4.122d), we get the change in wavelength, using the fact that $\lambda\nu = c$:

$$\Delta\lambda = \lambda_2 - \lambda_1 = \frac{h}{m_o c}(1 - \cos\theta), \tag{4.124}$$

where λ_1, λ_2 represent the wavelengths of the incident and the scattered X-rays.

The factor $\frac{h}{m_o c}$ appearing in the above equation is called the *Compton wavelength* and is denoted by the symbol λ_c. Its value can be computed:

$$\lambda_c = \frac{h}{m_o c} = \frac{6.63 \times 10^{-34}}{(9.11 \times 10^{-31}) \times (3 \times 10^8)} = 2.43 \times 10^{-12}\,\text{m} = 0.0243\,\text{Å}. \tag{4.125}$$

The kinetic energy T_e transferred to the ejected electron is the same as the change in the photon energy. To find this energy we first solve Eq. (4.123) for $h\nu_2$:

$$h\nu_2 = \frac{m_o c^2}{\frac{m_o c^2}{h\nu_1} + (1 - \cos\theta)} = \frac{h\nu_1}{1 + \left(\frac{1 - \cos\theta}{m_o c^2}\right) h\nu_1}. \tag{4.126a}$$

Therefore $$T_e = h\nu_1 - h\nu_2 = \frac{(1 - \cos\theta) h\nu_1}{\frac{m_o c^2}{h\nu_1} + (1 - \cos\theta)}. \tag{4.126b}$$

We shall now obtain a relation between the scattering angle θ and the electron ejection angle ϕ, from the momentum conservation equation (4.119):

$$\mathbf{p}_1 = p_1\mathbf{i}, \quad \mathbf{p}_2 = p_2(\cos\theta\mathbf{i} - \sin\theta\mathbf{j}),$$

$$\mathbf{p}_e = p_e(\cos\phi\mathbf{i} + \sin\phi\mathbf{j}), \tag{4.127a}$$

$$p_e\cos\phi = p_1 - p_2\cos\theta \quad \text{(conservation of } x\text{-momentum)}, \tag{4.127b}$$

$$p_e\sin\phi = p_2\sin\theta \quad \text{(conservation of } y\text{-momentum)}. \tag{4.127c}$$

$$\cot\phi = \frac{p_1/p_2 - \cos\theta}{\sin\theta} = \frac{h\nu_1/h\nu_2 - \cos\theta}{\sin\theta}$$

$$= \left(1 + \frac{h\nu_1}{m_o c^2}\right)\left(\frac{1-\cos\theta}{\sin\theta}\right). \tag{4.127d}$$

To obtain Eq. (4.127d) we used Eq. (4.126a), and made some algebraic manipulations.

We have plotted T_e vs. $h\nu_1$ and ϕ vs. θ in Fig. 4.11. It is seen from Fig. 4.11(b) that the electron is deflected by the angle $\pi/2$ at $\theta = 0$, whereas common sense would suggest that for a direct hit, the target particle should recoil in the same direction. It is now seen from Fig. 4.11(a), and also evident from Eq. (4.126b), that the electron's share of the incident γ-ray energy goes to zero as $\theta \to 0$. For a zero energy particle, the deflection angle can be anything. However, at back scattering of the γ-ray, i.e. at $\theta = \pi$, the deflection angle is zero, which is what is expected.

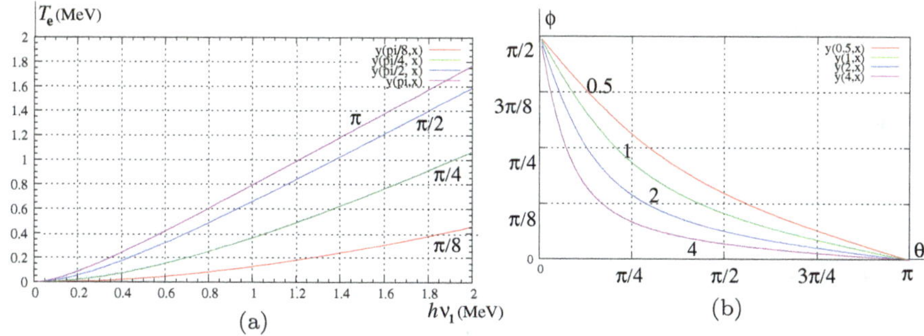

Fig. 4.11. Compton scattering. (a) Electron energy vs. incident photon energy for four values of photon scattering angle (indicated by the side of each graph). (b) Electron ejection angle vs. photon scattering angle for four values of photon energy in MeV (indicated by the side of each graph).

4.10. Summary of Important Formulas

Lorentz factor for boost. See Eq. (2.20).

$$\gamma \equiv \frac{1}{\sqrt{1-\beta^2}}, \quad \text{where } \beta \equiv \frac{u}{c}.$$

Lorentz transformation from S to S'. See Eq. (3.8).

$$\begin{aligned} ct' &= \gamma(ct - \beta x), \\ x' &= \gamma(x - \beta ct), \\ y' &= y, \\ z' &= z. \end{aligned}$$

Velocity addition formulas. See Eq. (4.6)

$$v'_x = \frac{v_x - u}{1 - \frac{v_x u}{c^2}}, \quad v'_y = \frac{v_y}{\gamma\left(1 - \frac{v_x u}{c^2}\right)}, \quad v'_z = \frac{v_z}{\gamma\left(1 - \frac{v_x u}{c^2}\right)}.$$

Other important formulas.

Dynamic Lorentz factor	$\Gamma \equiv \frac{1}{\sqrt{1-\beta^2}}, \quad \beta \equiv \frac{v}{c},$	(4.15)
Momentum	$\mathbf{p} = \Gamma m_0 \mathbf{v},$	(4.44)
Relativistic mass	$m \equiv \Gamma m_0 = \frac{m_0}{\sqrt{1 - \frac{v^2}{c^2}}},$	(4.45)
Equation of motion	$\mathbf{F} = \frac{d\mathbf{p}}{dt},$	(4.47)
Kinetic energy	$K = (\Gamma - 1) m_0 c^2,$	(4.77)
Total energy	$E = K + m_0 c^2 = \Gamma m_0 c^2 = mc^2,$	(4.79)
Relation between p and E	$E^2 = c^2 p^2 + m_0^2 c^4.$	(4.83)

4.11. Worked Out Problems II

Problem 4.1. A particle P decays into two smaller particles D_1 and D_2.[d]

$$P \to D_1 + D_2$$

[d]See [13].

(a famous example being $\Lambda \to p + \pi^-$; also see the example suggested in Example 4.5). Let M be the mass of the parent particle and let m_1, m_2 be the masses of the daughters ("mass" will mean "rest mass", unless otherwise stated). Energy consideration requires that $M > m_1 + m_2$, so that

$$Q \equiv (M - m_1 - m_2)c^2 \tag{4.128}$$

is the energy liberated in the reaction in the *rest frame* of P (or the disintegration energy).

(a) Let $\mathbf{p}_1$ and E_1 represent the momentum and energy of D_1 in the *rest frame* of P. Using energy and momentum conservation laws and Eq. (4.78), show that

$$p_1^2 = \left\{ \frac{(M^2 + m_2^2 - m_1^2)\, c}{2M} \right\}^2 - m_2^2 c^2, \tag{4.129}$$

$$E_1 = \left\{ \frac{M}{2} + \frac{m_1^2 - m_2^2}{2M} \right\} c^2. \tag{4.130}$$

(b) Note that $p_1 = p_2$. However, Eq. (4.129) does not exhibit explicit symmetry between m_1 and m_2. Therefore, recast Eq. (4.129) into the following symmetric form:

$$p_1 = p_2 = \frac{\sqrt{\{M^2 - (m_1 + m_2)^2\}\{M^2 - (m_1 - m_2)^2\}}}{2M} c. \tag{4.131}$$

(c) Hence, show that

$$Q = T_1 + T_2, \tag{4.132}$$

where T_1, T_2 are the kinetic energies of the daughters D_1, D_2, respectively, in the rest frame of P.

(d) Use the above equations to the observed decay of pi-meson (π^+) to mu-meson (μ^+) and neutrino (ν_μ). Take the masses, in unit of MeV/c^2 as follows: $M_\pi = 139.6$; $m_\mu = 105.7$; $m_\nu = 0$. Calculate the kinetic energies of μ^+ and ν_μ in the rest frame of π^+.

References: Jackson [13], Griffiths [14].

Solution. (a) Use energy–momentum conservation in the rest frame of P:

$$E_1 + E_2 = Mc^2, \quad p_1 + p_2 = 0. \tag{a}$$

$$\text{Hence, } E_1 = Mc^2 - E_2, \tag{b}$$

$$\text{or } \sqrt{p_1^2c^2 + m_1^2c^4} = Mc^2 - \sqrt{p_2^2c^2 + m_2^2c^4}$$

$$= Mc^2 - \sqrt{p_1^2c^2 + m_2^2c^4}. \tag{c}$$

$$\text{Squaring, simplifying, } \sqrt{p_1^2c^2 + m_2^2c^4} = \frac{(M^2 + m_2^2 - m_1^2)c^2}{2M}, \tag{d}$$

$$\text{or } p_1^2 = \left\{\frac{(M^2 + m_2^2 - m_1^2)c}{2M}\right\}^2 - m_2^2c^2. \tag{e}$$

$$\text{Also, } E_1^2 = p_1^2c^2 + m_1^2c^4$$

$$= \left\{\frac{(M^2 + m_2^2 - m_1^2)c}{2M}\right\}^2 c^2 - m_2^2c^4 + m_1^2c^4$$

$$= \left[\frac{c^2}{2M}\left(M^2 + m_2^2 - m_1^2\right)\right]^2. \tag{f}$$

$$\text{Hence, } E_1 = \left[\frac{M}{2} + \frac{m_2^2 - m_1^2}{2M}\right]c. \tag{g}$$

$$\text{Similarly, } E_2 = \left[\frac{M}{2} + \frac{m_1^2 - m_2^2}{2M}\right]c. \tag{h}$$

(b) By algebraic manipulation, we rewrite (e) as follows:

$$p_1^2 = \frac{c^2\left\{M^2 - (m_1 + m_2)^2\right\}\left\{M^2 - (m_1 - m_2)^2\right\}}{4M^2} = p_2^2$$

showing symmetry between m_1 and m_2.

(c)

$$Q = M - m_1 - m_2 = \text{mass excess.}$$

$$T_1 = E_1 - m_1c^2 = \frac{c^2}{2M}\left(M^2 + m_2^2 - m_1^2\right) - m_1c^2,$$

$$\text{or } \frac{2M}{c^2}T_1 = M^2 + m_1^2 - m_2^2 - 2Mm_1 = (M - m_1)^2 - m_2^2$$

$$= (M - m_1 - m_2)(M - m_1 + m_2) = Q(2M - 2m_1 - Q),$$

$$\text{or } T_1 = \frac{Q(2M - 2m_1 - Q)}{2M}c^2.$$

$$\text{Similarly, } T_2 = \frac{Q(2M - 2m_2 - Q)}{2M}c^2.$$

$$\text{Adding, } T_1 + T_2 = \frac{Q(2M + 2(M - m_1 - m_2 - Q)}{2M}c^2 = Qc^2.$$

(d) $M = 139.5\,\text{MeV}/c^2$; $m_1 = 105.7\,\text{MeV}/c^2$; $m_2 = 0\,\text{MeV}/c^2$;

$Q = 139.7 - 105.7 = 33.9\,\text{MeV}/c^2$.

$$T_1 = \frac{Q(2M - 2m_1 - Q)}{2M}c^2 = \frac{Q^2}{2M}c^2 = \frac{33.9^2}{2 \times 139.6} = 4.11\,\text{MeV}.$$

$$T_2 = \frac{Q(2M - 2m_2 - Q)}{2M}c^2 = \frac{Q(2M - Q)}{2M}c^2$$

$$= \frac{33.9 \times (2 \times 139.6 - 33.9)}{2 \times 139.6} = 29.78\,\text{MeV}.$$

Problem 4.2. A particle is moving with velocity $\mathbf{v}$ along the XY-plane, making an angle θ with the X-axis, as seen from the frame S.

(i) Find the magnitude of the velocity of this particle in the frame S' under the boost:$S(c\beta, 0, 0)S'$.
(ii) Show that the angle θ' of this velocity with the X'-axis is given by the *angle transformation formula*:

$$\tan\theta' = \frac{\nu \sin\theta}{\gamma(\nu\cos\theta - \beta)}. \tag{4.133}$$

(iii) Specialize to the case of a photon, and show that under the above boost, the magnitude of the velocity does not change. However, the angle of inclination changes to

$$\tan\theta' = \frac{\sin\theta}{\gamma(\cos\theta - \beta)}. \tag{4.134}$$

(iv) Modify the above equation to obtain the aberration angle ϕ' of a starlight which comes at an angle ϕ in the Sun's frame (frame of the "fixed stars"). See Fig. 4.12. Show that

$$\tan\phi' = \frac{\gamma(\sin\phi + \beta)}{\cos\phi}. \tag{4.135}$$

Hint: Set $\theta = -(\pi/2 + \phi)$.

Solution. The velocity of the particle in the frame S is $\mathbf{v} = v\,(\cos\theta\,\mathbf{e}_x + \sin\theta\,\mathbf{e}_y)$. Let $\mathbf{v}' = v'(\cos\theta'\,\mathbf{e}_x + \sin\theta'\,\mathbf{e}_y)$ be its velocity in the frame S'.

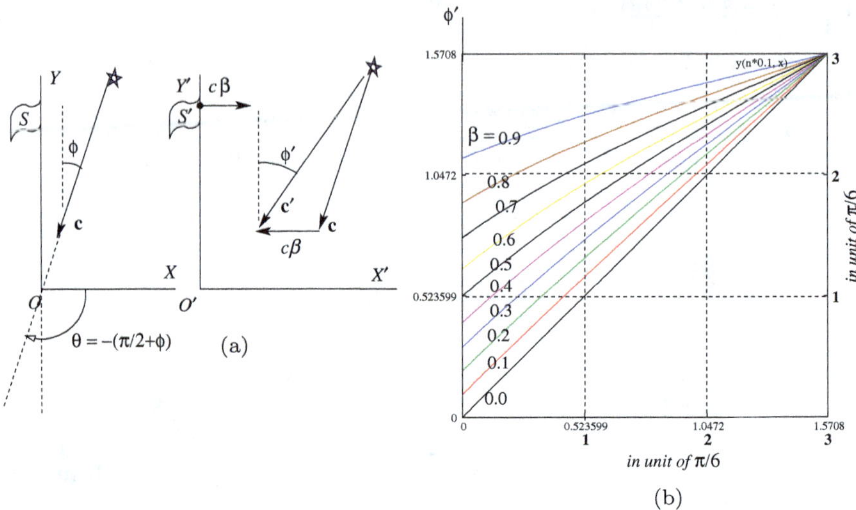

Fig. 4.12. Aberration of starlight. (a) The angles ϕ, ϕ' made by the starlight in the Sun's frame S and in the frame S' moving with speed $c\beta$ with respect to the Sun. (b) Plot of ϕ' vs. ϕ for values of $\beta = 0, 0.1, 0.2, \ldots, 0.9$.

Setting $v = c\nu$, and using velocity addition formula (4.23),

$$\nu' \cos\theta' = \frac{\nu\cos\theta - \beta}{1 - \beta\nu\cos\theta}, \tag{a}$$

$$\nu' \sin\theta' = \frac{\nu\sin\theta}{\gamma(1 - \beta\nu\cos\theta)}, \tag{b}$$

$$\nu'^2 = \left[\left\{\frac{\nu\cos\theta - \beta}{1 - \beta\nu\cos\theta}\right\}^2 + \left\{\frac{\nu\sin\theta}{\gamma(1 - \beta\nu\cos\theta)}\right\}^2\right] \tag{c}$$

$$= 1 + \frac{\beta^2(1 - \nu^2)}{(1 - \beta\nu\cos\theta)^2}. \tag{d}$$

$$\tan\theta' = \frac{\nu\sin\theta}{\gamma(\nu\cos\theta - \beta)}. \tag{f}$$

Answers to parts (i) and (ii) are given in Eqs. (d) and (f). That to part (iii) is obtained by setting $\nu = 1$.

We have explained the light aberration problem in Fig. 4.12(a). In Fig. 4.12(b), we have plotted ϕ' vs. ϕ, making tick marks on the axes at intervals of $\pi/6$.

Earth moves along its orbit with a speed of 30 km/s, which means $\beta = 10^{-4}$. This makes the $\phi' \approx \phi$. In actual terms, $\phi' \approx \beta = 10^{-4}$ radian, for $\phi = 0$, which is approximately equal to 20 arcseconds, when the star is at the zenith.

Problem 4.3. Let one of the daughters of Problem 4.1, say D_2, be a particle of zero rest mass (e.g. photon, neutrino). (One typical such reaction is the transition of an atom or nuclei from an excited state to the ground level, or any lower level, accompanied by the emission of a photon.) Show that for this case

$$p_1 c = \frac{(M^2 - m_1^2)}{2M} c^2 \tag{4.136a}$$

$$E_1 = \frac{(M^2 + m_1^2)}{2M} c^2. \tag{4.136b}$$

<u>*Solution.*</u> Use Eq. (4.130). Set $m_2 = 0$.

Problem 4.4. Let an atom (or a nucleus) make a radiative transition (i.e. the process is accompanied by the emission of a photon) from an excited state of energy ϵ (measured from the ground state) to the ground state. Let the mass of the atom (or the nucleus) at its *ground state* be M_0. Using Eq. (4.130), or starting from energy–momentum conservation equations, show that the frequency ν of the emitted photon is given by the relation

$$h\nu = \epsilon \left[1 - \frac{\epsilon}{2(\epsilon + M_0 c^2)}\right], \tag{4.137}$$

where h is Planck's constant.

<u>*Solution.*</u> Using Eq. (4.130), setting $m_1 = 0$, $E_1 = h\nu$, $M = M_0 + \varepsilon/c^2$, $m_2 = M_0$, we get the answer.

Problem 4.5. An object of mass m_1 and moving with momentum p_1 collides and coalesces with a stationary object of mass m_2. Show that the mass M of the compound object thus formed is given by the relation:

$$M^2 c^2 = m_1^2 c^2 + m_2^2 c^2 + 2m_2 \sqrt{p_1^2 c^2 + m_1^2 c^4}. \tag{4.138}$$

<u>*Solution.*</u> Let E_1, E_2 represent the energies of the objects before collision. Let P, E, M represent the momentum, energy and mass of the coalesced

object (after collision). Using Eq. (4.83), we get

$$E_1 = \sqrt{m_1^2c^4 + p_1^2c^2}; \quad E_2 = m_2c^2.$$

$$\text{Energy Conservation:} \quad E = E_1 + E_2 = \sqrt{m_1^2c^4 + p_1^2c^2} + m_2c^2.$$

$$\text{Momentum Conservation:} \quad P = p_1.$$

$$\begin{aligned}\text{Using Eq. (4.83) again:} \quad M^2c^2 &= E^2/c^2 - p_1^2 \\ &= m_1^2c^2 + m_2^2c^2 + 2m_2\sqrt{p_1^2c^2 + m_1^2c^4}.\end{aligned}$$

Problem 4.6. In a $\{p + {}^7\text{Li} \rightarrow {}^8\text{Be}^* \rightarrow \alpha + \alpha\}$ reaction, protons of energy 3.00 MeV fuse with a stationary ^{7}Li nucleus to form a compound nucleus ^{8}Be*, which is the nucleus of ^{8}Be in an excited state. The compound nucleus has a short life time $\sim 10^{-16}$ s. It breaks up into two α particles, moving in opposite directions.[e]

(a) Determine the mass, momentum and recoil velocity of the compound nucleus in the Lab frame. Is this velocity relativistic?
(b) What is the energy level of the compound nucleus (i.e. the energy of the nucleus, in MeV, above the ground level of ^{8}Be)?
(c) Find the kinetic energy, momentum and velocity of each α particle in the rest frame of the compound nucleus.

Solution.

Answers to Part (a). We have explained the nuclear reaction in Fig. 4.13

In the following, we shall use the symbol "u" to mean *atomic mass unit.*

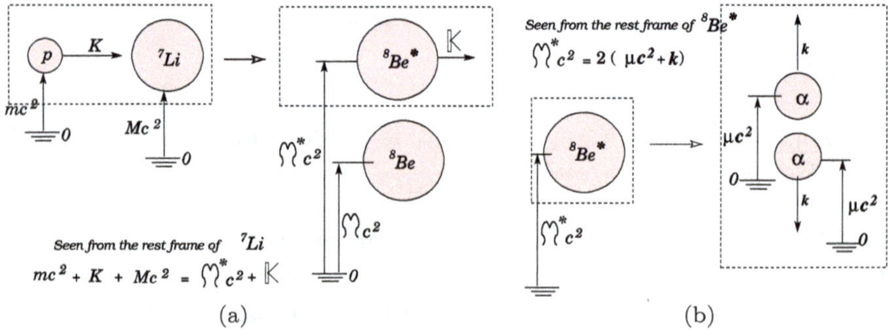

Fig. 4.13. Energy conservation in the nuclear reaction: $\{p + {}^7\text{Li} \rightarrow {}^8\text{Be}^* \rightarrow \alpha + \alpha\}$ reaction. (a) the reaction $\{p + {}^7\text{Li} \rightarrow {}^8\text{Be}^*\}$; (b) the reaction $\{{}^8\text{Be}^* \rightarrow \alpha + \alpha\}$.

[e]See [15, p. 203].

We shall use Eq. (4.138) in which particle # 1 will represent the proton, particle # 2, the Lithium nucleus ^{7}Li. The energy of the incident particle, i.e. the proton, is non-relativistic, being only 3 MeV compared the rest energy of the proton which is $\approx 1\,\text{u.c}^2 = 931.48\,\text{MeV}$.

We shall write (m, p) for the mass and the momentum of the proton, M for the mass of the target ^{7}Li nucleus, $\mathfrak{M}^*$ for the mass of the compound nucleus $^8\text{Be}^*$ in the resulting excited state, $\mathfrak{M}$ for the mass of the same nucleus in its ground state, and set

$$\delta M \equiv \mathfrak{M}^* - (M + m). \tag{4.139}$$

Equation (4.138) now becomes

$$\mathfrak{M}^{*2} = m^2 + M^2 + 2M\sqrt{\left(\frac{p}{c}\right)^2 + m^2} \tag{4.140a}$$

$$= (m + M)^2 - 2Mm + 2M\sqrt{\left(\frac{p}{c}\right)^2 + m^2}, \tag{4.140b}$$

$$\text{or} \quad [\mathfrak{M}^* + (m + M)][\mathfrak{M}^* - (m + M)] + 2mM$$

$$= 2M\sqrt{\left(\frac{p}{c}\right)^2 + m^2}, \tag{4.140c}$$

$$\text{or,} \quad [2(M + m) + \delta M]\delta M + 2mM = 2M\sqrt{\left(\frac{p}{c}\right)^2 + m^2}. \tag{4.140d}$$

Let K represent the K.E. of the proton. Then, in this N.R. case

$$\begin{aligned} K = \frac{p^2}{2m}; \Rightarrow \left(\frac{p}{c}\right)^2 = \frac{2mK}{c^2}. \\ \text{However,} \quad \left(\frac{p}{mc}\right)^2 = \frac{2K}{mc^2} \ll 1. \end{aligned} \tag{4.141}$$

Hence,

$$\sqrt{\left(\frac{p}{c}\right)^2 + m^2} = m\sqrt{1 + \left(\frac{p}{mc}\right)^2} \approx m\left(1 + \frac{1}{2}\left(\frac{p}{mc}\right)^2\right). \tag{4.142}$$

Therefore, from (4.140d)

$$[2(M + m) + \delta M]\,\delta M + 2mM = 2M\sqrt{\left(\frac{p}{c}\right)^2 + m^2}, \tag{4.143a}$$

$$\text{or,} \quad 2(M + m)\,\delta M + 2mM \approx 2Mm\left(1 + \frac{1}{2}\left(\frac{p}{mc}\right)^2\right), \tag{4.143b}$$

$$\text{or,}\quad 2(M+m)\,\delta M = Mm\left(\frac{p}{mc}\right)^2, \tag{4.143c}$$

$$\text{or,}\quad \delta M = \frac{Mm}{2(M+m)}\left(\frac{p}{mc}\right)^2 = \frac{Mm}{2(M+m)}\left(\frac{2K}{mc^2}\right)$$

$$= \frac{Mm}{M+m}\left(\frac{K}{mc^2}\right). \tag{4.143d}$$

Now we collect some mass values from [16].

$$\mathrm{u} = 1\,\text{atomic mass unit} = 931.481 \quad \mathrm{MeV}/c^2, \tag{4.144a}$$

$$1\,\mathrm{u}\times c^2 = 931.481\,\mathrm{MeV}, \tag{4.144b}$$

$$1\,\mathrm{MeV} = \frac{c^2}{931.481} = 1.07356\times 10^{-3}\times c^2 \quad \mathrm{u.(m/s)^2} \tag{4.144c}$$

$$= 9.662\times 10^{13} \quad \mathrm{u.(m/s)^2}, \tag{4.144d}$$

$$m \equiv m(\text{proton}) = 1.00782\,\mathrm{u}, \tag{4.144e}$$

$$M \equiv m(^7\mathrm{Li}) = 7.016004\,\mathrm{u}, \tag{4.144f}$$

$$\mathfrak{M} \equiv m(^8\mathrm{Be}) = 8.005305\,\mathrm{u}, \tag{4.144g}$$

$$\mu \equiv m(\alpha) = 4.0026\,\mathrm{u}. \tag{4.144h}$$

Go back to (4.143d), taking $K = 3\,\mathrm{MeV}$ as given:

$$\begin{aligned}
\frac{K}{mc^2} &= \frac{3}{1.00782\times 931.481} = 3.1957\times 10^{-3},\\
\frac{Mm}{M+m} &= \frac{7.016004\times 1.00782}{7.016004+1.00782} = \frac{7.0709}{8.02382} = 0.88124\,\mathrm{u},\\
\delta M &= 0.88124\times 3.1957\times 10^{-3} = 2.8162\times 10^{-3}\,\mathrm{u}.
\end{aligned} \tag{4.145}$$

Therefore, from (4.139), mass of the compound nucleus $^8\mathrm{Be}^*$:

$$\mathfrak{M}^* = (M+m) + \delta M = 8.02382 + 2.8162\times 10^{-3} = 8.02662\,\mathrm{u}. \tag{4.146}$$

Let us write $(\mathbb{E}, \mathbb{K}, \mathbb{P}, \mathbb{V})$ for the (total energy, kinetic energy, momentum, recoil velocity) of $^8\mathrm{Be}^*$, following the reaction. Since the energy of this nuclide is in the N.R. domain, we can write $\mathbb{E} = \mathbb{K} + \mathfrak{M}^*\,c^2$. See (4.80) and the comment following (4.83). Due to conservation of energy:

$$\begin{aligned} &K(p) + (M+m)c^2 = \mathbb{K} + \mathfrak{M}^* c^2, \\ &\text{or } \mathbb{K} = K(p) + \{(M+m) - \mathfrak{M}^*\}c^2 = K(p) - \delta M c^2 \\ &\quad = 3 - 2.8162 \times 10^{-3} \times 931.481 = 3 - 2.623 = 0.377\,\text{MeV}. \\ &\quad = 0.377 \times 9.662 \times 10^{13}\,\text{u.(m/s)}^2 = 3.6426 \times 10^{13}\,\text{u.(m/s)}^2 \\ &\quad [\text{used (4.144 d)}], \qquad (4.147) \\ &\mathbb{P} = \sqrt{2\,\mathfrak{M}^*\,\mathbb{K}} = \sqrt{2 \times 8.02662 \times 3.6426 \times 10^{13}} \\ &\quad = \sqrt{5.8475} \times 10^7 = 2.418 \times 10^7 \quad \text{u.m/s}, \\ &\mathbb{V} = \frac{\mathbb{P}}{\mathfrak{M}^*} = \frac{2.418}{8.02662} \times 10^7 \approx 3 \times 10^6 \quad \text{m/s} = \frac{c}{100}. \end{aligned}$$

The domain of relativistic dynamics starts at $\sim c/4$. See Example 1 in Sec. 9.7. Hence the recoil velocity $\mathbb{V}$ is non-relativistic.

Answer to Part (b). We shall find the difference between the mass $\mathfrak{M}^*$ of ^{8}Be* and the mass $\mathfrak{M}$ of ^{8}Be.

$$\begin{aligned} \delta M &\equiv \mathfrak{M}^* - (M+m) = 2.8162 \times 10^{-3}\,\text{u}, \\ \Delta M &\equiv (M+m) - \mathfrak{M} = (7.016004 + 1.00782) - 8.005305 \\ &= 18.519 \times 10^{-3}\,\text{u}, \\ (\mathfrak{M}^* - \mathfrak{M}) &= (\delta M + \Delta M) = (2.8162 + 18.519) \times 10^{-3} \\ &= 21.3352 \times 10^{-3}\,\text{u}, \\ (\mathfrak{M}^* - \mathfrak{M})c^2 &= 21.3352 \times 10^{-3} \times 931.481 = 19.87\,\text{MeV}, \qquad (4.148) \end{aligned}$$

which is the energy level of the compound nucleus ^{8}Be* above the ground state of ^{8}Be. This result concurs with the value of 19.9 MeV found in [15, p. 203].

Answer to Part (c). Let us compare the mass of two α-particles equal to 2μ with the mass of ^{8}Be, equal to $\mathfrak{M}$.

$$\begin{aligned} 2\mu - \mathfrak{M} &= 2 \times 4.0026 - 8.005305 = 8.0052 - 8.0053 = -0.00001\,\text{u} \\ &= -0.0931\,\text{MeV}/c^2. \end{aligned}$$

Compare this with the energy level (−0.096 MeV) for 2α shown in [15, p. 203]. The small discrepancy can be due to error margins of the mass values taken from [16].

Let us follow up by energy conservation of the decaying nucleus $^8\text{Be}^*$ in its rest frame. Let $k(\alpha)$ stand for the kinetic energy of each α particle (they are equal, as they move apart in opposite directions with equal and opposite momenta).

$$\begin{aligned}
\mathfrak{M}^* c^2 &= 2(\mu c^2 + k(\alpha)),\\
\text{or,}\quad 8.0266 &= 2 \times (4.0026 + k(\alpha)/c^2), \quad \text{from (4.146) and (4.144h)},\\
\text{or,}\quad k(\alpha)/c^2 &= \frac{1}{2} \times (8.0266 - 8.0052) = 0.0214/2 = 0.0107\,\text{u},\\
\text{or,}\quad k(\alpha) &= 0.0107 \times 9 \times 10^{16} = 9.63 \times 10^{14}\,\text{u.(m/s)}^2\\
&= \frac{9.63 \times 10^{14}}{9.662 \times 10^{13}} = 9.97\,\text{MeV} \quad \text{[we used (4.144d)]}
\end{aligned} \tag{4.149}$$

Let us compare $k(\alpha)$ with μc^2. From (4.144d) and (4.144f), $\mu c^2 = 4.0026\,\text{u}.c^2 = 4.0026 \times 931.481 = 3728\,\text{MeV}$. Thus, $k(\alpha) \ll \mu c^2$, and the N.R approximation is valid. See the comment following (4.147). Therefore, we can write

$$\begin{aligned}
p(\alpha) &= \sqrt{2\mu k(\alpha)} = \sqrt{2 \times 4.0026 \times 9.63 \times 10^{14}} = 8.78 \times 10^7\,\text{u.m/s}\\
v(\alpha) &= \frac{p(\alpha)}{\mu} = \frac{8.78 \times 10^7\,\text{u.m/s}}{4.0026\,\text{u}} = 2.19 \times 10^7\,\text{m/s} \approx \frac{2}{30}c.
\end{aligned} \tag{4.150}$$

The α particles emerging from the decay of $^8\text{Be}^*$ are far below the relativistic domain.

Problem 4.7. In this exercise, we shall compute the velocity of the zero momentum frame (Sec. 4.7) using three examples.

(a) Consider the first part of the reaction given in Problem 4.5, i.e. $\{p + {}^7\text{Li} \rightarrow {}^8\text{Be}^*\}$. Find the momentum and the velocity of p impinging on the ^7Li nucleus with kinetic energy 3 MeV.
(b) Find the velocity of the CM before and after the reaction, in the Lab frame.
(c) Consider the second part of the reaction given in Problem 4.6, i.e. $\{^8\text{Be}^* \rightarrow \alpha + \alpha\}$. Find the velocity of the CM before and after the reaction, in the ZM frame.

Answer to Part (a). In the following equation, the subscript "$_p$" represents proton.

$$\begin{aligned} p_p &= \sqrt{2m_pK_p} = \sqrt{2 \times 1.00782 \times 3 \times 9.662 \times 10^{13}} \\ &= 2.417 \times 10^7 \,\text{u.m/s}, \\ v_p &= \frac{p_p}{m_p} = \frac{2.417}{1.00782} \times 10^7 = 2.398 \times 10^7 \quad \text{m/s} \approx \frac{1}{10}c. \end{aligned} \tag{4.151}$$

Answer to Part (b). Let us first find the relativistic masses of $p, {}^7\text{Li}, {}^8\text{Be}^*$ in the atomic mass unit u.

$$\begin{aligned} \Gamma_p &= \frac{1}{\sqrt{1-\left(\frac{v_p}{c}\right)^2}} = \frac{1}{\sqrt{1-(0.1)^2}} = 1.01; \\ &\Gamma({}^7\text{Li}) = 1. \Gamma({}^8\text{Be}^*) \approx 1. \\ \therefore m_p &= 1.01 \times 1.00782 = 1.0179; \; m({}^7\text{Li}) \\ &= M = 7.016004; \quad m({}^8\text{Be}^*) = \mathfrak{M}^* = 8.02662. \end{aligned} \tag{4.152}$$

$$\begin{aligned} \therefore v_{\text{cm,before}} &= \frac{p_p + p_{\text{Li}}}{m_p + m_{\text{Li}}} = \frac{2.417 \times 10^7 + 0}{1.016004 + 7.01604} \\ &= 0.3 \times 10^7 \quad \text{m/s} \approx \frac{c}{100}. \\ v_{\text{cm,after}} &= \frac{\mathbb{P}}{\mathfrak{M}^*} \approx \frac{c}{100}. \quad \text{See Eq. (4.147).} \end{aligned}$$

Answer to Part (c). This is too trivial. The momentum of the parent nucleus is zero before the decay. The momenta of the daughter α-particles are equal and opposite. Hence $\mathbf{P} = 0$ before and after the decay process. Hence, $\mathbf{v}_{\text{cm}}$ is zero always.

4.12. Illustrative Numerical Examples II

Example 4.1. We observe two galaxies A and B moving in opposite directions with speeds $0.5c$ and $0.4c$, respectively. What is the velocity of B as seen from galaxy A?

Solution. The specified velocities $v_A = -0.5c$, $v_B = 0.4c$ are with respect to "our" frame S. Let S' be the frame of the galaxy A. Then the required velocity v'_B of B is obtained by applying the velocity addition formula (4.6).

$$v'_B = \frac{0.4 - (-0.5)}{1 - (0.4)(-0.5)}c = 0.75c.$$

Example 4.2. In the Lab frame a π^+ is moving in the positive X-direction with velocity $0.8c$ and a π^- is moving in the negative X-direction with velocity $0.9c$. Find (a) the velocity of π^- in the rest frame of π^+, (b) the velocity of π^+ in the rest frame of π^-.

Solution. This is similar to the previous problem. The given velocities $v_{\pi^+} = 0.8c$, $v_{\pi^-} = -0.9c$ are with respect to "our" frame S. Let us represent the rest frames of π^+ and π^- as S', and S'', respectively. We get the answers as follows.

$$\text{(a)} \quad v'_{\pi^-} = \frac{-0.9 - 0.8}{1 - (-0.9)(0.8)}c = -0.988c.$$

$$\text{(b)} \quad v''_{\pi^+} = \frac{0.8 - (-0.9}{1 - (0.8)(-0.9)}c = 0.988c.$$

Example 4.3. A rod of proper length $L_0 = 2\,\text{m}$ is moving in the $+X$-direction with velocity $0.6c$. An electron is moving in the $-X$-direction with velocity $0.8c$. Find the time T that the electron takes to traverse the length of the rod, as seen from the Lab frame S? Answer the question in the following three ways and see that the answers agree.

(a) View the motion from the frame S.
(b) View the motion from the electron's rest frame E, and convert the time thus obtained into Lab time T. (Hint: use velocity addition formula and Rules 2 and 3)
(c) View the motion from the rest frame of the meter-stick and convert back to Lab time T. (Hint: Use velocity addition formula and LT.)

Solution. (a) Let v_{ap} be the velocity with which the rod and the electron are approaching each other, as seen from the Lab frame S, and let L be the length of the rod in frame S which is related to the proper length L_0 by the length contraction formula (2.26). We need to calculate the

Lorentz factor γ for transformation from the Lab frame to the rest frame of the rod.

$$\begin{aligned}
v_{\mathrm{ap}} &= 0.6c + 0.8c = 1.4c,\\
\gamma &= \frac{1}{\sqrt{1-0.6^2}} = \frac{1}{0.8} = \frac{5}{4},\\
L &= \frac{L_0}{\gamma} = 2 \times 0.8 = 1.6\,\mathrm{m},\\
T &= \frac{L}{v_{\mathrm{ap}}} = \frac{1.6}{1.4c} = \frac{1.6}{1.4 \times 3 \times 10^8} = 0.38 \times 10^{-8}\,\mathrm{s}.
\end{aligned}$$

(b) We shall first find the time τ of traversing the length of the rod in the electron's rest frame E. For this we need to know the velocity $c\beta'$ and the length ℓ' of the rod in the frame E. We apply the velocity addition formula to get the first answer, and the length contraction formula (2.26) to get the second answer. For this second answer, we need the Lorentz factor γ' for transformation from the frame E to the rest frame of the rod:

$$\begin{aligned}
\beta' &= \frac{0.6-(-0.8)}{1-(0.6)(-0.8)} = 0.946, \quad \gamma' = \frac{1}{\sqrt{1-\beta'2}} = 3.04,\\
\text{Ł}' &= \frac{L_0}{\gamma'} = \frac{2}{3.04} = 0.658\,\mathrm{m},\\
\tau &= \frac{L'}{\beta' c} = \frac{0.658}{0.946 \times 3 \times 10^8} = 0.23 \times 10^{-8}\,\mathrm{s}.
\end{aligned}$$

We shall now convert this proper time into the Lab time T using the time dilation formula (2.23). For this, we need the Lorentz factor γ_e for transformation between the Lab frame S and the electron frame E:

$$\begin{aligned}
\gamma_e &= \frac{1}{\sqrt{1-0.8^2}} = \frac{1}{0.6} = \frac{5}{3},\\
T &= \gamma_e \tau = \frac{5}{3} \times 0.23 \times 10^{-8} = 0.38 \times 10^{-8}\,\mathrm{s}.
\end{aligned}$$

(c) This is left as an exercise for the reader.

Example 4.4. A proton is moving in the X-direction with velocity $0.999c$. The rest mass of a proton is $m_p = 1.0078$ u, 1u$= 1.66 \times 10^{-27}$ kg. Find (a) the energy equivalent of 1 u, denoted as $\mathcal{E}$, (b) the energy equivalent of the rest mass of the proton, to be denoted as E_0, (c) the Lorentz factor Γ for the moving proton, (d) the momentum p of the moving proton, (e) the energy E of the moving proton. Express energy in MeV, and momentum in MeV/c. Use the conversion factor $1\,\text{J} = 6.242 \times 10^{12}\,\text{MeV}$.

Solution. Let us denote 1 u as μ. Then $m_p = 1.0078\mu$.

$$\text{(a)}\quad \mathcal{E} = \mu c^2 = 1.66 \times 10^{-27} \times (3 \times 10^8)^2 = 14.94 \times 10^{-11}\,\text{J}$$
$$= 14.94 \times 10^{-11} \times 6.242 \times 10^{12}\,\text{MeV} = 932.56\,\text{MeV},$$

$$\text{(b)}\ E_0 = m_p c^2 = 1.0078\,\mathcal{E} = 939.83\,\text{MeV},$$

$$\text{(c)}\quad \Gamma = \frac{1}{\sqrt{1 - 0.999^2}} = 22.36,$$

$$\text{(d)}\quad p = \Gamma m_p v = \Gamma m_p c\beta = \Gamma m_p c^2 \times \tfrac{\beta}{c}.$$
$$= 22.36 \times 939.83 \times 0.999\,\text{MeV/c} = 20970\,\text{MeV/c},$$

$$\text{(e)}\quad E = \Gamma m_p c^2 = 22.36 \times 939.83 = 20991\,\text{MeV}.$$

4.13. Exercises for the Reader I

Listed below are values of some physical constants which the reader may need in working out some of the exercises. (The mass values are taken from [16].)

$c = 3 \times 10^5\,\text{km/sec.}$, $h =$ Planck's constant $= 6.63 \times 10^{-34}\,\text{J.sec.}$

Masses of e^- (i.e. electron), p (i.e. proton), n (i.e. neutron), α (i.e. alpha-particle), ^{7}Li nucleus are to be taken as $m_e = 5.49 \times 10^{-4}\,\text{u} = 0.511\,\text{MeV}/c^2$, $m_p = 1.0078\,\text{u}$, $m_n = 1.0087\,\text{u}$, $m_\alpha = 4.0026\,\text{u}$; $m\,(^7\text{Li}) = 7.0160\,\text{u}$; where "u" means "atomic mass unit (^{12}C scale)". $1\,\text{u} = 931.48\,\text{MeV}/c^2$. $1\,\text{eV} = 1.6021 \times 10^{-19}\,\text{J}$.

R1 We observe two galaxies A and B moving in opposite directions with speeds $0.5c$ and $0.4c$, respectively. What is the velocity of B as seen from galaxy A?

R2 In the Lab frame a π^+ is moving in the positive X-direction with velocity $0.8c$ and a π^- is moving in the negative X-direction with velocity $0.9c$. Find (a) the velocity of π^- in the rest frame of π^+, (b) the velocity of π^+ in the rest frame of π^-.

R3 A meter-stick (proper length 1 m) is moving in the $+X$-direction with velocity $0.6c$. An electron is moving in the $-X$-direction with velocity $0.8c$.

How long does the electron take to pass the meter stick, as measured in the Lab frame L? Answer the question in the following three ways and see that the answers agree.

(a) View the motion from the frame L.
(b) View the motion from the electron's rest frame E, and convert the time thus obtained into Lab time. (Hint: use velocity addition formula and Rules 2 and 3.)
(c) View the motion from the rest frame of the meter-stick and convert back to Lab time. (Hint: Use velocity addition formula and LT.)

R4 Two straight rods R_1 and R_2 of proper lengths $L_1 = 15\,\text{m}$ and $L_2 = 20\,\text{m}$ are moving in opposite directions with velocities $\beta = \frac{3}{5}$ and $\beta = \frac{4}{5}$ respectively as seen from the Lab frame. Find the time required by the rods to traverse the length of each other. Answer the question in three different ways and see that you get the same answer.

(a) View the motion from the Lab frame L.
(b) View the motion from the rest frame of R_1 and convert the time thus obtained to Lab time.
(c) View the motion from the rest frame of R_2 and convert the time thus obtained to Lab time.

R5 How much weight is gained or lost when 1 tonne ice melts? Take latent heat of fusion of water as $3.34 \times 10^5\,\text{J.kg}^{-1}$.

R6 A proton is moving in the X-direction with velocity $0.999c$ in the Lab frame. (a) Find the energy and momentum of the particle in the Lab frame. (b) Using energy–momentum transformation equations (4.83) determine the energy and momentum of the particle in the frame S' under the boost:$Lab(0.990, 0, 0)S'$.

R7 Answer the same questions, asked in Problem 4.6, for an electron of the same energy.

R8 Answer the same questions, asked in Problem 4.6, for an α particle of the same energy.

R9 Find the momenta and velocities of the following particles. (a) 1 MeV electron, (b) 1 MeV proton, (c) 1 MeV α particle, (d) 1 GeV electron, (e) 1 GeV proton, (f) 1 GeV α particle,

R10 A hydrogen atom decays from the level $n = 2$ to $n = 1$. The energy levels of the hydrogen atom is given by the formula $E_n = -\frac{13.6}{n^2}$ eV. The rest masses of a proton and an electron are $m_p = 938.3\,\text{MeV}$ and $m_e = 0.511\,\text{MeV}$. Using formulas (4.136) and (4.137) determine

(a) the momentum of the recoiling atom,
(b) the energy of the emitted photon.

R11 ^{7}Li nucleus consists of 7 nucleons of which three are protons and four are neutrons. Determine the energy in MeV required to dissociate this nucleus into its seven constituent nucleons. This energy is called the binding energy of the nucleus.

Answers to Selected Exercises

R1 $0.75c$. **R2** $-0.998c$. **R3** 0.19×10^{-8}s. **R4** $120/7c$. **R5** 0.37×10^{-5} gm.

R6(a) $E = 20{,}991\,\text{MeV}$; $pc = 20{,}970\,\text{MeV}$. **R6(b)** $E' = 1631\,\text{MeV}$; $p'c = 1335\,\text{MeV}$.

Part II
Amazing Power of Tensors

Chapter 5

Let Us Know Tensors

We shall now make a preparation to launch a vehicle that will take us from our mundane three-dimensional physical space, namely the *Euclidean space*, and denoted by the symbol E^3, which is encompassed by the three Cartesian axes X, Y, Z, or better still, spanned by the three unit space vectors $\mathbf{i}, \mathbf{j}, \mathbf{k}$, to a more esoteric and adventurous world of four-dimensions with the addition of just one more axis, the time axis cT. This new four-dimensional world will be called *Minkowski space–time*, and denoted by the symbol M^4. At the heart of this mathematical construct is Minkowski's assertion (which we are restating in a modern style) that any phenomenon in physics must be expressible in the form of a tensor equation in which both sides must be a tensor of the same rank, and the same sequence of contravariant and covariant indices. Such a statement, without proper clarification, will scare the reader, as they have done to umpteen ordinary persons, who have been fed with stories and myths of a Relativity demon, living in a lofty mountain, beyond the range of ordinary vision.

Our objective is to demystify this demon. Nothing that Einstein or Minkowski said can be above the level of a student who has studied calculus and the basic principles of electromagnetism. In fact, the basic impulse behind Einstein's construction of Special Relativity, his puzzle and how he overcame it, were worked out in a series of exercises, all of which can be a standard set of homework problems in a serious course in *Electromagnetism* at the undergraduate level.

We have presented an exposition of the original papers of Einstein and Minkowsi in two articles which can be downloaded from the website of this author (see Preface).

In this part of this book, we shall make a special effort to explain what is a tensor [17]. We encounter this strange object as *stress tensor* in engineering mechanics (sometimes without being aware of it). We shall find out how the same object makes a rebirth in M^4 as a 4-tensor. It is our hope that the step-by-step approach we have undertaken in the following sections will equip the reader with the necessary gear for climbing the Relativity mountain.

5.1. Introduction to Tensor

5.1.1. *Vector–tensor analogy*

It may help the reader get an intuitive impression of a tensor if we tell him what is common between a vector and a tensor. In fact going by the gradation of tensor, a scalar quantity, like the electric charge of a particle (e.g., an electron), is a tensor of rank 0. A vector quantity like the momentum of a particle is a tensor of rank 1. The inertia tensor of a rigid body, from which its moment of inertia about any axis can be obtained is a tensor of rank 2.

In this chapter, we shall use the term tensor to mean a tensor of rank 2, even though one can build a tensor of arbitrary rank $n > 2$, in which we are not interested.

In Fig. 5.1(b) we have drawn a vector $\mathbf{V}$, by which we mean a *directed* straight line segment of a *measured length.* We have chosen a set of Cartesian axes X, Y, Z, and projected the vector on these axes, by dropping perpendicular straight lines on these axes, with intercepts V_x, V_y, V_z. We call these intercepts the *scalar-components* of $\mathbf{V}$ associated with the respective axes, their directions represented by the unit vectors (also called the base vectors) $\mathbf{e}_x, \mathbf{e}_y, \mathbf{e}_z$. We can then write $\mathbf{V}$ either as a column matrix, or as a linear superposition of the base vectors with (V_x, V_y, V_z) as the coefficients.

$$\mathbf{V} = \begin{pmatrix} V_x \\ V_y \\ V_z \end{pmatrix} = V_x\mathbf{e}_x + V_y\mathbf{e}_y + V_z\mathbf{e}_z. \tag{5.1}$$

We can project $\mathbf{V}$ on any arbitrary direction represented by the unit vector $\mathbf{n}$ and get the intercept V_n, and call it the scalar component of $\mathbf{V}$ in the direction of $\mathbf{n}$. Mathematically, we obtain V_n by taking a dot product

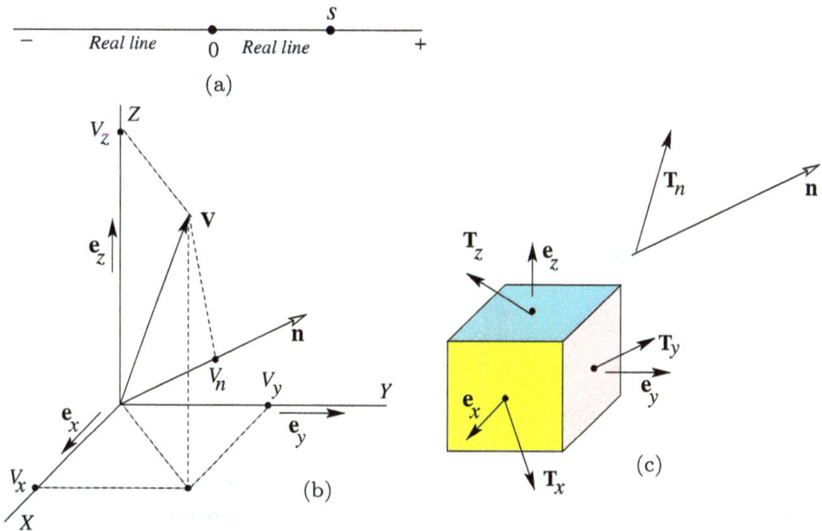

Fig. 5.1. Scalar, vector and tensor as geometrical objects.

of $\mathbf{V}$ with $\mathbf{n}$. Let $\mathbf{n} = n_x\mathbf{e}_x + n_y\mathbf{e}_y + n_z\mathbf{e}_z$. Then

$$V_n = \mathbf{V} \cdot \mathbf{n} = \begin{pmatrix} V_x \\ V_y \\ V_z \end{pmatrix} (n_x\, n_y\, n_z) = V_x n_x + V_y n_y + V_z n_z. \tag{5.2}$$

By analogy, we can think of a mathematical object $\widehat{\mathbf{T}}$ having *vector-components* $\mathbf{T}_x, \mathbf{T}_y, \mathbf{T}_z$ associated with the axes X, Y, Z, as shown in Fig. 5.1(c), and call it a *tensor*. We can then write

$$\widehat{\mathbf{T}} = \begin{pmatrix} \mathbf{T}_x \\ \mathbf{T}_y \\ \mathbf{T}_z \end{pmatrix} = \mathbf{T}_x\mathbf{e}_x + \mathbf{T}_y\mathbf{e}_y + \mathbf{T}_z\mathbf{e}_z. \tag{5.3}$$

We can take a dot product of $\widehat{\mathbf{T}}$ with $\mathbf{n}$, and call it the vector-component of $\widehat{\mathbf{T}}$ associated with the direction $\mathbf{n} = n_x\mathbf{e}_x + n_y\mathbf{e}_y + n_z\mathbf{e}_z$, and represent it by $\mathbf{T}_n$:

$$\mathbf{T}_n = \widehat{\mathbf{T}} \cdot \mathbf{n} = \begin{pmatrix} \mathbf{T}_x \\ \mathbf{T}_y \\ \mathbf{T}_z \end{pmatrix} (n_x\, n_y\, n_z) = \mathbf{T}_x n_x + \mathbf{T}_y n_y + \mathbf{T}_z n_z. \tag{5.4}$$

Note the most distinguishing features of a vector and a tensor. The scalar components (V_x, V_y, V_z) uniquely determine the vector $\mathbf{V}$. The vector components $(\mathbf{T}_x, \mathbf{T}_y, \mathbf{T}_z)$ uniquely determine the tensor $\widehat{\mathbf{T}}$.

Scalars, vectors, tensors are all geometrical objects[a] illustrated in Figs. 5.1(a)–5.1(c). They can be represented, respectively, as a point (on the real line), a measured and directed straight line $\mathbf{V}$, as a triplet of straight lines $(\mathbf{T}_x, \mathbf{T}_y, \mathbf{T}_z)$.

We shall now take a rigorous look at tensor, using a language which can be somewhat abstract.

5.1.2. *Linear operator in a vector space*

We shall begin by explaining what we mean by *linear operator* in a vector space.

By the three-dimensional linear vector space $\mathcal{V}$ we mean the set of all vectors $\mathbf{A}, \mathbf{B}, \mathbf{C}, \ldots$ we can think of and all such vectors we can construct by combining them linearly, e.g. $\eta\mathbf{A} + \lambda\mathbf{B}$ where η, λ are real numbers.

Let us think of two vectors $\mathbf{C}$ and $\mathbf{D}$ having Cartesian components (C_x, C_y, C_z) and (D_x, D_y, D_z) and related to each other in such a way that the values of the former determine the values of the latter. This means that $\mathbf{C}$ is an independent vector and $\mathbf{D}$ is a dependent one. In other words, $\mathbf{D}$ is a function of $\mathbf{C}$. Let us further assume that $\mathbf{D}$ is proportional to $\mathbf{C}$. That is, if for example we double $\mathbf{C}$, then $\mathbf{D}$ is doubled. These two vectors, however, may or may not be in the same direction. In that case, we say that a *linear operator* $\widehat{\mathcal{O}}$ transforms $\mathbf{C}$ into $\mathbf{D}$. We may like to write this transformation symbolically as

$$\widehat{\mathcal{O}}(\mathbf{C}) = \mathbf{D}. \tag{5.5}$$

The property of linearity means that

$$\text{If } \widehat{\mathcal{O}}(\mathbf{C}) = \mathbf{D} \text{ and } \widehat{\mathcal{O}}(\mathbf{E}) = \mathbf{F}, \text{ then } \widehat{\mathcal{O}}(a\mathbf{C} + b\mathbf{E}) = a\mathbf{D} + b\mathbf{F}, \tag{5.6}$$

where a, b are two arbitrary scalar constants.

In Fig. 5.2 we have shown two simple examples of how the operation $\widehat{\mathcal{O}}$ can take place. In Fig. 5.2(a), we have shown a particle of constant mass m in arbitrary motion along some trajectory Γ. At some instant of time t it has velocity $\mathbf{v}$. Therefore, its momentum at the same instant is $\mathbf{p} = m\mathbf{v}$. We can therefore think of the operator $\widehat{\mathcal{O}}$ transforming velocity

[a] As stressed by Misner *et al.* [5, Sec. 2.2].

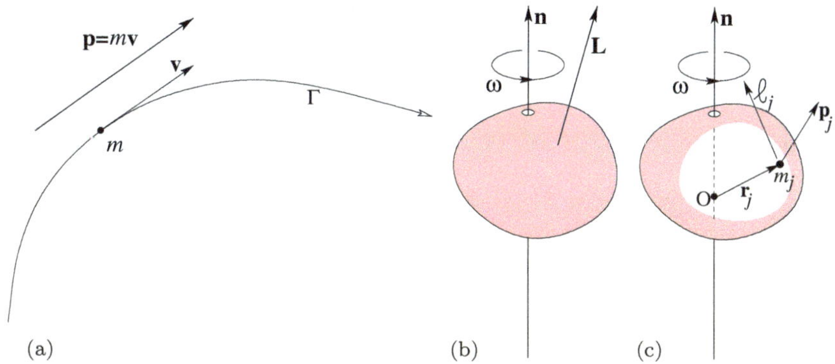

Fig. 5.2. Two examples of how a linear operator $\widehat{\mathcal{O}}$ transforms a vector into another vector: (a) $\widehat{\mathcal{O}}$ acting on $\mathbf{v}$ yields $\mathbf{p}$; (b,c) $\widehat{\mathcal{O}}$ acting on $\boldsymbol{\omega}$ yields $\mathbf{L}$.

$\mathbf{v}$ into momentum $\mathbf{p}$ by scaling the length of the former by the factor m without changing its direction.

In Fig. 5.2(b), we have shown a rigid body rotating about some axis pointing in the direction of the unit vector $\mathbf{n}$ with angular speed ω, so that its angular velocity is $\boldsymbol{\omega} = \omega\mathbf{n}$. Its angular momentum is $\mathbf{L}$, which (in general) does not coincide with the direction of $\boldsymbol{\omega}$. In this case, the operator $\widehat{\mathcal{O}}$ transforms the angular velocity $\boldsymbol{\omega}$ into angular momentum $\mathbf{L}$ by changing the length as well as the direction. The linear operator $\widehat{\mathcal{O}}$ in this case is the inertia tensor $\widehat{\mathcal{I}}$ about which we shall give some more insight in Sec. 5.1.5.

For our immediate purpose, we shall look upon a tensor $\widehat{\mathbf{T}}$ as a linear operator. The linear operation mentioned above suggests that $\widehat{\mathbf{T}}$ can be represented by a matrix, and the "tensor operation" can be represented as a matrix multiplication. This will become evident in the next section.

5.1.3. *Tensor as a dyadic*

Two arbitrary vectors $\mathbf{A}, \mathbf{B}$ can be combined in three types of "multiplication operation", the first two of which the reader is familiar with, namely, (1) the dot product $\mathbf{A} \cdot \mathbf{B}$ which is a scalar; (2) the cross product $\mathbf{A} \times \mathbf{B}$ which is a vector. Now comes (3) the third type, namely the *dyadic product* $\mathbf{AB}$, which is a simple juxtaposition of the vectors, without any dot or cross in between, which we shall call a *dyad*.[b]

[b]See a good book on intermediate mechanics, e.g. Ref. [18, Chapter 10], which also discusses Inertia Tensor in detail.

We define the *dyad* $\mathbf{AB}$ to be a *linear operator* which converts any vector $\mathbf{C}$ to another vector $\mathbf{D}$ and this conversion can be done in either of the following two ways:

$$\begin{aligned} &\text{operating } \textit{on the right}\text{: } \mathbf{AB}\cdot\mathbf{C} \stackrel{\text{def}}{=} \mathbf{A}(\mathbf{B}\cdot\mathbf{C}) = \eta\mathbf{A} \\ &\quad \text{where } \eta = \mathbf{B}\cdot\mathbf{C} = \text{scalar;} \end{aligned} \tag{5.7a}$$

$$\begin{aligned} &\text{operating } \textit{on the left}\text{: } \mathbf{C}\cdot\mathbf{AB} \stackrel{\text{def}}{=} (\mathbf{C}\cdot\mathbf{A})\mathbf{B} = \lambda\mathbf{B} \\ &\quad \text{where } \lambda = \mathbf{C}\cdot\mathbf{A} = \text{scalar.} \end{aligned} \tag{5.7b}$$

The linearity property follows from the operation defined in (5.7). Also note that in general, $\mathbf{AB} \neq \mathbf{BA}$.

We shall write the sum of two dyads $\mathbf{AB}$ and $\mathbf{EF}$ as $\mathbf{AB}+\mathbf{EF}$ and define it by the distributive property:

$$\begin{aligned} (\mathbf{AB}+\mathbf{EF})\cdot\mathbf{C} &\stackrel{\text{def}}{=} \mathbf{AB}\cdot\mathbf{C}+\mathbf{EF}\cdot\mathbf{C} = \mathbf{A}(\mathbf{B}\cdot\mathbf{C})+\mathbf{E}(\mathbf{F}\cdot\mathbf{C}), \\ \mathbf{C}\cdot(\mathbf{AB}+\mathbf{EF}) &\stackrel{\text{def}}{=} \mathbf{C}\cdot\mathbf{AB}+\mathbf{C}\cdot\mathbf{EF} = (\mathbf{C}\cdot\mathbf{A})\mathbf{B}+(\mathbf{C}\cdot\mathbf{E})\mathbf{F}. \end{aligned} \tag{5.8}$$

It should be a simple exercise to show from Eq. (5.7) that the dyadic product is distributive, i.e. if $\mathbf{E}, \mathbf{F}, \mathbf{C}$ are three arbitrary vectors, then

$$\begin{aligned} (\mathbf{E}+\mathbf{F})\mathbf{C} &= \mathbf{EC}+\mathbf{FC}, \\ \mathbf{C}(\mathbf{E}+\mathbf{F}) &= \mathbf{CE}+\mathbf{CF}. \end{aligned} \tag{5.9}$$

As a corollary,

$$(\mathbf{A}+\mathbf{B})(\mathbf{E}+\mathbf{F}) = \mathbf{AE}+\mathbf{AF}+\mathbf{BE}+\mathbf{BF}. \tag{5.10}$$

A sum of dyads can be called a *dyadic*. We shall prefer to use the term "dyadic" as a general name for sums of dyads as well as individual dyads.

We shall frequently use the symbols $\mathbf{e}_x, \mathbf{e}_y, \mathbf{e}_z$ to represent unit vectors in the directions of the X-, Y-, Z-axes, for which we had used $\mathbf{i}, \mathbf{j}, \mathbf{k}$ earlier in this chapter. As we progress, we shall use another set of symbols $\mathbf{e}_1, \mathbf{e}_2, \mathbf{e}_3$ to mean the same unit vectors. This transition $(\mathbf{i}, \mathbf{j}, \mathbf{k}) \to (\mathbf{e}_x, \mathbf{e}_y, \mathbf{e}_z) \to (\mathbf{e}_1, \mathbf{e}_2, \mathbf{e}_3)$, side by side with $(x, y, z) \to (x_1, x_2, x_3)$ will restore symmetry and help us use Einstein's summation convention (following Eq. (5.16).

The unit vectors $(\mathbf{e}_1, \mathbf{e}_2, \mathbf{e}_3)$, in both Cartesian and spherical coordinate systems, form an *orthogonal right handed triple* and this property is expressed as

$$\begin{aligned} \mathbf{e}_1 \cdot \mathbf{e}_2 &= \mathbf{e}_2 \cdot \mathbf{e}_3 = \mathbf{e}_3 \cdot \mathbf{e}_1 = 0, \\ \mathbf{e}_1 \cdot \mathbf{e}_1 &= \mathbf{e}_2 \cdot \mathbf{e}_2 = \mathbf{e}_3 \cdot \mathbf{e}_3 = 1, \\ \mathbf{e}_1 \times \mathbf{e}_2 &= \mathbf{e}_3; \quad \mathbf{e}_2 \times \mathbf{e}_3 = \mathbf{e}_1; \quad \mathbf{e}_3 \times \mathbf{e}_1 = \mathbf{e}_2; \end{aligned} \tag{5.11}$$

or, more compactly as

$$\begin{aligned} \mathbf{e}_i \cdot \mathbf{e}_j &= \delta_{ij}, \\ \mathbf{e}_i \times \mathbf{e}_j &= \varepsilon_{ijk}\mathbf{e}_k, \end{aligned} \tag{5.12}$$

where δ_{ij}, called *Kronecker delta,* and ε_{ijk}, called *Levi-Civita Symbol,* are defined as

$$\delta_{ij} \equiv \delta^{ij} \equiv \delta^i_j = \begin{cases} 1, & \text{if } i = j, \\ 0, & \text{if } i \neq j. \end{cases} \tag{5.13}$$

$$\varepsilon_{ijk} = \begin{cases} 1 & \text{if } ijk = 123, 231, 312, \\ -1 & \text{if } ijk = 213, 321, 132, \\ 0 & \text{if } i = j, \text{ or } j = k, \text{ or, } k = i. \end{cases} \tag{5.14}$$

Let us now consider the set of 12 dyads: $\{\mathbf{e}_x\mathbf{e}_x, \mathbf{e}_x\mathbf{e}_y, \mathbf{e}_x\mathbf{e}_z, \ldots, \mathbf{e}_z\mathbf{e}_z\}$. Using them we can construct the following dyadic

$$\begin{aligned} \widehat{\mathbf{T}} &= T_{xx}\mathbf{e}_x\mathbf{e}_x + T_{yx}\mathbf{e}_y\mathbf{e}_x + \cdots + T_{yz}\mathbf{e}_y\mathbf{e}_z + T_{zz}\mathbf{e}_z\mathbf{e}_z \\ &= \sum_{i=1}^{3}\sum_{j=1}^{3} T_{ij}\mathbf{e}_i\mathbf{e}_j \equiv T_{ij}\mathbf{e}_i\mathbf{e}_j, \end{aligned} \tag{5.15}$$

where the subscripts $(1, 2, 3)$ represent (x, y, z), respectively. That is

$$\begin{aligned} \mathbf{e}_1 \equiv \mathbf{e}_x; \quad \mathbf{e}_2 &\equiv \mathbf{e}_y; \quad \mathbf{e}_3 \equiv \mathbf{e}_z; \\ \text{and,} \quad T_{11} \equiv T_{xx}; \quad T_{12} &\equiv T_{xy}; \cdots; \\ T_{32} \equiv T_{zy}; \quad T_{33} &= T_{zz} \end{aligned} \tag{5.16}$$

are arbitrary real numbers.

In the second line of Eq. (5.15), we have introduced *Einstein's summation convention*: *sum over repeated index,* without explicitly inserting the sum symbol $\sum$. The subscript "$_i$" appears twice, implying a sum over i. The subscript "$_j$" appears twice, implying one more sum, this time over j.

The mathematical object $\widehat{\mathbf{T}}$ appearing in Eq. (5.15) is what we shall call a *tensor* for all purposes in this book. The set of dyads $\{\mathbf{e}_x\mathbf{e}_x, \mathbf{e}_x\mathbf{e}_y, \mathbf{e}_x\mathbf{e}_z, \ldots, \mathbf{e}_z\mathbf{e}_z\}$ can be looked upon as a complete set of base dyads forming a *basis* $\widehat{\mathcal{B}}$ in the tensor space $\mathfrak{T}$ of $\widehat{\mathbf{T}}$. This is analogous to the way that the vectors $\{\mathbf{e}_x, \mathbf{e}_y, \mathbf{e}_z\}$ form a *basis* $\mathcal{B}$ in the vector space $\mathfrak{V}$ of $\mathbf{V}$. Any arbitrary vector $\mathbf{V}$ can be written as a linear superposition of the base vectors as

$$\mathbf{V} = V_x\mathbf{e}_x + V_y\mathbf{e}_y + V_z\mathbf{e}_z, \tag{5.17a}$$

$$\text{where } V_x = \mathbf{V}\cdot\mathbf{e}_x,\ V_y = \mathbf{V}\cdot\mathbf{e}_y,\ V_z = \mathbf{V}\cdot\mathbf{e}_z, \tag{5.17b}$$

are the Cartesian (scalar) components of $\mathbf{V}$ in the basis $\mathcal{B}$. In the same way any arbitrary tensor $\widehat{\mathbf{T}}$ can be written as a linear superposition of the base dyads, as in Eq. (5.15), where the nine quantities $\{T_{xx}, T_{xy}, \ldots T_{zy}, T_{zz}\}$ are to be interpreted as the Cartesian (scalar) *components* of $\widehat{\mathbf{T}}$ with respect to this basis $\widehat{\mathcal{B}}$.

From the definition of dyad given in (5.7), and the orthogonality of the base vectors $\{\mathbf{e}_x, \mathbf{e}_y, \mathbf{e}_z\}$, i.e.

$$\mathbf{e}_j\cdot\mathbf{e}_k = \delta_{jk}, \quad j,k = 1,2,3 = x,y,z, \tag{5.18}$$

it should be apparent that the base dyads operating on any arbitrary vector $\mathbf{V}$ will yield the following vectors:

$$\begin{aligned}
&\mathbf{e}_x\mathbf{e}_x\cdot\mathbf{V} = \mathbf{e}_xV_x; \quad \mathbf{e}_x\mathbf{e}_y\cdot\mathbf{V} = \mathbf{e}_xV_y; \cdots; \mathbf{e}_z\mathbf{e}_y\cdot\mathbf{V} = \mathbf{e}_zV_y;\\
&\mathbf{V}\cdot\mathbf{e}_x\mathbf{e}_x = V_x\mathbf{e}_x; \quad \mathbf{V}\cdot\mathbf{e}_x\mathbf{e}_y = V_x\mathbf{e}_y; \cdots; \mathbf{V}\cdot\mathbf{e}_z\mathbf{e}_y = V_z\mathbf{e}_y.
\end{aligned} \tag{5.19}$$

Hence, if $\mathbf{A} = A_x\mathbf{e}_x + A_y\mathbf{e}_x + A_z\mathbf{e}_z$ and $\mathbf{B} = B_x\mathbf{e}_x + B_y\mathbf{e}_x + B_z\mathbf{e}_z$ are two arbitrary vectors, then, $\mathbf{A}\cdot\widehat{\mathbf{T}}\cdot\mathbf{B} \stackrel{\text{def}}{=} \mathbf{A}\cdot(\widehat{\mathbf{T}}\cdot\mathbf{B}) = A_iT_{ij}B_j = (\mathbf{A}\cdot\widehat{\mathbf{T}})\cdot\mathbf{B}$.

$$\mathbf{A}\cdot\widehat{\mathbf{T}}\cdot\mathbf{B} = A_iT_{ij}B_j \tag{5.20a}$$

$$\text{Special case: } \mathbf{e}_i\cdot\widehat{\mathbf{T}}\cdot\mathbf{e}_j = T_{ij}. \tag{5.20b}$$

If the nine components $\{T_{ij}\}$ of a tensor $\widehat{\mathbf{T}}$ are given, the tensor can be constructed using Eq. (5.15). Conversely, if a tensor $\widehat{\mathbf{T}}$ is given in the form of a mathematical relation, its nine components T_{ij} can be retrieved by means of Eq. (5.20b).

Using the distributive property given in (5.10) it is seen that the dyadic product of $\mathbf{A}$ and $\mathbf{B}$ has the following dyadic representation:

$$\begin{aligned}\mathbf{AB} &= A_xB_x\mathbf{e}_x\mathbf{e}_x + A_xB_y\mathbf{e}_x\mathbf{e}_y + \cdots + A_zB_y\mathbf{e}_z\mathbf{e}_y + A_zB_z\mathbf{e}_z\mathbf{e}_z \\ &= A_iB_j\mathbf{e}_i\mathbf{e}_j. \end{aligned} \tag{5.21}$$

Hence, if we write

$$\widehat{\mathbf{T}} = \mathbf{AB}, \quad \text{then} \quad T_{ij} = A_iB_j. \tag{5.22}$$

Using Eq. (5.7a), the operation of the tensor $\widehat{\mathbf{T}}$ on the vector $\mathbf{C} = C_k\mathbf{e}_k$ placed on the *right* works out as follows:

$$\begin{aligned}\widehat{\mathbf{T}} \cdot \mathbf{C} &= (T_{ij}\mathbf{e}_i\mathbf{e}_j) \cdot (C_k\mathbf{e}_k) \\ &= T_{ij}C_k\mathbf{e}_i(\mathbf{e}_j \cdot \mathbf{e}_k) \\ &= \mathbf{e}_i(T_{ij}C_j). \end{aligned} \tag{5.23}$$

We have used the orthogonality relation (5.18) to get to the last line.

In a similar way, using Eq. (5.7b), the operation of the tensor $\widehat{\mathbf{T}}$ on the vector $\mathbf{C} = C_k\mathbf{e}_k$ placed on the *left* works out as follows.

$$\begin{aligned}\mathbf{C} \cdot \widehat{\mathbf{T}} &= (C_k\mathbf{e}_k) \cdot (T_{ij}\mathbf{e}_i\mathbf{e}_j) \\ &= C_kT_{ij}(\mathbf{e}_k \cdot \mathbf{e}_i)\mathbf{e}_j \\ &= (C_kT_{kj})\mathbf{e}_j. \end{aligned} \tag{5.24}$$

The above two equations suggest that if we write $\mathbf{D} = \widehat{\mathbf{T}} \cdot \mathbf{C}$ and $\mathbf{F} = \mathbf{C} \cdot \widehat{\mathbf{T}}$, then the Cartesian components (D_1, D_2, D_3) of $\mathbf{D}$ and (F_1, F_2, F_3) of $\mathbf{F}$ can be obtained from matrix multiplications:

$$\begin{pmatrix} D_1 \\ D_2 \\ D_3 \end{pmatrix} = \begin{pmatrix} T_{11} & T_{12} & T_{13} \\ T_{21} & T_{22} & T_{23} \\ T_{31} & T_{32} & T_{33} \end{pmatrix} \begin{pmatrix} C_1 \\ C_2 \\ C_3 \end{pmatrix}, \tag{5.25a}$$

$$\begin{pmatrix} F_1 & F_2 & F_3 \end{pmatrix} = \begin{pmatrix} C_1 & C_2 & C_3 \end{pmatrix} \begin{pmatrix} T_{11} & T_{12} & T_{13} \\ T_{21} & T_{22} & T_{23} \\ T_{31} & T_{32} & T_{33} \end{pmatrix}. \tag{5.25b}$$

In the above equations, starting from Eq. (5.7), we have used a dot $(\cdot)$ to separate the tensor from the vector on which it is operating. We shall frequently refer to a tensor operation as a *dot product* between the tensor and the vector. Equations (5.25a) and (5.25b) show that a *dot product actually involves a matrix multiplication.* A tensor is to be represented as a *square matrix*, and a vector either as a *column matrix* or a *row matrix*, depending on whether the tensor operation is on the right or on the left.[c]

$$\widehat{\mathbf{T}} = \begin{pmatrix} T_{11} & T_{12} & T_{13} \\ T_{21} & T_{22} & T_{23} \\ T_{31} & T_{32} & T_{33} \end{pmatrix} = [T], \quad \mathbf{C} = \begin{pmatrix} C_1 \\ C_2 \\ C_3 \end{pmatrix} = \{C\},$$
$$\mathbf{F} = \begin{pmatrix} F_1 & F_2 & F_3 \end{pmatrix} = (F). \tag{5.26}$$

In the above equations, we have adopted the convention of indicating a 3×3 square matrix by [ ], a 3×1 column matrix by { }, and a 1×3 row matrix by ( ). Hence, Eqs. (5.25a) and (5.25b) can be written as

$$\{D\} = [T]\{C\}, \quad (F) = (C)[T]. \tag{5.27}$$

It follows from Eq. (5.21) that the matrix representation of the dyadic **AB** is

$$\mathbf{AB} = \begin{pmatrix} A_1B_1 & A_1B_2 & A_1B_3 \\ A_2B_1 & A_2B_2 & A_2B_3 \\ A_3B_1 & A_3B_2 & A_3B_3 \end{pmatrix}. \tag{5.28}$$

We shall define the dot product of two tensors $\widehat{\mathbf{S}}$ and $\widehat{\mathbf{T}}$ as the tensor $\widehat{\mathbf{R}} = \widehat{\mathbf{S}} \cdot \widehat{\mathbf{T}}$ by its operation on an arbitrary vector $\mathbf{C}$ *on the right* in the following way:

$$(\widehat{\mathbf{S}} \cdot \widehat{\mathbf{T}}) \cdot \mathbf{C} \overset{\text{def}}{=} \widehat{\mathbf{S}} \cdot (\widehat{\mathbf{T}} \cdot \mathbf{C}). \tag{5.29}$$

From this, it follows that the matrix representing $\widehat{\mathbf{R}}$ is given by the product of the matrices representing $\widehat{\mathbf{S}}$ and $\widehat{\mathbf{T}}$. That is,

$$[R] = [S][T], \quad \text{implying: } R_{ij} = S_{ik}T_{kj}. \tag{5.30}$$

It is then obvious that, in general, $\widehat{\mathbf{S}} \cdot \widehat{\mathbf{T}} \neq \widehat{\mathbf{T}} \cdot \widehat{\mathbf{S}}$.

[c]In Quantum Mechanics (QM), a clear distinction is made between a vector **A** on left and a vector **B** on right, as in the scalar product $\mathbf{A} \cdot \mathbf{B}$. The former is called a *bra* vector and the latter a *ket* vector, and together, in the scalar product, they constitute a *bra-ket*: $\mathbf{A} \to \langle A|$; $\mathbf{B} \to |B\rangle$; $\mathbf{A} \cdot \mathbf{B} \to \langle A|B\rangle$. However, these vectors are in general infinite dimensional, their components are complex numbers, and the components of the bra vector $\langle A|$ are complex conjugates of the respective components of the ket vector $|A\rangle$.

Using the matrix representation as given in Eq. (5.30), and the tensor operation on the left as found out in (5.24), we can now see how the product tensor $\widehat{\mathbf{R}} = \widehat{\mathbf{S}} \cdot \widehat{\mathbf{T}}$ will act *on the left.*

$$\begin{aligned} \mathbf{C} \cdot \widehat{\mathbf{R}} &= (C_k R_{kj})\mathbf{e}_j = (C_k S_{km} T_{mj})\mathbf{e}_j \\ &= (C_k S_{km})(T_{mj}\mathbf{e}_j), \\ \text{or} \quad \mathbf{C} \cdot (\widehat{\mathbf{S}} \cdot \widehat{\mathbf{T}}) &= (\mathbf{C} \cdot \widehat{\mathbf{S}}) \cdot \widehat{\mathbf{T}}. \end{aligned} \tag{5.31}$$

We can extend the definition of matrix product to any number of tensors, by writing the matrix representation of the product tensor as the product of the representative matrices of the component tensors. For example,

$$\text{if } \widehat{\mathbf{R}} = \widehat{\mathbf{A}} \cdot \widehat{\mathbf{B}} \cdot \widehat{\mathbf{C}}, \quad \text{then } [R] = [A]\,[B]\,[C]. \tag{5.32}$$

At this point we shall add a word of caution. A tensor is not the same as a square matrix, just as a vector is not the same as a column matrix or a row matrix. The row matrix shown in Eq. (5.26), for example, gives the components of the vector $\mathbf{F}$ in a given coordinate system XYZ. As the coordinates are changed from (x, y, z) to (x', y', z'), the components will transform from (F_1, F_2, F_3) to (F_1', F_2', F_3'). However, the vector $\mathbf{F}$ itself is a "geometrical object" (a straight line of measured length pointing in an assigned direction) which remains invariant under all coordinate transformations. In the same way, the tensor $\widehat{\mathbf{T}}$ is a geometrical object, which remains invariant under all coordinate transformations, even though its components will change from the square matrix $[T_{ij}]$ to another square matrix $[T_{ij}']$ under the same coordinate transformation.

Yes, the components of all tensors will transform, except the components of the identity tensor which we shall introduce in the next section. They will remain the same, the same as in (5.34), following any coordinate transformations.

5.1.4. *Identity tensor, completeness relation, components of a tensor in the spherical coordinate system*

In matrix multiplication one needs the *identity matrix* $\widehat{1}$ which in the present context, is the matrix representation of the *identity tensor,* also known by the alternative name *idemfactor.* It will be recognized by the symbol $\widehat{\mathbf{1}}$. Its sole property is that when it operates on any vector $\mathbf{V}$, either on the right, or on the left, it gives back the same vector.

$$\widehat{\mathbf{1}} \cdot \mathbf{V} \stackrel{\text{def}}{=} \mathbf{V}, \quad \mathbf{V} \cdot \widehat{\mathbf{1}} \stackrel{\text{def}}{=} \mathbf{V}. \tag{5.33}$$

Such a tensor must have $\widehat{1}$ for its matrix representation. The dyadic representation (shown below) follows from the above property and the orthogonality relation (5.18).

$$\widehat{\mathbf{1}} = \widehat{1} = \begin{pmatrix} 1 & 0 & 0 \\ 0 & 1 & 0 \\ 0 & 0 & 1 \end{pmatrix}, \tag{5.34a}$$

$$\widehat{\mathbf{1}} = \mathbf{e}_x\mathbf{e}_x + \mathbf{e}_y\mathbf{e}_y + \mathbf{e}_z\mathbf{e}_z = \mathbf{e}_i\mathbf{e}_i. \tag{5.34b}$$

Equation (5.34a) gives the Matrix representation, and Eq. (5.34b) the dyadic representation.

It will be advantageous to write the tensor $\widehat{\mathbf{T}}$ in a curvilinear coordinate system, in particular, spherical coordinate system. For this purpose, we shall write down the transformation equations for the coordinates and the base vectors:

$$\begin{aligned} x &= r \sin\theta \cos\phi, \quad [0 \le r < \infty], \\ y &= r \sin\theta \sin\phi, \quad [0 \le \theta \le \pi], \\ z &= r \cos\theta, \quad [0 \le \phi < 2\pi]. \end{aligned} \tag{5.35}$$

In the above equations, we have indicated the "ranges" of the three coordinates within the [ ] brackets.

$$\begin{aligned} \mathbf{e}_r &= \sin\theta(\cos\phi\mathbf{e}_x + \sin\phi\mathbf{e}_y) + \cos\theta\mathbf{e}_z, \\ \mathbf{e}_\theta &= \cos\theta(\cos\phi\mathbf{e}_x + \sin\phi\mathbf{e}_y) - \sin\theta\mathbf{e}_z, \\ \mathbf{e}_\phi &= -\sin\phi\mathbf{e}_x + \cos\phi\mathbf{e}_y. \end{aligned} \tag{5.36}$$

Using these equations (and remembering that $\mathbf{e}_r\mathbf{e}_\theta \neq \mathbf{e}_\theta\mathbf{e}_r$, for example), it should be a simple exercise to show that

$$\mathbf{e}_r\mathbf{e}_r + \mathbf{e}_\theta\mathbf{e}_\theta + \mathbf{e}_\phi\mathbf{e}_\phi = \mathbf{e}_x\mathbf{e}_x + \mathbf{e}_y\mathbf{e}_y + \mathbf{e}_z\mathbf{e}_z = \widehat{\mathbf{1}}. \tag{5.37}$$

If we have three unit vectors $\{\boldsymbol{a}, \boldsymbol{b}, \boldsymbol{c}\}$ which are mutually orthogonal at every point in space and such that

$$\boldsymbol{aa} + \boldsymbol{bb} + \boldsymbol{cc} = \widehat{\mathbf{1}}, \tag{5.38}$$

then we say that these three vectors form a *complete orthogonal set*, and hence a basis, so that any arbitrary vector $\mathbf{V}$ can be represented as a linear superposition of these three vectors.[d] This should be clear from the following:

$$\begin{aligned} &\mathbf{V} = \mathbf{V}\cdot\hat{1} = \mathbf{V}\cdot(\boldsymbol{aa}+\boldsymbol{bb}+\boldsymbol{cc}) = V_a\boldsymbol{a}+V_b\boldsymbol{b}+V_c\boldsymbol{c}, \\ \text{where}\quad &V_a = \mathbf{V}\cdot\boldsymbol{a},\ V_b = \mathbf{V}\cdot\boldsymbol{b},\ V_c = \mathbf{V}\cdot\boldsymbol{c}, \end{aligned} \tag{5.39}$$

are the components of $\mathbf{V}$ in the directions of $\{\boldsymbol{a},\boldsymbol{b},\boldsymbol{c}\}$, respectively. Using the completeness property, it can be advantageous to write a tensor in the following style.

$$\begin{aligned} \widehat{\mathbf{T}} &= \hat{1}\cdot\widehat{\mathbf{T}}\cdot\hat{1} = (\boldsymbol{aa}+\boldsymbol{bb}+\boldsymbol{cc})\cdot\widehat{\mathbf{T}}\cdot(\boldsymbol{aa}+\boldsymbol{bb}+\boldsymbol{cc}) \\ &= T_{aa}\boldsymbol{aa}+T_{ab}\boldsymbol{ab}+T_{ac}\boldsymbol{ac}+\cdots+T_{cb}\boldsymbol{cb}+T_{cc}\boldsymbol{cc}, \quad \text{where} \\ T_{aa} &= \boldsymbol{a}\cdot\widehat{\mathbf{T}}\cdot\boldsymbol{a},\ T_{ab} = \boldsymbol{a}\cdot\widehat{\mathbf{T}}\cdot\boldsymbol{b},\ldots,T_{cb} = \boldsymbol{c}\cdot\widehat{\mathbf{T}}\cdot\boldsymbol{b},\ T_{cc} = \boldsymbol{c}\cdot\widehat{\mathbf{T}}\cdot\boldsymbol{c} \end{aligned} \tag{5.40}$$

are the components of $\widehat{\mathbf{T}}$ with respect to the basis $\{\boldsymbol{a},\boldsymbol{b},\boldsymbol{c}\}$.

We shall illustrate the operation shown in Eq. (5.40) by writing the tensor $\widehat{\mathbf{T}}$ in Cartesian and spherical coordinate systems:

$$\begin{aligned} \widehat{\mathbf{T}} &= (\mathbf{e}_x\mathbf{e}_x+\mathbf{e}_y\mathbf{e}_y+\mathbf{e}_z\mathbf{e}_z)\cdot\widehat{\mathbf{T}}\cdot(\mathbf{e}_x\mathbf{e}_x+\mathbf{e}_y\mathbf{e}_y+\mathbf{e}_z\mathbf{e}_z) \\ &= T_{xx}\mathbf{e}_x\mathbf{e}_x+T_{xy}\mathbf{e}_x\mathbf{e}_y+T_{xz}\mathbf{e}_x\mathbf{e}_z+\cdots+T_{zx}\mathbf{e}_z\mathbf{e}_y+T_{zz}\mathbf{e}_z\mathbf{e}_z, \end{aligned} \tag{5.41a}$$

$$T_{xx} = \mathbf{e}_x\cdot\widehat{\mathbf{T}}\cdot\mathbf{e}_x,\ T_{xy} = \mathbf{e}_x\cdot\widehat{\mathbf{T}}\cdot\mathbf{e}_y,\ldots,\ T_{zy} = \mathbf{e}_z\cdot\widehat{\mathbf{T}}\cdot\mathbf{e}_y, \tag{5.41b}$$

$$T_{zz} = \mathbf{e}_z\cdot\widehat{\mathbf{T}}\cdot\mathbf{e}_z, \tag{5.41c}$$

$$\begin{aligned} \widehat{\mathbf{T}} &= (\mathbf{e}_r\mathbf{e}_r+\mathbf{e}_\theta\mathbf{e}_\theta+\mathbf{e}_\phi\mathbf{e}_\phi)\cdot\widehat{\mathbf{T}}\cdot(\mathbf{e}_r\mathbf{e}_r+\mathbf{e}_\theta\mathbf{e}_\theta+\mathbf{e}_\phi\mathbf{e}_\phi) \\ &= T_{rr}\mathbf{e}_r\mathbf{e}_r+T_{r\theta}\mathbf{e}_r\mathbf{e}_\theta+T_{r\phi}\mathbf{e}_r\mathbf{e}_\phi+\cdots+T_{\phi\theta}\mathbf{e}_\phi\mathbf{e}_\theta+T_{\phi\phi}\mathbf{e}_\phi\mathbf{e}_\phi, \end{aligned} \tag{5.41d}$$

$$T_{rr} = \mathbf{e}_r\cdot\widehat{\mathbf{T}}\cdot\mathbf{e}_r,\ T_{r\theta} = \mathbf{e}_r\cdot\widehat{\mathbf{T}}\cdot\mathbf{e}_\theta,\ldots,T_{\phi\theta} = \mathbf{e}_\phi\cdot\widehat{\mathbf{T}}\cdot\mathbf{e}_\theta, \tag{5.41e}$$

$$T_{\phi\phi} = \mathbf{e}_\phi\cdot\widehat{\mathbf{T}}\cdot\mathbf{e}_\phi. \tag{5.41f}$$

Equations (5.41a)–(5.41c) represent the tensor $\widehat{\mathbf{T}}$ in a Cartesian coordinate system, and Eqs. (5.41d)–(5.41f) in a spherical coordinate system.

---

[d]In QM, the completeness of a set of orthonormal vectors $\{|u_i\rangle;\ i = 1,2,\ldots,\infty\}$ is expressed through the statement $\sum_i |u_i\rangle\langle u_i| = 1$. This relation is used to change the representation of a Hermitean operator $\widehat{\mathbf{T}}$, the equivalent of the tensors we are considering here. The (i,j) component of $\widehat{\mathbf{T}}$ will then be written as $T_{ij} = \langle u_i|\widehat{\mathbf{T}}|u_j\rangle$, the equivalent of Eq. (5.20b).

We can then write the components of $\widehat{\mathbf{T}}$ in the following matrix forms:

$$\widehat{\mathbf{T}} \overset{\text{(Cart)}}{\longrightarrow} \begin{pmatrix} T_{xx} & T_{xy} & T_{xz} \\ T_{yx} & T_{yy} & T_{yz} \\ T_{zx} & T_{zy} & T_{zz} \end{pmatrix}, \quad \widehat{\mathbf{T}} \overset{\text{(sphr)}}{\longrightarrow} \begin{pmatrix} T_{rr} & T_{r\theta} & T_{r\phi} \\ T_{\theta r} & T_{\theta\theta} & T_{\theta\phi} \\ T_{\phi r} & T_{\phi\theta} & T_{\phi\phi} \end{pmatrix}. \tag{5.42}$$

The first matrix gives the Cartesian components, and the second one the spherical components.

Using the transformation of the base vectors (5.36), and the completeness relations (5.37), one can transform the Cartesian components to spherical components, for both vectors and tensors, as we shall show. For this purpose, we shall temporarily denote the spherical base vectors with a prime, i.e. $\{\mathbf{e}'_i : i = r, \theta, \phi\}$ and make a table of *transformation coefficients* $\{c_{ij}\}$:

$$\begin{aligned} \mathbf{e}'_i &= \mathbf{e}'_i \cdot \mathbf{e}_j \mathbf{e}_j = c_{ij}\mathbf{e}_j, \\ \text{where} \quad c_{ij} &\equiv \mathbf{e}'_i \cdot \mathbf{e}_j : \; i = r, \theta, \phi; \; j = x, y, z. \\ &= \begin{pmatrix} \sin\theta\cos\phi & \sin\theta\sin\phi & \cos\theta \\ \cos\theta\cos\phi & \cos\theta\sin\phi & -\sin\theta \\ -\sin\phi & \cos\phi & 0 \end{pmatrix} \end{aligned} \tag{5.43}$$

Now, let $\mathbf{V}$ be a vector and $\widehat{\mathbf{T}}$ be a tensor with Cartesian components $[\{V_j\}, \{T_{ij}\}, i, j = x, y, z]$, respectively. Then the spherical components of the same vector and tensor, namely, $[\{V'_j\}, \{T'_{ij}\}, i, j = r, \theta, \phi]$ will be obtained in the following ways[e]:

$$V'_j = \mathbf{V} \cdot \mathbf{e}'_j = \mathbf{V} \cdot \mathbf{e}_k \mathbf{e}_k \cdot \mathbf{e}'_j = c_{jk} V_k, \tag{5.44a}$$

$$\begin{aligned} T'_{ij} &= \mathbf{e}'_i \cdot \widehat{\mathbf{T}} \cdot \mathbf{e}'_j = \mathbf{e}'_i \cdot \mathbf{e}_k \mathbf{e}_k \cdot \widehat{\mathbf{T}} \cdot \mathbf{e}_l \mathbf{e}_l \cdot \mathbf{e}'_j \\ &= c_{ik} c_{jl} T_{kl}. \end{aligned} \tag{5.44b}$$

Note that we have used the summation convention: sum over k in (5.44a), sum over k, l in (5.44b).

[e]In Tensor analysis, the primary language of the theory of relativity, the rule of transformation has different forms for contravariant and covariant vectors, and for contravariant, covariant and mixed tensors. The rules we are establishing here are different from them. The components of vectors, tensors we are using may be called *physical components*, in contrast to their contravariant and covariant components for which a more elegant transformation rule is used.

We shall illustrate the transformation formulas (5.44) with two examples, i.e., $V_r \equiv V_1'$ and $T_{r\theta} \equiv T_{12}'$.

$$\begin{aligned} V_r &= \sin\theta\cos\phi V_x + \sin\theta\sin\phi V_y + \cos\theta V_z, \\ T_{r\theta} &= \sin\theta\cos\phi(\cos\theta\cos\phi T_{xx} + \cos\theta\sin\phi T_{xy} - \sin\theta T_{xz}) \\ &\quad + \sin\theta\sin\phi(\cos\theta\cos\phi T_{yx} + \cos\theta\sin\phi T_{yy} - \sin\theta T_{yz}) \\ &\quad + \cos\theta(\cos\theta\cos\phi T_{zx} + \cos\theta\sin\phi T_{zy} - \sin\theta T_{zz}). \end{aligned} \tag{5.45}$$

5.1.5. *Example: Inertia tensor*

We shall illustrate the tensor concept by showing two important examples, namely (1) the inertia tensor and (2) the stress tensor. We shall take up a short discussion of the first example in this section leaving the second example, which needs a more detailed coverage, to the next section.

In Sec. 5.1.2 we talked about the tensor operation converting the angular velocity $\boldsymbol{\omega}$ into angular momentum $\mathbf{L}$. The corresponding operator is the *inertia tensor* $\widehat{\mathcal{I}}$ of the rigid body. Its dot product with the angular velocity $\boldsymbol{\omega}$ gives the *angular momentum* $\mathbf{L}$ of the rigid body. That is,

$$\mathbf{L} = \widehat{\mathcal{I}} \cdot \boldsymbol{\omega}. \tag{5.46}$$

We shall find an expression for the vector angular momentum $\mathbf{L}$ of a rigid body which is rotating about a point O (which can be a moving point, e.g. the CM) with angular velocity $\boldsymbol{\omega} = \omega\mathbf{n}$ about the axis pointing in the direction of the unit vector $\mathbf{n}$. Let j be one of the constituent particles, having mass m_j, and located at the radius vector $\mathbf{r}_j$ with respect to O, as shown in Fig. 5.2(c). The velocity of this point is $\mathbf{v}_j = \boldsymbol{\omega} \times \mathbf{r}_j$. Therefore, this particle has an angular momentum with respect to the point O, equal to

$$\boldsymbol{\ell}_j = \mathbf{r}_j \times \mathbf{p}_j = \mathbf{r}_j \times m_j\mathbf{v}_j = m_j\mathbf{r}_j \times (\boldsymbol{\omega} \times \mathbf{r}_j) = m_j[r_j^2\boldsymbol{\omega} - (\mathbf{r}_j \cdot \boldsymbol{\omega})\mathbf{r}_j]. \tag{5.47}$$

Assuming that the rigid body is made of N particles (which is a very large number), we add the angular momentum of each particle to obtain the angular momentum of the rigid body about the point O, given as

$$\mathbf{L}_O = \sum_{j=0}^{N} m_j[r_j^2\boldsymbol{\omega} - (\mathbf{r}_j \cdot \boldsymbol{\omega})\mathbf{r}_j]. \tag{5.48}$$

We can write the quantity within square brackets as

$$[r_j^2\boldsymbol{\omega} - (\mathbf{r}_j \cdot \boldsymbol{\omega})\mathbf{r}_j] = [r_j^2\widehat{\mathbf{1}} - \mathbf{r}_j\mathbf{r}_j] \cdot \boldsymbol{\omega}, \tag{5.49}$$

and construct the inertia tensor as the dyadic (sum of infinitely small dyads)

$$\widehat{\mathcal{I}} = \sum_{j=0}^{N} m_j [r_j^2 \widehat{\mathbf{1}} - \mathbf{r}_j \mathbf{r}_j]. \tag{5.50}$$

Then, we get the angular momentum as the dot product

$$\mathbf{L}_O = \widehat{\mathcal{I}} \cdot \boldsymbol{\omega}. \tag{5.51}$$

We have thus derived Eq. (5.46), and along with it have found an expression for the inertia tensor in Eq. (5.50). Note that the expression within the square brackets is the difference of two dyadics, namely, the identity dyadic $\widehat{\mathbf{1}}$ multiplied by the scalar r_j^2, and the dyadic product of $\mathbf{r}_j$ with itself.

For further clarification we shall write down the components of the tensor. Assuming that the rigid body has uniform mass density ρ distributed over its volume V, the sum in Eq. (5.50) becomes the integral:

$$\widehat{\mathcal{I}} = \rho \iiint_V [r^2 \widehat{\mathbf{1}} - \mathbf{r}\mathbf{r}] d^3 r. \tag{5.52}$$

Some of its components are

$$\begin{aligned} \mathcal{I}_{xx} &= \rho \iiint_V [r^2 - x^2] d^3 r = \rho \iiint_V (y^2 + z^2) d^3 r; \\ \mathcal{I}_{xy} &= -\rho \iiint_V (xy)\, d^3 r; \quad \text{etc.} \end{aligned} \tag{5.53}$$

It is now seen that the *inertia tensor is a symmetric tensor*, i.e.

$$\mathcal{I}_{xy} = \mathcal{I}_{yx}; \quad \mathcal{I}_{yz} = \mathcal{I}_{zy}; \quad \mathcal{I}_{zx} = \mathcal{I}_{xz}. \tag{5.54}$$

This symmetry property is preserved under all coordinate transformations.

5.2. Stress in a Medium

5.2.1. *Stress vector*

By (mechanical) stress we mean *internal forces* (in the form of intermolecular forces) called into play when bulk matter, either in the form of solid, liquid or gas, is subjected to external forces. These internal forces exist throughout the bulk matter and its mathematical expression is given by a stress tensor field $\widehat{\mathcal{T}}(x, y, z)$.

For simplicity we shall consider a solid block in Fig. 5.3(a). It has been cut into two parts, the upper block $\mathcal{U}$ and the lower block $\mathcal{L}$, by an imaginary plane Σ, leaving a trace Γ of its boundary. This plane is identified by the unit normal vector $\mathbf{n}$ pointing from the lower block to the upper block.

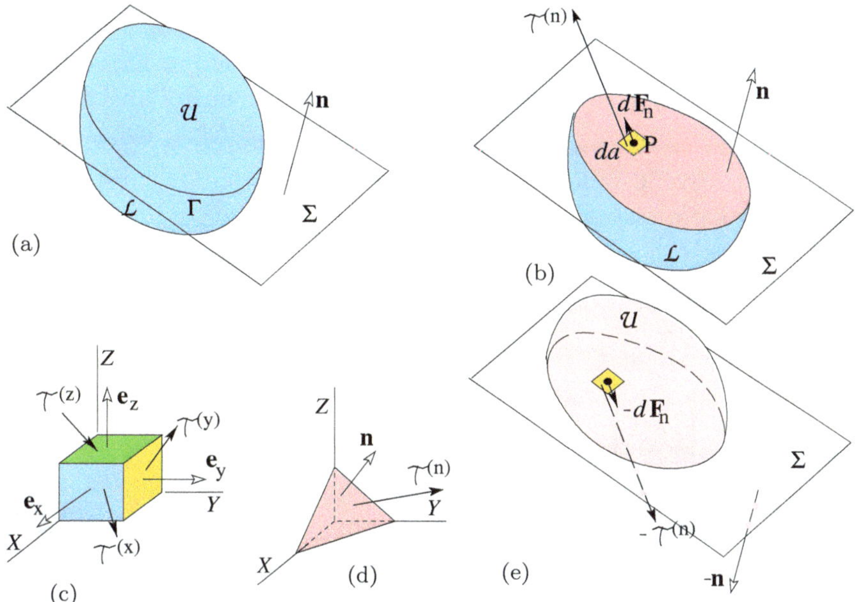

Fig. 5.3. Explaining the stress tensor.

In Fig. 5.3(b), we have shown the lower block $\mathcal{L}$ with the plane of separation Σ exposed. Let us consider a small area da at the point $\mathrm{P}(x, y, z)$ inside the solid, but lying on this plane. Then the *stress vector* $\widehat{\mathcal{T}}^n(x, y, z)$ is defined to be the force per unit area at $\mathrm{P}(x, y, z)$, exerted by the atoms of the upper block $\mathcal{U}$ on the atoms of the lower block $\mathcal{L}$ across the plane $\mathbf{n}$. The infinitesimal force acting on the area da is then

$$d\mathbf{F}^{(n)} = \widehat{\mathcal{T}}^n(x, y, z)\, da. \tag{5.55}$$

Note that in general the direction of the stress vector $\widehat{\mathcal{T}}^n(x, y, z)$ is different from the direction of the normal $\mathbf{n}$. If, however, $\widehat{\mathcal{T}}^n(x, y, z) \parallel \mathbf{n}$ (i.e. perpendicular to the plane), the stress (vector) is called *normal stress.* If $\widehat{\mathcal{T}}^n(x, y, z) \perp \mathbf{n}$ (i.e. parallel to the plane), it is called *shear stress.*

5.2.2. *Stress tensor*

Let us go back to the matrix representation of $\widehat{\mathbf{T}}$ given in Eq. (5.42), and be specific that we are considering only the Cartesian components. Take the dot product of $\widehat{\mathbf{T}}$ with $\mathbf{e}_x$, using the dyadic form (5.15), and call it the vector component of $\widehat{\mathbf{T}}$ associated with the X-direction, and write it as $\mathbf{T}^{(x)}$:

$$\mathbf{T}^{(x)} = \widehat{\mathbf{T}} \cdot \mathbf{e}_x = T_{j\ell}\mathbf{e}_j\mathbf{e}_l \cdot \mathbf{e}_1 = T_{j1}\mathbf{e}_j = T_{xx}\mathbf{e}_x + T_{yx}\mathbf{e}_y + T_{zx}\mathbf{e}_z. \tag{5.56}$$

Comparing with (5.42) we notice that the columns 1, 2, 3 of the Cartesian matrix $\widehat{\mathbf{T}}$ represent three vectors $\mathbf{T}^{(x)}, \mathbf{T}^{(y)}, \mathbf{T}^{(z)}$, respectively, each as a column matrix.

$$\widehat{\mathbf{T}} = \begin{pmatrix} T_{xx} & T_{xy} & T_{xz} \\ T_{yx} & T_{yy} & T_{yz} \\ T_{zx} & T_{zy} & T_{zz} \end{pmatrix} = (\mathbf{T}^{(x)}\ \mathbf{T}^{(y)}\ \mathbf{T}^{(z)}), \tag{5.57}$$

$$\mathbf{T}^{(x)} = \begin{pmatrix} T_{xx} \\ T_{yx} \\ T_{zx} \end{pmatrix}; \quad \mathbf{T}^{(y)} = \begin{pmatrix} T_{xy} \\ T_{yy} \\ T_{zy} \end{pmatrix}; \quad \mathbf{T}^{(z)} = \begin{pmatrix} T_{xz} \\ T_{yz} \\ T_{zz} \end{pmatrix}. \tag{5.58}$$

Now let $\mathbf{n} = n_x\mathbf{e}_x + n_x\mathbf{e}_x + n_x\mathbf{e}_z$ be a unit vector, representing some direction in space. Let us construct the dot product of $\widehat{\mathbf{T}}$ with $\mathbf{n}$

$$\mathbf{T}^{(n)} = \widehat{\mathbf{T}} \cdot \mathbf{n}. \tag{5.59}$$

We shall call $\mathbf{T}^{(n)}$ the vector component of the tensor $\widehat{\mathbf{T}}$ associated with the direction $\mathbf{n}$.

The above dot product operation can be represented as the matrix multiplication:

$$\mathbf{T}^{(n)} = \begin{pmatrix} T_x^{(n)} \\ T_y^{(n)} \\ T_z^{(n)} \end{pmatrix} = \begin{pmatrix} T_{xx} & T_{xy} & T_{xz} \\ T_{yx} & T_{yy} & T_{yz} \\ T_{zx} & T_{zy} & T_{zz} \end{pmatrix} \begin{pmatrix} n_x \\ n_y \\ n_z \end{pmatrix}, \tag{5.60}$$

or, more compactly as

$$\mathbf{T}^{(n)} = (\mathbf{T}^{(x)}\ \mathbf{T}^{(y)}\ \mathbf{T}^{(z)}) \begin{pmatrix} n_x \\ n_y \\ n_z \end{pmatrix} = \mathbf{T}^{(x)}n_x + \mathbf{T}^{(y)}n_y + \mathbf{T}^{(z)}n_z. \tag{5.61}$$

Let us specialize $\widehat{\mathbf{T}}$ to stress tensor $\widehat{\boldsymbol{\mathcal{T}}}$.

$$\widehat{\boldsymbol{\mathcal{T}}} = \begin{pmatrix} \boldsymbol{\mathcal{T}}^{(x)} & \boldsymbol{\mathcal{T}}^{(y)} & \boldsymbol{\mathcal{T}}^{(z)} \\ \Downarrow & \Downarrow & \Downarrow \\ \mathcal{T}_{xx} & \mathcal{T}_{xy} & \mathcal{T}_{xz} \\ \mathcal{T}_{yx} & \mathcal{T}_{yy} & \mathcal{T}_{yz} \\ \mathcal{T}_{zx} & \mathcal{T}_{zy} & \mathcal{T}_{zz} \end{pmatrix}. \tag{5.62}$$

Note from the above equation that in $\mathcal{T}_{ij}$ the second index j is the "surface index" (indicating the direction of the surface on which stands the stress vector $\mathcal{T}^{(j)}$) and the first index i the "component index" (indicating x, y, z components of $\mathcal{T}^{(j)}$).

Now rewrite Eq. (5.61) as

$$\mathcal{T}^{(n)} = \mathcal{T}^{(x)} n_x + \mathcal{T}^{(y)} n_y + \mathcal{T}^{(z)} n_z. \tag{5.63}$$

The above equality involving the stress components can be proved using Newton's second law of motion applied to a fluid in motion or a solid under deformation. In other words, the stress vectors on three perpendicular surfaces determine the stress on any other surface pointing in any arbitrary direction. Therefore, stress $\mathcal{T}$ is a tensor as per the qualification written in Sec. 5.1.1

In Fig. 5.3(e) we have shown the upper part of the solid of Fig. 5.3(a), and the same area da as in Fig. 5.3(b), but now on the upper block $\mathcal{U}$. The normal vector now is $-\mathbf{n}$, and the stress vector is

$$\mathcal{T}^{(-n)}(x, y, z) = \widehat{\mathcal{T}}(x, y, z) \cdot (-\mathbf{n}) = -\mathcal{T}^{(n)}(x, y, z), \tag{5.64}$$

so that the force exerted by the atoms of the lower block $\mathcal{L}$ on the atoms of the upper block $\mathcal{U}$ across the same area da is $d\mathbf{F}'^{(n)} = -\mathcal{T}^{(n)} da = -d\mathbf{F}^{(n)}$. Which is in conformity with Newton's third law of motion.

In obtaining the last equality in Eq. (5.58) we have used the linearity property of the tensor as stipulated in (5.6). In this case $\widehat{\mathcal{T}} \cdot (a\mathbf{n}) = a\widehat{\mathcal{T}} \cdot \mathbf{n}$ where $a = -1$.

Like the inertia tensor, the stress tensor is a *symmetric tensor*, i.e.

$$\mathcal{T}_{xy} = \mathcal{T}_{yx}; \quad \mathcal{T}_{yz} = \mathcal{T}_{zy}; \quad \mathcal{T}_{zx} = \mathcal{T}_{xz}. \tag{5.65}$$

which can be proved using the equation of motion of the angular momentum.

5.2.3. *Diagonalization of a symmetric tensor*

A symmetric tensor can be always diagonalized. By this we mean the following. Let $\widehat{\mathbf{T}}$ be a symmetric tensor. Let its components with respect to some axes (XYZ) be $\{T_{ij}\}$ and that $T_{ij} = T_{ji}$. By a suitable rotation of the axes (XYZ), one can arrive at another set of axes $(X_0Y_0Z_0)$ such that $T_{ij} = 0$ if $i \neq j$.

We can express this formally in the form of the following equation:

$$\widehat{\mathbf{T}} = \begin{pmatrix} T_{xx} & T_{xy} & T_{xz} \\ T_{yx} & T_{yy} & T_{yz} \\ T_{zx} & T_{zy} & T_{zz} \end{pmatrix}_{XYZ} \rightarrow \widehat{\mathbf{T}} = \begin{pmatrix} T_1 & 0 & 0 \\ 0 & T_2 & 0 \\ 0 & 0 & T_3 \end{pmatrix}_{X_0 Y_0 Z_0} . \tag{5.66}$$

The axes $(X_0 Y_0 Z_0)$ are called the *principal axes*, and the diagonal components (T_1, T_2, T_3) are called the *principal moments of inertia* in the case of Inertia Tensor, and the *principal stresses* in the case of Stress Tensor.[f]

5.2.4. *Gauss's divergence theorem for a tensor field*

When we say tensor field, we mean a physical quantity represented by a tensor $\widehat{\mathbf{T}}(x, y, z)$ whose nine components $T_{xx}(x, y, z), T_{xy}(x, y, z), \ldots, T_{zz}(x, y, z)$ are defined at every coordinate point (x, y, z). We assume that these nine components are all differentiable functions of the coordinates x, y, z. For such a tensor field, we define its *divergence* to be the formal dot product of the grad operator $\boldsymbol{\nabla}$ with the tensor $\widehat{\mathbf{T}}(x, y, z)$, it being assumed that $\boldsymbol{\nabla}$ will appear on the left.

Let us write the tensor $\widehat{\mathbf{T}}$ by the dyadic representation

$$\widehat{\mathbf{T}} = \mathbf{T}^{(x)}\mathbf{e}_x + \mathbf{T}^{(y)}\mathbf{e}_y + \mathbf{T}^{(z)}\mathbf{e}_z, \tag{5.67}$$

as in Eq. (5.3). Then

$$\begin{aligned} \operatorname{div} \widehat{\mathbf{T}} &\equiv \boldsymbol{\nabla} \cdot \widehat{\mathbf{T}} = \boldsymbol{\nabla} \cdot (\mathbf{T}^{(x)}\mathbf{e}_x + \mathbf{T}^{(y)}\mathbf{e}_y + \mathbf{T}^{(z)}\mathbf{e}_z) \\ &\stackrel{\text{def}}{=} (\boldsymbol{\nabla} \cdot \mathbf{T}^{(x)})\mathbf{e}_x + (\boldsymbol{\nabla} \cdot \mathbf{T}^{(y)})\mathbf{e}_y + (\boldsymbol{\nabla} \cdot \mathbf{T}^{(z)})\mathbf{e}_z. \end{aligned} \tag{5.68}$$

Note that $\boldsymbol{\nabla} \cdot \mathbf{T}^{(x)}, \boldsymbol{\nabla} \cdot \mathbf{T}^{(y)}, \boldsymbol{\nabla} \cdot \mathbf{T}^{(z)}$ are the familiar scalar divergences of the vector fields $\mathbf{T}^{(x)}, \mathbf{T}^{(y)}, \mathbf{T}^{(z)}$ respectively,

$$\begin{aligned} \boldsymbol{\nabla} \cdot \mathbf{T}^{(x)} &= \frac{\partial T_{xx}}{\partial x} + \frac{\partial T_{yx}}{\partial y} + \frac{\partial T_{zx}}{\partial z}, \\ \boldsymbol{\nabla} \cdot \mathbf{T}^{(y)} &= \frac{\partial T_{xy}}{\partial x} + \frac{\partial T_{yy}}{\partial y} + \frac{\partial T_{zx}}{\partial z}, \\ \boldsymbol{\nabla} \cdot \mathbf{T}^{(z)} &= \frac{\partial T_{xz}}{\partial x} + \frac{\partial T_{yz}}{\partial y} + \frac{\partial T_{zz}}{\partial z}, \end{aligned} \tag{5.69}$$

[f]See [18, Sec. 10-4].

and constitute three (scalar) components of the vector $\boldsymbol{\nabla}\cdot\widehat{\mathbf{T}}$ along the X-, Y- and Z-axes, respectively. Combining (5.68) and (5.69), we get

$$\boldsymbol{\nabla}\cdot\widehat{\mathbf{T}} = \sum_{j=1}^{3}\sum_{i=1}^{3}\frac{\partial T_{ij}}{\partial x_i}\mathbf{e}_j \equiv \frac{\partial T_{ij}}{\partial x_i}\mathbf{e}_j. \tag{5.70}$$

In the second equality, we have employed Einistein's summation convention (introduced on p. 147).

The divergence of a vector field is sometimes interpreted as "outflux per unit volume". This association of divergence with outflux is due to Gauss's divergence theorem briefly recalled in Sec. 11.2.2. Applying the divergence theorem, to the three vector fields $\mathbf{T}^{(x)}, \mathbf{T}^{(y)}, \mathbf{T}^{(z)}$ separately, we get the following three equivalence relations:

$$\iiint_V \boldsymbol{\nabla}\cdot\mathbf{T}^{(x)}(\mathbf{r})d^3r = \iint_S \mathbf{T}^{(x)}(\mathbf{r})\cdot\mathbf{n}(\mathbf{r})\,da, \tag{5.71a}$$

$$\iiint_V \boldsymbol{\nabla}\cdot\mathbf{T}^{(y)}(\mathbf{r})d^3r = \iint_S \mathbf{T}^{(y)}(\mathbf{r})\cdot\mathbf{n}(\mathbf{r})\,da, \tag{5.71b}$$

$$\iiint_V \boldsymbol{\nabla}\cdot\mathbf{T}^{(z)}(\mathbf{r})d^3r = \iint_S \mathbf{T}^{(z)}(\mathbf{r})\cdot\mathbf{n}(\mathbf{r})\,da. \tag{5.71c}$$

Multiplying either side of Eqs. (5.71a)–(5.71c) with $\mathbf{e}_x, \mathbf{e}_y, \mathbf{e}_z$ respectively, and adding, we get

$$\begin{aligned}&\iiint_V \boldsymbol{\nabla}\cdot(\mathbf{T}^{(x)}\mathbf{e}_x + \mathbf{T}^{(y)}\mathbf{e}_y + \mathbf{T}^{(z)}\mathbf{e}_z)d^3r\\ &\quad= \iint_S (\mathbf{T}^{(x)}\mathbf{e}_x + \mathbf{T}^{(y)}\mathbf{e}_y + \mathbf{T}^{(z)}\mathbf{e}_z)\cdot\mathbf{n}\,da.\end{aligned} \tag{5.72}$$

Identifying the dyadic within the parentheses as the tensor $\widehat{\mathbf{T}}$, we obtain the divergence theorem for the tensor field:

$$\boxed{\iiint_V \boldsymbol{\nabla}\cdot\widehat{\mathbf{T}}(\mathbf{r})d^3r = \iint_S \widehat{\mathbf{T}}(\mathbf{r})\cdot\mathbf{n}\,da.} \tag{5.73}$$

We shall find this theorem to be crucial for constructing Maxwell's stress tensor in the next chapter.

Since the stress tensor $\widehat{\mathcal{T}}$ is symmetric, we can write its divergence as follows:

$$\boldsymbol{\nabla}\cdot\widehat{\mathcal{T}} = \frac{\partial \mathcal{T}_{ij}}{\partial x_i}\mathbf{e}_j = \mathbf{e}_j\frac{\partial \mathcal{T}_{ji}}{\partial x_i} = \mathbf{e}_j\mathcal{T}_{ji,i}. \tag{5.74}$$

In the last equality, we have adopted the convention $\Phi_{,i} \equiv \frac{\partial \Phi}{\partial x_i}$. The jth Cartesian component of the divergence theorem (5.73) can therefore be written as follows:

$$\iiint_V \frac{\partial \mathcal{T}_{ji}}{\partial x_i} d^3r = \iint_S \mathcal{T}_{ji} n_i \, da. \tag{5.75}$$

We shall find this practice useful while writing covariant equations in the context of the Special Theory of Relativity.

Most authors prefer the expression (5.74) for the divergence. However, following our chain of discourse leading to the construction of the stress tensor, the expression (5.70) seems to be most natural.

5.2.5. *Volume force density in a stress tensor field*

Figure 5.4 shows an imaginary rectangular box *abcdefgh* of infinitesimal dimensions $\delta x, \delta y, \delta z$ inside a medium under stress (which may be matter, or field). The centre P of this box is located at the coordinates (x, y, z). Let us assume that the stress in the medium is given by the tensor field

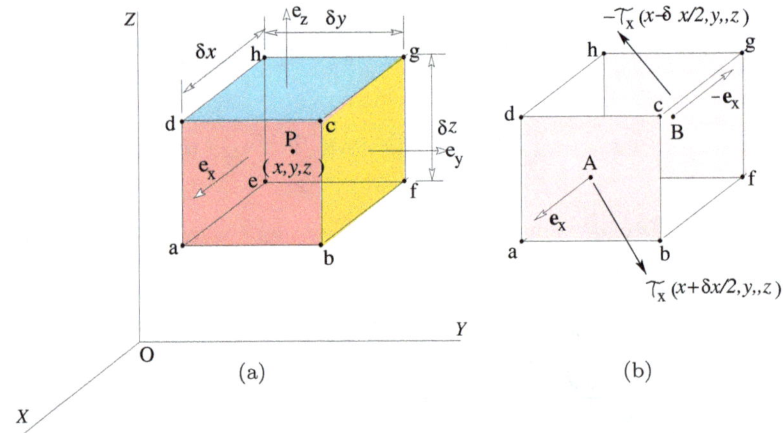

Fig. 5.4. Stress force on a volume element.

$\widehat{\mathcal{T}}(x, y, z)$, whose components are differentiable functions of the coordinates. We shall find the total force on this box due to this stress.

We have shown in Fig. 5.4(a) the outward normal vectors $(\mathbf{e}_x, \mathbf{e}_y, \mathbf{e}_z)$ on the three faces of the box that are exposed to our view. The outward normals on the other faces which are hidden from our view are $(-\mathbf{e}_x, -\mathbf{e}_y, -\mathbf{e}_z)$. We shall identify each one of the six surfaces of the box by their outward normal vectors.

Let us consider the opposite faces *abcd* and *efgh*, recognized by the normals $(\mathbf{e}_x)$ and $(-\mathbf{e}_x)$. The locations of their centres are $(x + \frac{\delta x}{2}, y, z)$ and $(x - \frac{\delta x}{2}, y, z)$, respectively. The stress forces on these two faces are

$$\begin{aligned}
\delta \mathbf{F}_{+x} &= \widehat{\mathcal{T}}\left(x + \frac{\delta x}{2}, y, z\right) \cdot (+\mathbf{e}_x)\delta y \delta z = \mathcal{T}^{(x)}\left(x + \frac{\delta x}{2}, y, z\right)\delta y \delta z \\
&= \left[\mathcal{T}^{(x)}(x, y, z) + \frac{\partial \mathcal{T}^{(x)}}{\partial x}\frac{\delta x}{2}\right]\delta y \delta z, \\
\delta \mathbf{F}_{-x} &= \widehat{\mathcal{T}}\left(x - \frac{\delta x}{2}, y, z\right) \cdot (-\mathbf{e}_x)\delta y \delta z = -\mathcal{T}^{(x)}\left(x - \frac{\delta x}{2}, y, z\right)\delta y \delta z \\
&= -\left[\mathcal{T}^{(x)}(x, y, z) - \frac{\partial \mathcal{T}^{(x)}}{\partial x}\frac{\delta x}{2}\right]\delta y \delta z,
\end{aligned}$$

$$\delta \mathbf{F}_{+x} + \delta \mathbf{F}_{-x} = \frac{\partial \mathcal{T}^{(x)}}{\partial x}\delta x \delta y \delta z = \frac{\partial \mathcal{T}^{(x)}}{\partial x}\delta V, \tag{5.76}$$

where $\delta V = \delta x \delta y \delta z$ is the volume of the infinitesimal box. In the same way, we find the forces on the other four faces of the block. Adding the stress forces on all the six surfaces, we get

$$\delta \mathbf{F}_s = \left[\frac{\partial \mathcal{T}^{(x)}}{\partial x} + \frac{\partial \mathcal{T}^{(y)}}{\partial y} + \frac{\partial \mathcal{T}^{(z)}}{\partial z}\right]\delta V \tag{5.77}$$

as the total stress force on the box. The volume force density $\mathbf{f}_s$, which gives the stress force acting per unit volume of the media under stress, is then given as

$$\begin{aligned}
\mathbf{f}_s &= \frac{\partial \mathcal{T}^{(x)}}{\partial x} + \frac{\partial \mathcal{T}^{(y)}}{\partial y} + \frac{\partial \mathcal{T}^{(z)}}{\partial z} = \frac{\partial(\mathbf{e}_x \cdot \widehat{\mathcal{T}})}{\partial x} + \frac{\partial(\mathbf{e}_y \cdot \widehat{\mathcal{T}})}{\partial y} + \frac{\partial(\mathbf{e}_z \cdot \widehat{\mathcal{T}})}{\partial z} \\
&= \left[\mathbf{e}_x \frac{\partial}{\partial x} + \mathbf{e}_x \frac{\partial}{\partial x} + \mathbf{e}_x \frac{\partial}{\partial x}\right] \cdot \widehat{\mathcal{T}},
\end{aligned} \tag{5.78}$$

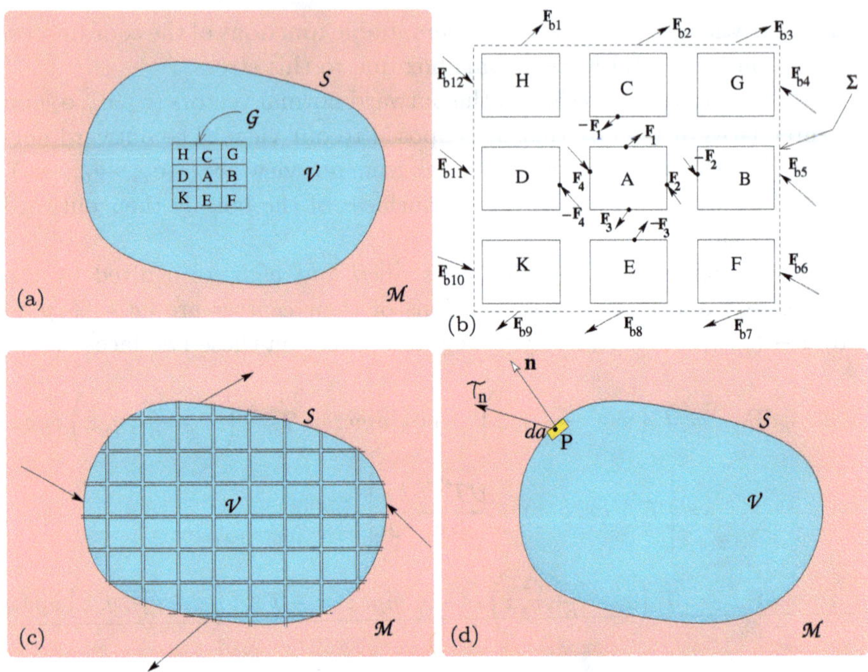

Fig. 5.5. Stress forces on a bulk volume.

or

$$\boxed{\mathbf{f}_s = \boldsymbol{\nabla} \cdot \widehat{\mathcal{T}}.} \tag{5.79}$$

One may conclude that total stress force $\mathbf{F}_s$ on a bulk volume $\mathcal{V}$ carved out inside a medium $\mathcal{M}$, as shown in Fig. 5.5(a), is the volume integral of the force density $\mathbf{f}_s$ carried out over the entire volume $\mathcal{V}$. We shall carefully analyze the forces inside the medium before jumping into this conclusion.

Let us consider a two-dimensional view of nine tiny, imaginary neighbouring blocks lying inside the medium and forming a group $\mathcal{G}$. We have marked the blocks as A, B, C, D, E, F, G, H, K, with A at the centre. In Fig. 5.4(b) we have shown the forces on the four sides of A as $\mathbf{F}_1, \mathbf{F}_2, \mathbf{F}_3, \mathbf{F}_4$. The force $\mathbf{F}_1$ comes from the neighbour B, and by Newton's third law of motion, A applies an equal and opposite force $-\mathbf{F}_1$ on B. Similarly, the forces $\mathbf{F}_2, \mathbf{F}_3, \mathbf{F}_4$ come from the neighbours C, D, E. And A applies equal and opposite forces $-\mathbf{F}_2, -\mathbf{F}_3, -\mathbf{F}_4$ on them. It *may* then appear that these internal forces, when added together, get cancelled out and there should not be any stress force on the group $\mathcal{G}$ at all.

A close examination will disprove this judgement. We have surrounded $\mathcal{G}$ by an imaginary boundary surface Σ. It is now seen that even though the action–reaction forces cancel out in the interior of the group $\mathcal{G}$, they survive on the boundary surface Σ. These surface forces $\mathbf{F}_{b1}, \mathbf{F}_{b2}, \ldots, \mathbf{F}_{b12}$, when added together constitute the total force $\mathbf{F}_s$ on the group $\mathcal{G}$.

In Fig. 5.5(c), we have divided the volume $\mathcal{V}$ into an infinite number of infinitesimal blocks. The interior stress forces between adjoining blocks will cancel out. However, the forces on the boundary surface, some of which we have shown as $\mathbf{F}_{b1}, \mathbf{F}_{b2}, \mathbf{F}_{b3}, \mathbf{F}_{b4}$, will survive and add together to constitute the net stress force $\mathbf{F}_s$ on the volume $\mathcal{V}$.

We now get a clue of how to find the net stress force $\mathbf{F}_s$ on the volume $\mathcal{V}$. In Fig. 5.5(d), we have shown the volume $\mathcal{V}$ once again. At a certain point P on this surface we have pictured a tiny patch of area da, on which we have drawn a unit outward normal $\mathbf{n}$. The stress force on this patch is $d\mathbf{f}_s = \mathcal{T}^{(n)}\, da = \widehat{\mathcal{T}} \cdot \mathbf{n}\, da$. Integrate this force over the entire boundary to get $\mathbf{F}_s$. We shall perform this integration and convert the surface integral into volume integral by applying Gauss's Divergence Theorem as derived in Eq. (5.73):

$$\boxed{\mathbf{F}_s = \iint_S \widehat{\mathcal{T}}(\mathbf{r}) \cdot \mathbf{n}\, da = \iiint_V \boldsymbol{\nabla} \cdot \widehat{\mathcal{T}}(\mathbf{r})\, d^3r.} \tag{5.80}$$

which reconfirms Eq. (5.79)

The simplest example of stress field in matter is provided by a *perfect fluid*, which by definition, does not support shear stress. Since tensile stress is also ruled out in a fluid, the stress field inside a perfect fluid is left with only *normal compressive stress*, which is known more familiarly by the name *pressure*, to be written as $p(x, y, z)$. It is then obvious that the *stress field* in a perfect fluid is the "pressure tensor", having only three equal diagonal elements $p(x, y, z)$:

$$\widehat{\mathcal{T}}(\mathbf{r}) \equiv \hat{\mathbf{p}}(\mathbf{r}) = -p(x, y, z)\widehat{\mathbf{1}} = -\begin{pmatrix} p(x, y, z) & 0 & \\ 0 & p(x, y, z) & 0 \\ 0 & 0 & p(x, y, z) \end{pmatrix}. \tag{5.81}$$

According to Eq. (5.79) the volume force density inside a pressure field is given as

$$\mathbf{f}_p(x, y, z) = -\boldsymbol{\nabla} \cdot [p(x, y, z)\widehat{\mathbf{1}}] = -\boldsymbol{\nabla} p(x, y, z), \tag{5.82}$$

and the stress vector on a surface $\mathbf{n}$ as

$$\boldsymbol{\mathcal{T}}^{(\mathbf{n})} = \widehat{\boldsymbol{\mathcal{T}}} \cdot \mathbf{n} = -p\,\mathbf{n}. \tag{5.83}$$

Chapter 6

Maxwell's Stress Tensor

6.1. Introduction

'Action at a distance' (AAD) was an enigma to natural philosophers, from Rene Descartes[a] (1596–1650) to James Clerk Maxwell (1831–1879). We find an account of the evolution of physical concepts in [19]. According to Descartes, space was a plenum, a medium called *aether*, capable of transmitting force on material bodies. "It was to be regarded as the solitary tenant of the universe, save for that infinitesimal fraction of space which is occupied by ordinary matter."

Subsequent theoretical physicists and mathematicians, Robert Hooke (1635–1703), Isaac Newton (1642–1727), Reimann (1826–1866), W. Thomson (1824–1907), Maxwell and others lent their support to this view. Implicit in their belief was the assumption that *force cannot be transmitted except by actual pressure or impact.* AAD was a taboo, as abhorrent as witchcraft: I wave my hand here and a fire is ignited there. In order to support their faith in aether they contrived every possible idea, any possible mechanical model, to make aether viable.

According to Newton "All space is pervaded by an elastic medium or aether, which is capable of propagating vibrations in the same way as air propagates the vibrations of sound. This aether pervades the pores of all material bodies, and is the cause of their cohesion; its density varies from one body to another, being greatest in the interplanetary space."

[a]The Cartesian coordinate system is associated with his name

Maxwell inherited this legacy. We shall quote a few passages from his celebrated paper *A Dynamical Theory of the Electromagnetic Field* read to the Royal Society of London on December 8, 1864 [20].

> "(1) In this way mathematical theories of statical electricity, of magnetism, of the mechanical action between conductors carrying currents, and of the induction if currents have been formed. *In these theories the force acting between two bodies is treated with reference only to the condition of the bodies and their relative position, and without reference to the surrounding medium.*"
>
> "(2) The *mechanical* difficulties, however, which are involved in the assumption of particles acting at a distance with forces which depend on their velocities are such as to prevent me from considering this theory as an ultimate one, though it may have been, and may yet be useful to the coordination of phenomena."
>
> "(3) The theory I propose may therefore be called a theory of the *Electromagnetic Field*, because *it has to do with the space in the neighbourhood of the electric and magnetic bodies*, and it may be called a *Dynamical Theory*, because it assumes that in that space there is matter in motion, by which the observed electromagnetic phenomena are produced."
>
> "(4) The electromagnetic field is that part of space which contains and surrounds bodies in electric and magnetic conditions. ... It may contain any kind of matter, or *we may render it empty of all gross matter*, as in the case of Geissler's Tubes and other so-called *vacua*.
>
> *There is always, however, enough matter to receive and transmit the undulations of light and heat*, and it is because of the transmission of these radiations is not greatly altered when transparent bodies of measurable densities are substituted for the so-called vacuum, that we are obliged to admit that the *undulations are those of aetherial substance*, and not of the gross matter, the presence of which merely modifies in some way the motion of the aether.
>
> We have therefore some reason to believe, from the phenomena of light and heat, that *there is an aetherial medium filling space and permeating bodies, capable of being set in motion and of transmitting that motion from one part to another, and communicating that motion to gross matter so as to heat it and affect it in various ways.*"

One aspect of the mechanical model Maxwell built up to present a complete picture of the electromagnetic field was the proposition that space, i.e. aether, can sustain stress, and a force is transmitted from one body (electrified or magnetized) to another by means of stress, in the same way a force is transmitted from one end of a cable to the other by means of tensile stress, and from one part of a beam to another by means of shear stress.

In his two-volume book *A treatise on Electricity and Magnetism* Maxwell presents a complete formulation of the stress in the field (read aether) by constructing the *Stress Tensor* for the *Static Electric Field* [21] and for the *Static Magnetic Field*, in terms of the field potentials.

We have derived the stress tensors for electrostatic field, magnetostatic field and time varying electromagnetic field in terms of the electric field **E**, magnetic field **B** in a unified manner exploiting the useful identity given in Eq. (6.7).

Einstein's formulation of the Special Theory of Relativity saw the demise of the Luminiferous (i.e. light carrying) Aether. Light travels in empty space, electric and magnetic forces also propagate from one body to another (with the speed of light) in empty space. Is there then any place for Maxwell's stress tensor? Is it only for historical reason that we are writing this long article? We shall attempt to provide the answer in four steps.

First, it is indeed an amazing thing that the force acting on an isolated body A (which may consist of electric charges and currents), due to the presence of charges and currents elsewhere, can be computed *exactly* by drawing a boundary surface $\mathcal{S}$ of our convenience surrounding A, as in Fig. 6.1(a), finding the "*stress*" all over this surface, and by integrating this stress. In other words, there *is* stress even in vacuum. The purpose of this chapter is to articulate how this stress is to be found out. Also it should be noted with interest that even empty space is not a true vacuum. When loaded with the electric and magnetic fields, space comes under stress. Empty space is always buzzing with emission and absorption of virtual particles, with the virtual photons mediating the interaction among electrified and magnetized objects. Aren't these virtual photons the new *avatar* of the aether?

Secondly, calculating the force on an isolated object A requires *exact* knowledge of the **E** or **B** field in which A is immersed. In recognizing these fields, one has to be very careful that these **E**, **B** fields do not contain any trace of the fields contributed by A itself. This is sometimes a challenging task. Consider for example the force acting on the surface of a conductor

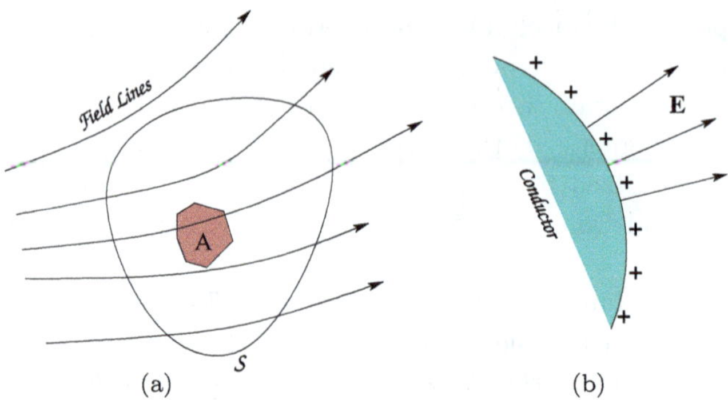

Fig. 6.1. Electrified object in $\mathbf{E}$ field.

carrying a surface charge density σ, as in Fig. 6.1(b). The electric field just outside the surface is $\mathbf{E} = (\sigma/\varepsilon_0)\mathbf{n}$ where $\mathbf{n}$ is a unit normal to the surface. One may be tempted to conclude that the force per unit area of the surface is $\mathbf{F}' = \sigma\mathbf{E} = (\sigma^2/\varepsilon_0)\mathbf{n}$, forgetting the fact that an infinitesimal area *da* on the surface contributes the same $\mathbf{E}$ field perpendicular to the surface as the rest of the surface, so that the true force is

$$\mathbf{F} = \frac{1}{2}\mathbf{F}' = (\sigma^2/2\varepsilon_0)\mathbf{n} = (\varepsilon_0 E^2/2)\mathbf{n}. \tag{6.1}$$

The stress tensor approach, which uses the total field $\mathbf{E}_{\text{total}}$, making no distinction between the test object and the source object, will give the right result without creating any confusion, as we shall show following Eq. (6.15).

Thirdly, it is always advisable to arrive at the same answer through several alternative routes, if available, just to make sure that we have not made any mistakes. The stress tensor provides that valuable alternative route.

And fourthly, *Maxwell's stress tensor*, which we shall denote by the symbol $\widehat{\mathcal{T}}$, is needed for understanding conservation and flow of *momentum* in the electromagnetic field, which we shall present in Sec. 6.5. When one goes deeper into the theory of relativity the same tensor appears as the most important component of the energy–momentum tensor required not only for presenting a four dimensional and unified view of the conservation of energy and momentum, but also for building up the source term in formulating Einstein's field equation for the gravitational field, in his General Theory of Relativity.

6.2. Maxwell's Stress Tensor for the Electrostatic Field

6.2.1. *Volume force density in terms of the field*

We shall now construct the stress tensor for the electrostatic field.

We shall call this tensor *Maxwell's Stress Tensor* and represent it by the symbol $\widehat{\mathcal{T}}^{(\mathrm{e})}$, where the superscript $^{(\mathrm{e})}$ implies electric field.

Figure 6.2 shows a system of electric charges $\mathcal{S}$ placed in an electric field $\mathbf{E}(\mathbf{r})$. In Fig. 6.2(a), the system consists of discrete charges $q_1, q_2, q_3, \ldots$ placed at the radius vectors $\mathbf{r}_1, \mathbf{r}_2, \mathbf{r}_3, \ldots$. In Fig. 6.2(b), the system is a continuous distribution characterized by a smooth charge density function $\rho(\mathbf{r})$ confined within a volume. Our intention is to write the total electric force $\mathbf{F}$ on this system.

The force on the discrete system shown in Fig. 6.2(a) is given as follows:

$$\mathbf{F} = \sum_j q_j \mathbf{E}^{(\mathrm{ext})}(\mathbf{r}_j). \tag{6.2}$$

Here the sum is over all the charges in the system, and $\mathbf{E}^{(\mathrm{ext})}(\mathbf{r}_j)$ is the *external electric field* at the radius vector $\mathbf{r}_j$ caused by the presence of all *other* charges lying outside the system $\mathcal{S}$.

For the case of continuous distribution, shown in Fig. 6.2(b), the individual charges become infinitesimal elementary charges, i.e. $q_j \to \rho(\mathbf{r})d^3r$, and the sum becomes the integral

$$\mathbf{F} = \iiint_V \rho(\mathbf{r})\mathbf{E}^{(\mathrm{ext})}(\mathbf{r})\, d^3r. \tag{6.3}$$

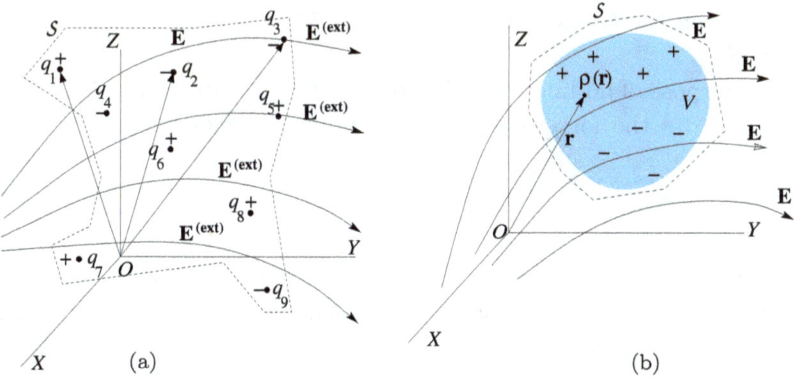

Fig. 6.2. Forces on charges in an electric field.

What about the force from the charges inside the system $\mathcal{S}$. They are *internal* forces, and cancel due to Newton's third law of motion.

Let $\mathbf{E}_i^{(\text{int})}(\mathbf{r}_j)$ be the "internal" field caused at $\mathbf{r}_j$ by a member particle i lying within the system $\mathcal{S}$. Then $\mathbf{F}_{ij} = q_j\mathbf{E}_i^{(\text{int})}(\mathbf{r}_j)$ is the force that the member particle i exerts on the member particle j. By Newton's third law of motion, $q_j\mathbf{E}_i^{(\text{int})}(\mathbf{r}_j) + q_i\mathbf{E}_j^{(\text{int})}(\mathbf{r}_i) = \mathbf{0}$. Adding together over all pairs for the discrete distribution, and integrating over the entire distribution for the continuous distribution we get

$$\begin{aligned}&\text{For discrete: } \sum_{j=1}^{N} q_j {\sum_{i=1}^{N}}' \mathbf{E}_i^{(\text{int})}(\mathbf{r}_j) = \sum_{j=1}^{N} q_j\mathbf{E}^{(\text{int})}(\mathbf{r}_j) = \mathbf{0}.\\ &\text{For continuous: } \iiint_V \rho(\mathbf{r})\,\mathbf{E}^{(\text{int})}(\mathbf{r}) = \mathbf{0}.\end{aligned} \tag{6.4}$$

In the first equation, the sum symbol $\sum'$ means that while summing over i, the term $i = j$ (corresponding to the "self field" of the member j) is to be avoided. The "internal field" $\mathbf{E}^{(\text{int})}(\mathbf{r}_j)$ is the field at the location of the member j caused by "all other members" in the system $\mathcal{S}$. In the second equation $\mathbf{E}^{(\text{int})}(\mathbf{r})$ is the "internal field" at the radius vector $\mathbf{r}$, as sensed by a tiny volume element d^3r at this point.

We shall add the null contribution shown in the second line of Eq. (6.4) to the right-hand side of Eq. (6.3) and write

$$\mathbf{F} = \iiint_V \rho(\mathbf{r})\mathbf{E}\,(\mathbf{r})\,d^3r. \tag{6.5}$$

Here $\mathbf{E}\,(\mathbf{r})$ is the actual field at the point $\mathbf{r}$, being the sum of two contributions, from the (i) external sources, and (ii) the internal sources of the system $\mathcal{S}$.

The purpose of adding the null integral of Eq. (6.4b) to Eq. (6.3) is that when we write the force density $\mathbf{f}$, the internal forces need to be added. That is,

$$\mathbf{f}(\mathbf{r}) = \rho(\mathbf{r})\mathbf{E}\,(\mathbf{r}) \tag{6.6}$$

is the force on unit volume of the charge distribution at $\mathbf{r}$, in which $\mathbf{E}\,(\mathbf{r})$ is necessarily the *total* field at this location, caused by *both* external and internal sources. Now we manipulate the right-hand side of Eq. (6.6) so as to convert $\rho\mathbf{E} \to \boldsymbol{\nabla}\cdot\widehat{\mathcal{T}}^{(\text{e})}$, as suggested in Eq. (5.79). This new tensor field $\widehat{\mathcal{T}}^{(\text{e})}(\mathbf{r})$ would represent "stress" in the electrostatic field.

Construction of the stress tensor for electrostatic field, magnetostatic field and time varying electromagnetic field will be facilitated by the following identity [22]:

$$\nabla \cdot \left[\mathbf{AA} - \frac{1}{2}A^2\hat{\mathbf{1}}\right] = (\nabla \cdot \mathbf{A})\mathbf{A} - \mathbf{A} \times (\nabla \times \mathbf{A}). \tag{6.7}$$

Before establishing the above identity we shall need a standard formula (see, for example, *vector identities* compiled in Griffiths, 4th edn).

$$\begin{aligned}\nabla(\mathbf{A} \cdot \mathbf{B}) &= \mathbf{A} \times (\nabla \times \mathbf{B}) + \mathbf{B} \times (\nabla \times \mathbf{A}) \\ &\quad + (\mathbf{A} \cdot \nabla)\mathbf{B} + (\mathbf{B} \cdot \nabla)\mathbf{A}.\end{aligned} \tag{6.8}$$

By setting $\mathbf{B} = \mathbf{A}$ in the above formula, we get

$$\nabla\left(\frac{1}{2}A^2\right) = \mathbf{A} \times (\nabla \times \mathbf{A}) + (\mathbf{A} \cdot \nabla)\mathbf{A}. \tag{6.9}$$

We shall now prove the identity (6.7).

Proof.

$$\begin{aligned}\nabla \cdot (\mathbf{AA}) &= \left(\mathbf{e}_l \frac{\partial}{\partial x_l}\right) \cdot (\mathbf{e}_i \mathbf{e}_j A_i A_j), \\ &= \frac{\partial}{\partial x_i}(A_i A_j)\mathbf{e}_j \\ &= \left\{\left(\frac{\partial A_i}{\partial x_i}\right) A_j + \left(A_i \frac{\partial}{\partial x_i}\right) A_j\right\} \mathbf{e}_j \\ &= (\nabla \cdot \mathbf{A})\mathbf{A} + (\mathbf{A} \cdot \nabla)\mathbf{A}, \qquad \text{(a)} \\ \nabla \cdot \left(\frac{1}{2}A^2\hat{\mathbf{1}}\right) &= \left(\mathbf{e}_l \frac{\partial}{\partial x_l}\right) \cdot \left(\frac{1}{2}\mathbf{e}_i \mathbf{e}_i A^2\right) \\ &= \frac{1}{2}\mathbf{e}_i \frac{\partial A^2}{\partial x_i} = \nabla\left(\frac{1}{2}A^2\right) \\ &= \mathbf{A} \times (\nabla \times \mathbf{A}) + (\mathbf{A} \cdot \nabla)\mathbf{A}, \text{ by (6.9).} \qquad \text{(b)}\end{aligned}$$

The identity (6.7) follows when we subtract line (b) from line (a). □

Note that we have used Einstein's summation convention introduced on p. 149. That is, $\mathbf{e}_l \frac{\partial}{\partial x_l} \equiv \sum_{l=1}^{3} \mathbf{e}_l \frac{\partial}{\partial x_l}$; $\mathbf{e}_i \mathbf{e}_j A_i A_j \equiv \sum_{i=1}^{3} \sum_{j=1}^{3} \mathbf{e}_i \mathbf{e}_j A_i A_j$, etc.

The stress tensor for the electrostatic field follows when we set $\mathbf{E}$ for $\mathbf{A}$ in (6.7), and use the field equations: $\boldsymbol{\nabla}\cdot\mathbf{E} = \rho/\varepsilon_0$; $\boldsymbol{\nabla}\times\mathbf{E} = \mathbf{0}$:

$$\boxed{\begin{aligned} \mathbf{f}^{(\mathrm{e})} &= \rho\mathbf{E} = \boldsymbol{\nabla}\cdot\widehat{\mathcal{T}}^{(\mathrm{e})}, \quad \text{(a)} \\ \text{where } \widehat{\mathcal{T}}^{(\mathrm{e})} &= \varepsilon_0[\mathbf{E}\mathbf{E} - \tfrac{1}{2}E^2\widehat{\mathbf{1}}]. \ \text{(b)} \end{aligned}} \tag{6.10}$$

It will be a simple exercise to write the Cartesian components of this tensor:

$$\begin{aligned} \widehat{\mathcal{T}}^{(\mathrm{e})} &= (\widehat{\mathcal{T}}^{(\mathrm{e})}\cdot\mathbf{e}_x \quad \widehat{\mathcal{T}}^{(\mathrm{e})}\cdot\mathbf{e}_y \quad \widehat{\mathcal{T}}^{(\mathrm{e})}\cdot\mathbf{e}_z) \\ &= \varepsilon_0\begin{pmatrix} \frac{1}{2}(E_x^2 - E_y^2 - E_z^2) & E_xE_y & E_xE_z \\ E_yE_x & \frac{1}{2}(E_y^2 - E_z^2 - E_x^2) & E_yE_z \\ E_zE_x & E_zE_y & \frac{1}{2}(E_z^2 - E_y^2 - E_z^2) \end{pmatrix}. \end{aligned} \tag{6.11}$$

6.2.2. *Example 1: Stress vector on a plane as a function of the angle of inclination*

The stress tensor (6.10) will remain abstract and obscure unless the reader works out a few examples. We shall provide two examples of which the first one is depicted in Fig. 6.3. A *uniform* electric field $\mathbf{E} = E\mathbf{e}_x$ exists in a certain region of space. The stress tensor is then given by the following expression:

$$\widehat{\mathcal{T}}^{(\mathrm{e})} = \frac{\varepsilon_0}{2}E^2(\mathbf{e}_x\mathbf{e}_x - \mathbf{e}_y\mathbf{e}_y - \mathbf{e}_z\mathbf{e}_z) = \frac{\varepsilon_0}{2}\begin{pmatrix} E^2 & 0 & 0 \\ 0 & -E^2 & 0 \\ 0 & 0 & -E^2 \end{pmatrix}. \tag{6.12}$$

Imagine a plane running parallel to the Z-axis, but inclined to the X-axis by an angle θ (Fig. 6.3(a)). The normal vector is then given as

$$\mathbf{n} = \mathbf{e}_x\sin\theta + \mathbf{e}_y\cos\theta = \begin{pmatrix} \sin\theta \\ \cos\theta \\ 0 \end{pmatrix}. \tag{6.13}$$

The stress vector $\mathcal{T}^{(n)}$ on this plane is then

$$\mathcal{T}^{(n)} = \widehat{\mathcal{T}}^{(\mathrm{e})}\cdot\mathbf{n} = \frac{\varepsilon_0}{2}E^2(\mathbf{e}_x\sin\theta - \mathbf{e}_y\cos\theta) = \frac{\varepsilon_0}{2}E^2\begin{pmatrix} \sin\theta \\ -\cos\theta \\ 0 \end{pmatrix}. \tag{6.14}$$

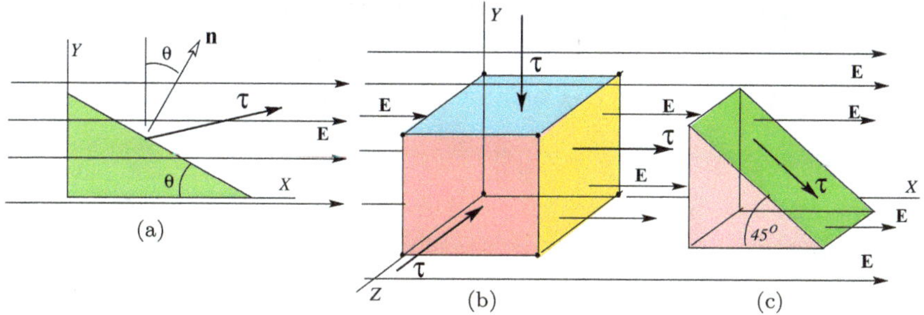

Fig. 6.3. Stress vector on an inclined plane placed in a uniform electric field.

Let us consider some special cases:

$$\boldsymbol{\mathcal{T}}^{(x)} = \frac{\varepsilon_0}{2}E^2\mathbf{e}_x \quad (\text{by setting}\,\theta = \pi/2), \tag{6.15a}$$

$$\boldsymbol{\mathcal{T}}^{(y)} = -\frac{\varepsilon_0}{2}E^2\mathbf{e}_y \quad (\text{by setting}\,\theta = 0), \tag{6.15b}$$

$$\boldsymbol{\mathcal{T}}^{(z)} = -\frac{\varepsilon_0}{2}E^2\mathbf{e}_z \quad (\text{same as}\,\widehat{\boldsymbol{\mathcal{T}}}^{(\mathrm{e})}\cdot\mathbf{e}_z), \tag{6.15c}$$

$$\boldsymbol{\mathcal{T}}^{(45^\circ)} = \frac{\varepsilon_0}{2}E^2\frac{1}{\sqrt{2}}(\mathbf{e}_x - \mathbf{e}_y). \tag{6.15d}$$

Equations (6.15a)–(6.15c) give the stress vectors on the planes identified by the normal vectors $\mathbf{e}_x, \mathbf{e}_y, \mathbf{e}_z$, and Eq. (6.15d) gives the stress vector on a plane making an angle of 45° with X-axis. We have illustrated these points in Figs. 6.3(b) and 6.3(c). We have shown the stress vectors with thick arrows, and labelled them with the bold Greek letter $\boldsymbol{\mathcal{T}}$. We draw the following conclusion.

Conclusions:

(a) If the field is *perpendicular* to the plane, the stress vector is *normal* and *outward* (tensile stress), and equal to $\frac{\varepsilon_0}{2}E^2$.
(b) If the field is *tangential* to the plane, the stress vector is *normal* and *inward* (compressive stress), and equal to $\frac{\varepsilon_0}{2}E^2$.
(c) If the field makes angle 45° to the plane, the stress vector is *tangential* (shear stress), and equal to $\frac{\varepsilon_0}{2}E^2$.

Case (a) applies to a conductor in an electric field **E**. The field is perpendicular to the surface. The surface force density is the same as the stress vector. We get back the same answer as in Eq. (6.1) using the stress tensor, without laboring to find out what is the "external field".

6.2.3. *Example 2: Force transmitted between two charged particles across a spherical boundary*

We shall obtain the familiar Coulomb force between two charged particles using Maxwell's stress tensor.

We shall use spherical coordinate system. The reader must have used the spherical coordinate system to construct vectors in situations of spherical symmetry, as in the case of central forces in mechanics. This example, and two more the next sections, will give an opportunity to deploy the same coordinate system to construct a tensor, the stress tensor to be specific, to get at the answer with less expenditure of time.

We shall first obtain an expression for the **E** field at any arbitrary point P (r, θ, ϕ) located on the spherical surface Σ. The point P is at the displacement vector $\boldsymbol{\eta}$ from A and $\mathbf{r}$ from O (Fig. 6.4(a)). In order to avoid repeated appearance of the constant $\frac{1}{4\pi\epsilon_0}$, we shall set $\mathbf{E} = \frac{1}{4\pi\epsilon_0}\boldsymbol{\mathcal{E}}$. Note that

$$\boldsymbol{\eta} = \mathbf{r} - \boldsymbol{a} = \mathbf{r} - a\mathbf{e}_z, \tag{6.16a}$$

$$\text{so that} \quad \eta^2 = r^2 + a^2 - 2ra\cos\theta, \tag{6.16b}$$

$$\text{and} \quad \mathbf{e}_z = \cos\theta\mathbf{e}_r - \sin\theta\mathbf{e}_\theta. \tag{6.16c}$$

Then

$$\boldsymbol{\mathcal{E}} = \frac{Q\mathbf{r}}{r^3} + \frac{q\boldsymbol{\eta}}{\eta^3} \tag{6.17a}$$

$$= \frac{Q\mathbf{e}_r}{r^2} + \frac{q(\mathbf{r} - a\mathbf{e}_z)}{(r^2 + a^2 - 2ra\cos\theta)^{3/2}}. \tag{6.17b}$$

Therefore,

$$\boldsymbol{\mathcal{E}} = \mathcal{E}_r\mathbf{e}_r + \mathcal{E}_\theta\mathbf{e}_\theta, \tag{6.18a}$$

$$\text{where} \quad \mathcal{E}_r = \frac{Q}{r^2} + \frac{q(r - a\cos\theta)}{(r^2 + a^2 - 2ra\cos\theta)^{3/2}}, \tag{6.18b}$$

$$\mathcal{E}_\theta = \frac{qa\sin\theta}{(r^2 + a^2 - 2ra\cos\theta)^{3/2}}. \tag{6.18c}$$

From Eq. (6.10), the stress tensor is

$$\widehat{\mathcal{T}}^{(\mathrm{e})} = \varepsilon_0(\mathbf{EE} - \frac{1}{2}E^2\widehat{\mathbf{1}}) = \frac{1}{16\pi^2\varepsilon_0}(\boldsymbol{\mathcal{E}}\boldsymbol{\mathcal{E}} - \frac{1}{2}\mathcal{E}^2\widehat{\mathbf{1}}) = \frac{1}{16\pi^2\varepsilon_0}\tilde{\mathcal{T}}^{(\mathrm{e})}, \tag{6.19}$$

$$\text{where} \quad \tilde{\mathcal{T}}^{(\mathrm{e})} = \boldsymbol{\mathcal{E}}\boldsymbol{\mathcal{E}} - \frac{1}{2}\mathcal{E}^2\widehat{\mathbf{1}},$$

which we may refer to as the "reduced stress tensor".

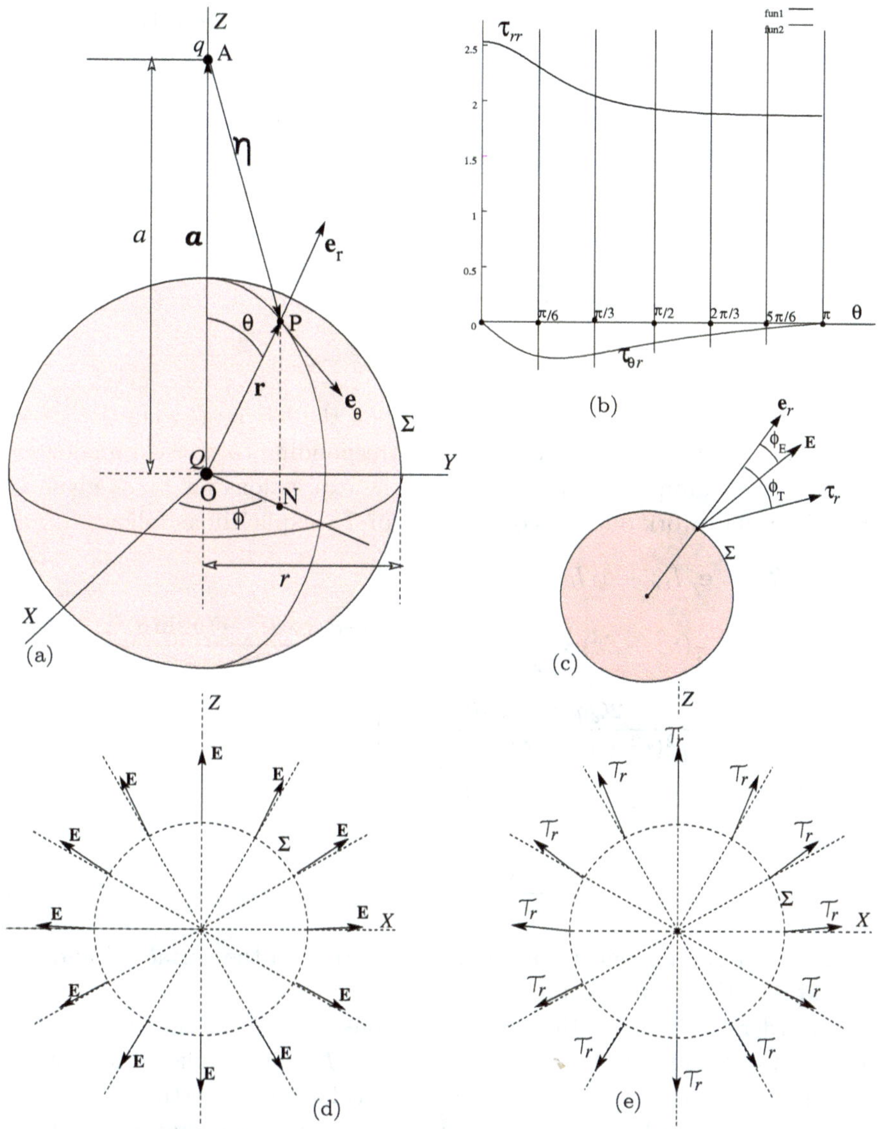

Fig. 6.4. Electric stress vector on a spherical surface.

Since we have invoked the spherical coordinate system to write the expression for the $\boldsymbol{\mathcal{E}}$ field, the components of the tensor $\widehat{\boldsymbol{\mathcal{T}}}^{(\mathrm{e})}$ will have to be written in this coordinate system. Since only r and θ components of $\boldsymbol{\mathcal{E}}$ are non-zero, the non-zero components of this tensor are $\mathcal{T}_{rr}, \mathcal{T}_{r\theta}, \mathcal{T}_{\theta r}, \mathcal{T}_{\theta\theta}$, as seen from (6.19). Therefore $\mathcal{E}^2 = \mathcal{E}_r^2 + \mathcal{E}_\theta^2$, and we write this tensor as

$$\tilde{\mathcal{T}}^{(\mathrm{e})} = \begin{pmatrix} \mathcal{T}_{rr} & \mathcal{T}_{r\theta} & 0 \\ \mathcal{T}_{\theta r} & \mathcal{T}_{\theta\theta} & 0 \\ 0 & 0 & 0 \end{pmatrix}, \quad \text{where}$$

$$\begin{aligned} \mathcal{T}_{rr} &= \mathcal{E}_r^2 - \frac{1}{2}\mathcal{E}^2 = \frac{1}{2}(\mathcal{E}_r^2 - \mathcal{E}_\theta^2), \\ \mathcal{T}_{r\theta} &= \mathcal{T}_{\theta r} = \mathcal{E}_r \mathcal{E}_\theta, \\ \mathcal{T}_{\theta\theta} &= \mathcal{E}_\theta^2 - \frac{1}{2}\mathcal{E}^2 = \frac{1}{2}(\mathcal{E}_\theta^2 - \mathcal{E}_r^2). \end{aligned} \tag{6.20}$$

The first column in the square matrix on the left represents the stress vector $\boldsymbol{\mathcal{T}}_r$ on the spherical surface Σ (corresponding to $\mathbf{n} = \mathbf{e}_r$, analogous to the first column in Eq. (5.62). Using the expressions for $\mathcal{E}_r, \mathcal{E}_\theta$ given in (6.18) we shall work out the components of $\boldsymbol{\mathcal{T}}_r$ explicitly as follows:

$$\begin{aligned} \boldsymbol{\mathcal{T}}_r &= \mathbf{e}_r \mathcal{T}_{rr} + \mathbf{e}_\theta \mathcal{T}_{\theta r}, \\ \mathcal{T}_{rr} &= \frac{1}{2}(\mathcal{E}_r^2 - \mathcal{E}_\theta^2) = \frac{1}{2}\left[\frac{Q^2}{r^4} + \frac{q^2[(r - a\cos\theta)^2 - (a\sin\theta)^2]}{(r^2 + a^2 - 2ra\,\cos\theta)^3}\right. \\ &\quad \left. + \frac{2Qq(r - \cos\theta)}{r^2(r^2 + a^2 - 2ra\,\cos\theta)^{3/2}}\right], \\ \mathcal{T}_{\theta r} &= \mathcal{E}_r \mathcal{E}_\theta = \frac{Qqa\sin\theta}{r^2(r^2 + a^2 - 2ra\,\cos\theta)^{3/2}} \\ &\quad + \frac{q^2 a\sin\theta(r - a\cos\theta)}{(r^2 + a^2 - 2ra\,\cos\theta)^3}. \end{aligned} \tag{6.21}$$

The first component $\mathcal{T}_{rr}$ is the normal stress on the surface Σ and the second one $\mathcal{T}_{\theta r}$ the tangential (or, the shear) stress.

In order to illustrate the above equations, and to see how the electric field vector **E** and the Maxwell's stress vector $\boldsymbol{\mathcal{T}}_r$ vary on the surface of the imaginary sphere Σ, we shall make a numerical example, setting $Q = 2, q = -1, a = 3, r = 1$ in Eqs. (6.18) and (6.21). The expressions we now get are functions of the polar angle θ only. We have plotted $\mathcal{T}_{rr}, \mathcal{T}_{\theta r}$ in Fig. 6.4(b), using Maxima.

In order to show how the field vector $\boldsymbol{\mathcal{E}}$ and the stress vector $\boldsymbol{\mathcal{T}}_r$ vary on the surface of the sphere Σ we have prepared Table 3.1 after evaluating the corresponding quantities in the columns 1–9, using Maxima. The angles $\phi_{\mathrm{E}}, \phi_{\mathrm{T}}$ appearing in columns 5 and 9 have been explained in Fig. 6.4(c).

Table 3.1. $\boldsymbol{\mathcal{E}}$ and $\boldsymbol{\mathcal{T}}_r$ vectors on the surface of the sphere.

1	2	3	4	5	6	7	8	9
θ	$\mathcal{E}_r$	$\mathcal{E}_\theta$	$\mathcal{E}$	ϕ_{E}	$\mathcal{T}_{rr}$	$\mathcal{T}_{\theta r}$	$\mathcal{T}_r$	ϕ_{T}
0°	2.25	0	2.25	0°	2.53	0	2.53	0°
30°	2.15	−0.14	2.16	−3.8°	2.30	−0.31	2.33	−7.6°
60°	2.03	−0.14	2.03	−4°	2.04	−0.28	2.06	−7.9°
90°	1.97	−0.10	1.97	−2.8°	1.93	−0.19	1.94	−5.5°
120°	1.95	−0.05	1.95	−1.6°	1.89	−0.11	1.90	−3.3°
150°	1.94	−0.02	1.94	−0.8°	1.88	−0.05	1.88	−1.5°
180°	1.94	0	1.94	0°	1.88	0	1.88	0°

The first one is the angle between the normal $\mathbf{e}_r$ to the surface Σ and the electric field $\boldsymbol{\mathcal{E}}$ at the surface, and the second one is the angle between $\mathbf{e}_r$ and the stress vector $\boldsymbol{\mathcal{T}}_r$ on the surface.

$$\begin{aligned} \mathcal{E} &= \sqrt{\mathcal{E}_r^2 + \mathcal{E}_\theta^2}; \quad \tan\phi_{\mathrm{E}} = \frac{\mathcal{E}_\theta}{\mathcal{E}_r}; \\ \mathcal{T}_r &= \sqrt{\mathcal{T}_{rr}^2 + \mathcal{T}_{\theta r}^2}; \quad \tan\phi_{\mathrm{T}} = \frac{\mathcal{T}_{\theta r}}{\mathcal{T}_{rr}}. \end{aligned} \tag{6.22}$$

We have drawn the field vectors $\mathbf{E}$ and the stress vectors $\boldsymbol{\mathcal{T}}_r$ on the sphere Σ in Figs. 6.4(d) and 6.4(e) (using two different scales for the two sets of vectors).

All this tedious work will have been fruitful if we could show that the surface force density, when integrated over the entire surface Σ, will give us back the familiar Coulomb force between the two charges. The surface force density is the same as the stress vector on this surface. We shall work with the "reduced" surface force density, same as $\boldsymbol{\mathcal{T}}_r$.

The Coulomb force of attraction (if Q, q are of opposite signs) or repulsion (if they are of the same sign) will be along the line OA joining the two charges. Since this line coincides with the Z-axis, we shall integrate the Z component of $\boldsymbol{\mathcal{T}}_r$, which we shall denote as $\tilde{f}_z$. We go back to Eqs. (6.16) and (6.21) to compute this force, and get the following results after some simplification:

$$\begin{aligned} \tilde{f}_z &= \mathbf{e}_z \cdot \boldsymbol{\mathcal{T}}_r \\ &= (\cos\theta\mathbf{e}_r - \sin\theta\mathbf{e}_\theta) \cdot (\mathbf{e}_r\mathcal{T}_{rr} + \mathbf{e}_\theta\mathcal{T}_{\theta r}) \end{aligned} \tag{6.23a}$$

$$= \cos\theta \mathcal{T}_{rr} - \sin\theta \mathcal{T}_{\theta r} \tag{6.23b}$$

$$= \tilde{f}_z(Q^2) + \tilde{f}_z(Qq) + \tilde{f}_z(q^2), \text{ where} \tag{6.23c}$$

$$\tilde{f}_z(Q^2) = \frac{1}{2}\frac{Q^2}{r^4}\cos\theta, \tag{6.23d}$$

$$\tilde{f}_z(Qq) = \frac{Qq[r\cos\theta - a]}{r^2(r^2+a^2-2ra\,\cos\theta)^{3/2}}, \tag{6.23e}$$

$$\tilde{f}_z(q^2) = \frac{1}{2}\frac{q^2[(r^2+a^2)\cos\theta - 2ra]}{(r^2+a^2-2ra\,\cos\theta)^3}. \tag{6.23f}$$

The expressions in lines (6.23d) and (6.23f), involving Q^2 and q^2, are "self-terms", whereas the expression in (6.23e) involving Qq is the "interaction term". The reader should complete the steps leading from (6.23b) to these equations. We shall soon show that the self-terms will vanish upon integration, leaving the integrated stress force entirely a function of Qq.

The "reduced" force transmitted across the surface Σ, and hence acting on the charge Q, is the surface integral of $\tilde{f}_z$. Let us denote this integral as $\tilde{F}$. An area element on Σ is $da = r^2 \sin\theta\, d\theta\, d\phi$. Therefore,

$$\tilde{F} = \iint_\Sigma \tilde{f}_z\, r^2 \sin\theta\, d\theta\, d\phi$$

$$= 2\pi r^2 \int_0^\pi \tilde{f}_z \sin\theta\, d\theta \tag{6.24a}$$

$$= 2\pi r^2[\mathcal{I}(Q^2) + \mathcal{I}(Qq) + \mathcal{I}(q^2)], \tag{6.24b}$$

$$\text{where } \mathcal{I}(Q^2) = \int_0^\pi \tilde{f}_z(Q^2)\sin\theta\, d\theta = 0, \tag{6.24c}$$

$$\mathcal{I}(Qq) = \int_0^\pi \tilde{f}_z(Qq)\sin\theta\, d\theta = -\frac{2Qq}{a^2r^2}, \tag{6.24d}$$

$$\mathcal{I}(q^2) = \int_0^\pi \tilde{f}_z(q^2)\sin\theta\, d\theta = 0. \tag{6.24e}$$

$$\text{Hence,} \quad \tilde{F} = -\frac{4\pi Qq}{a^2}. \tag{6.24f}$$

The integral given in (6.24c) is easy to evaluate. The other integrals have been worked out in the Sec. B.1. They can be worked out more easily using Maxima with a computer.

To get the true force we go back to (6.19), multiply $\tilde{F}$ with the factor $\frac{1}{16\pi^2\varepsilon_0}$, and get the force $\mathbf{F}_Q$ acting on the charge Q:

$$\mathbf{F}_Q = \frac{1}{16\pi^2\varepsilon_0}\tilde{F}\mathbf{e}_z = -\frac{Qq}{4\pi\varepsilon_0 a^2}\mathbf{e}_z. \tag{6.25}$$

This force is the familiar Coulomb force on the charge Q located at the origin, exerted on it by another charge q located at a distance a on the positive Z-axis. It is repulsive, i.e. towards the negative Z-axis, if Qq is positive, and attractive, i.e. towards the positive Z-axis, if Qq is negative.

6.3. Maxwell's Stress Tensor for the Magnetostatic Field

This section is the magnetostatic analogue of the electrostatic stress tensor presented in Sec. 6.2.3. The steps are parallel, so that we shall avoid detailed explanation.

6.3.1. *Volume force density in terms of the field*

We shall construct Maxwell's stress tensor for the magnetostatic field, represent it by the symbol $\widehat{\mathcal{T}}^{(m)}$. The volume force density in a magnetic field is $\mathbf{f}^{(m)} = \mathbf{J} \times \mathbf{B}$. Therefore, we need to construct the tensor $\widehat{\mathcal{T}}^{(m)}$ under the specification

$$\nabla \cdot \widehat{\mathcal{T}}^{(m)} \equiv \mathbf{f}^{(m)} = \mathbf{J} \times \mathbf{B}. \tag{6.26}$$

This is now an easy task, thanks to the identity (6.7) we had established in Sec. 6.2. We set $\mathbf{B}$ for $\mathbf{A}$ in that equation, and use the field equations: $\nabla \cdot \mathbf{B} = 0$; $\nabla \times \mathbf{B} = \mu_0 \mathbf{J}$, leading to:

$$\boxed{\begin{aligned} \mathbf{f}^{(m)} = \mathbf{J} \times \mathbf{B} = \nabla \cdot \widehat{\mathcal{T}}^{(m)}, \quad &\text{(a)} \\ \text{where } \widehat{\mathcal{T}}^{(m)} = \tfrac{1}{\mu_0}\left[\mathbf{BB} - \tfrac{1}{2}B^2\widehat{\mathbf{1}}\right]. \quad &\text{(b)} \end{aligned}} \tag{6.27}$$

Note the similarity between the stress tensor $\widehat{\mathcal{T}}^{(m)}$ written above and the stress tensor $\widehat{\mathcal{T}}^{(e)}$ written in Eq. (6.10). The former converts into the latter if we replace $\mathbf{E}$ with $\mathbf{B}$ and ε_0 with $\frac{1}{\mu_0}$. In the same way the matrix form given in Eq. (6.11) converts to the matrix form of $\widehat{\mathcal{T}}^{(m)}$. Consequently, the stress vector changes from normal outward, to tangential, to normal inward, as the angle between the plane and the direction of the $\mathbf{B}$ field changes from 90° to 45° to 0°, as shown in (6.15) and illustrated in Fig. 6.3, and

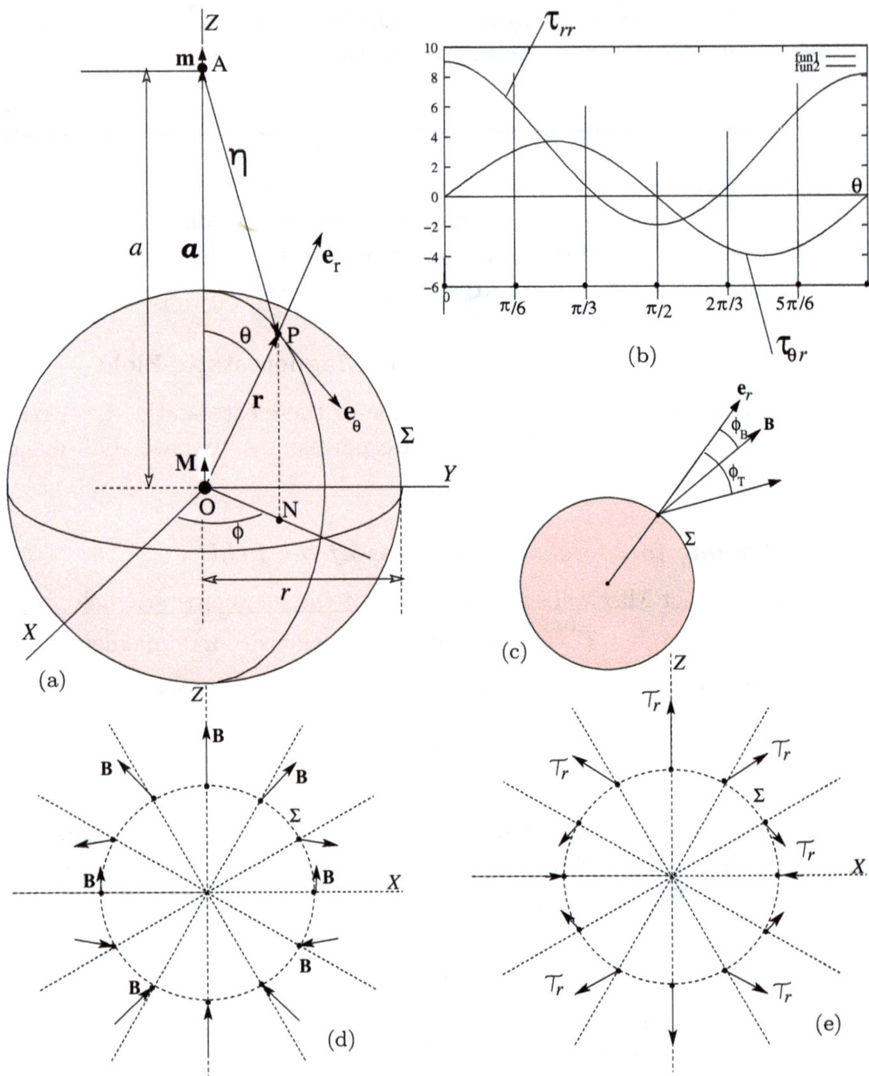

Fig. 6.5. Magnetic stress vector on a spherical surface.

the "Conclusion" written on p. 175 carries over to the case of a magnetic field without any change. Each point in the conclusion is well illustrated in Fig. 6.5 (see next section) if the reader compares the direction of the field vector **B** in Fig. 6.5(d) with the direction of stress vector $\boldsymbol{\mathcal{T}}_r$ in Fig. 6.5(e).

6.3.2. *Example 3: Force transmitted between two magnetic dipoles across a spherical boundary*

The smallest denomination of the source of a magnetic field is a magnetic dipole, consisting of a tiny current loop. We shall therefore think of the force between two magnetic dipoles. We have placed these dipoles along the Z-axis, oriented them in the positive direction of this axis. Figure 6.5(a) shows the geometry of this configuration. The dipoles are shown by tiny spherical blobs with an arrow pointing in the direction of this vector. As in the electrostatic example, we shall illustrate Maxwell's stress tensor $\widehat{\boldsymbol{\mathcal{T}}}^{(\mathrm{m})}$ by finding the stress vector on the surface of an imaginary sphere Σ of radius r surrounding the point magnetic dipole $\mathbf{M}$ which is placed at a distance a from the other point magnetic dipole $\mathbf{m}$ such that $r < a$, and then integrate this stress vector over the spherical surface to obtain the force $\mathbf{F}_{\mathrm{m}}$ on $\mathbf{M}$ exerted by $\mathbf{m}$.

We shall first obtain the $\mathbf{B}$ field at any arbitrary point P (r, θ, ϕ) located on the spherical surface Σ, at the displacement vector $\boldsymbol{\eta}$ from A and $\mathbf{r}$ from O. In order to avoid repeated appearance of the constant $\frac{\mu_0}{4\pi}$, we shall set $\mathbf{B} = \frac{\mu_0}{4\pi}\boldsymbol{\mathcal{B}}$, and use Eq. (6.16).

Let $\boldsymbol{\mathcal{B}}^{(M)}(r, \theta, \phi)$, and $\boldsymbol{\mathcal{B}}^{(m)}(r, \theta, \phi)$ be the fields[b] produced by the dipoles $\mathbf{M}$ and $\mathbf{m}$ respectively, at any coordinate point (r, θ, ϕ). Adding them we get the total field $\boldsymbol{\mathcal{B}}(r, \theta, \phi)$:

$$\boldsymbol{\mathcal{B}}(r, \theta, \phi) = \boldsymbol{\mathcal{B}}^{(M)}(r, \theta, \phi) + \boldsymbol{\mathcal{B}}^{(m)}(r, \theta, \phi), \tag{6.28a}$$

$$\boldsymbol{\mathcal{B}}^{(M)}(r, \theta, \phi) = \frac{3(\mathbf{M} \cdot \mathbf{r})\,\mathbf{r} - \mathbf{M}r^2}{r^5} = \mathcal{B}_r^{(M)}\mathbf{e}_r + \mathcal{B}_\theta^{(M)}\mathbf{e}_\theta, \tag{6.28b}$$

$$\text{where} \quad \mathcal{B}_r^{(M)} = \frac{2M\,\cos\theta}{r^3}, \quad \mathcal{B}_\theta^{(M)} = \frac{M\,\sin\theta}{r^3}, \tag{6.28c}$$

$$\boldsymbol{\mathcal{B}}^{(m)}(r, \theta, \phi) = \frac{3(\mathbf{m} \cdot \boldsymbol{\eta})\,\boldsymbol{\eta} - \mathbf{m}\eta^2}{\eta^5} = \mathcal{B}_r^{(m)}\mathbf{e}_r + \mathcal{B}_\theta^{(m)}\mathbf{e}_\theta, \tag{6.28d}$$

$$\text{where} \quad \mathcal{B}_r^{(m)} = \frac{m[2(r^2 + a^2)\cos\theta - (3 + \cos^2\theta)ar]}{\eta^5}, \tag{6.28e}$$

$$\mathcal{B}_\theta^{(m)} = \frac{m(r^2 - 2a^2 + ar\cos\theta)\sin\theta}{\eta^5}. \tag{6.28f}$$

[b] See [14, Eq. (3.89)].

For future convenience, we write

$$\begin{aligned}
&\boldsymbol{\mathcal{B}} = \mathcal{B}_r \mathbf{e}_r + \mathcal{B}_\theta \mathbf{e}_\theta, \quad \text{where} \\
&\mathcal{B}_r = \left(\frac{M}{r^3}\right)\alpha + \left(\frac{m}{\eta^5}\right)\beta, \\
&\mathcal{B}_\theta = \left(\frac{M}{r^3}\right)\gamma + \left(\frac{m}{\eta^5}\right)\delta.
\end{aligned} \tag{6.29}$$

From (6.28): $\alpha = 2\cos\theta, \quad \beta = 2(r^2 + a^2)\cos\theta - (3 + \cos^2\theta)ar,$

$$\gamma = \sin\theta, \quad \delta = (r^2 - 2a^2 + ar\cos\theta)\sin\theta.$$

From Eq. (6.27), the stress tensor is

$$\begin{aligned}
\widehat{\mathcal{T}}^{(\mathrm{m})} &= \frac{1}{\mu_0}\left(\mathbf{BB} - \frac{1}{2}B^2\widehat{\mathbf{1}}\right) = \frac{\mu_0}{16\pi^2}\left(\boldsymbol{\mathcal{B}}\boldsymbol{\mathcal{B}} - \frac{1}{2}\mathcal{B}^2\widehat{\mathbf{1}}\right) \\
&= \frac{\mu_0}{16\pi^2}\tilde{\mathcal{T}}^{(\mathrm{m})},
\end{aligned} \tag{6.30a}$$

$$\text{where} \quad \tilde{\mathcal{T}}^{(\mathrm{m})} = \boldsymbol{\mathcal{B}}\boldsymbol{\mathcal{B}} - \frac{1}{2}\mathcal{B}^2\widehat{\mathbf{1}}, \tag{6.30b}$$

which we may refer to as the "reduced stress tensor". The non-zero components of this tensor needed by us are

$$\mathcal{T}_{rr} = \mathcal{B}_r^2 - \frac{1}{2}\mathcal{B}^2 = \frac{1}{2}(\mathcal{B}_r^2 - \mathcal{B}_\theta^2); \quad \mathcal{T}_{r\theta} = \mathcal{T}_{\theta r} = \mathcal{B}_r\mathcal{B}_\theta. \tag{6.31}$$

In order to illustrate the above equations, and to see how the magnetic field vector **B** and the Maxwell's stress vector $\widehat{\mathcal{T}}^{(\mathrm{m})}$ look like on the surface of the imaginary sphere surrounding the charge Q, we shall make a numerical example, setting $M = 2, m = 1, a = 3, r = 1$ in Eqs. (6.29) and (6.31). For this purpose, we have prepared Table 3.2, after evaluating the corresponding quantities in the columns 1–9 using Maxima. The angles $\phi_{\mathrm{B}}, \phi_{\mathrm{T}}$ appearing in this table have been explained in Fig. 6.5(c). See also Eq. (6.21).

We have plotted $\mathcal{T}_{rr}, \mathcal{T}_{\theta r}$ as functions of the polar angle θ in Figs. 6.5(b), using Maxima, and have drawn the vectors **B** and $\boldsymbol{\mathcal{T}}_r$ on the sphere Σ in Figs. 6.5(d) and 6.5(e) (using two different scales for the two sets of vectors).

All this tedious work will have been fruitful if we could show that the surface force density, when integrated over the entire surface Σ, will yield the same force between the two dipoles that we can calculate using the standard formulas of magnetostatics. Let us then first apply the "standard formula".

Table 3.2. $\boldsymbol{\mathcal{B}}$ and $\boldsymbol{\mathcal{T}}_r$ vectors on the surface of the sphere.

1	2	3	4	5	6	7	8	9
θ	$\mathcal{B}_r$	$\mathcal{B}_\theta$	$\mathcal{B}$	ϕ_B	$\mathcal{T}_{rr}$	$\mathcal{T}_{\theta r}$	$\mathcal{T}$	ϕ_T
0°	4.25	0	4.25	0°	9.03	0	9.03	0°
30°	3.58	0.86	3.69	13.5°	6.05	3.09	6.80	26.9°
60°	2.00	1.64	2.59	39.3°	0.66	3.28	3.34	78.5°
90°	−0.03	1.96	1.96	−89.4°	−1.91	−0.06	1.91	1.7°
120°	−2.03	1.71	2.66	−40.1°	0.60	−3.48	3.53	−76.8°
150°	−3.5	0.99	3.63	−16.0°	5.62	−3.47	6.60	−31.5°
180°	−4.03	0	4.03	0°	8.13	0	8.13	0°

The force $\mathbf{F}_\mathrm{m}$ on $\mathbf{m}$ is given by the formula $\mathbf{F} = (\mathbf{m}\cdot\boldsymbol{\nabla})\mathbf{B}$, in which $\mathbf{B}$ is the field created by $\mathbf{M}$. The $\mathbf{m}$ vector is in the Z-direction. Therefore, $\mathbf{m}\cdot\boldsymbol{\nabla} = m\frac{\partial}{\partial z}$, which means that we can treat the (x, y) coordinates as constant and equal to zero. Therefore,

$$\mathbf{F}_{\mathrm{m}} = m\frac{\partial \mathbf{B}}{\partial z}\bigg|_{x=y=0,z=a},$$

$$\text{where,}\quad \mathbf{B}(0,0,z) = \frac{\mu_0 M}{4\pi}\left[\frac{3z^2 - z^2}{z^5}\right]\mathbf{e}_z,$$

$$\frac{\partial \mathbf{B}}{\partial z}\bigg|_{x=y=0,z=a} = -\frac{3\mu_0 M}{2\pi}\frac{1}{a^4}\mathbf{k}. \tag{6.32}$$

$$\text{Hence,}\quad \mathbf{F}_{\mathrm{m}} = -\frac{3\mu_0 mM}{2\pi a^4}\mathbf{e}_z.$$

By Newton's third law of motion,

$$\mathbf{F}_\mathrm{m} = -\mathbf{F}_{\mathrm{m}} = \frac{3\mu_0 Mm}{2\pi a^4}\mathbf{e}_z. \tag{6.33}$$

Now we shall calculate the same force using the stress tensor. The surface force density is the same as the stress vector on this surface. We shall work with the "reduced" surface force density, same as $\boldsymbol{\mathcal{T}}_r$.

The force of attraction between the dipoles will be along the line OA joining them, which lies on the Z-axis. Therefore, we need the Z component of the surface force density $\tilde{f}_z$:

$$\tilde{f}_z = \mathbf{e}_z\cdot\boldsymbol{\mathcal{T}}_r = (\cos\theta\mathbf{e}_r - \sin\theta\mathbf{e}_\theta)\cdot(\mathbf{e}_r\mathcal{T}_{rr} + \mathbf{e}_\theta\mathcal{T}_{\theta r}) = \cos\theta\mathcal{T}_{rr} - \sin\theta\mathcal{T}_{\theta r}. \tag{6.34}$$

We shall break up this force density into three components: (1) $\tilde{f}_z(M^2)$ representing self-term for $\mathbf{M}$, (2) $\tilde{f}_z(Mm)$ representing interaction term between **M** and **m**, (3) $\tilde{f}_z(m^2)$ representing self-term for **m**. From Eqs. (6.29), (6.31) and (6.34):

$$\tilde{f}_z(M^2) = \left[(\alpha^2 - \gamma^2)\cos\theta - 2\alpha\gamma\sin\theta\right]\frac{M^2}{2r^6}, \tag{6.35a}$$

$$\tilde{f}_z(Mm) = [(\alpha\beta - \gamma\delta)\cos\theta - (\alpha\delta + \beta\gamma)\sin\theta]\frac{Mm}{r^3\eta^5}, \tag{6.35b}$$

$$\tilde{f}_z(m^2) = \left[(\beta^2 - \delta^2)\cos\theta - 2\beta\delta\sin\theta\right]\frac{m^2}{2\eta^{10}}. \tag{6.35c}$$

The "reduced" force $\tilde{F}$ transmitted across the surface Σ, and hence acting on the dipole **M**, is the surface integral of $\tilde{f}_z$, which is the sum of the integrals of $\tilde{f}_z(M^2), \tilde{f}_z(Mm)$, and $\tilde{f}_z(m^2)$. Each integral is difficult to evaluate, because $\alpha, \beta, \gamma, \delta$ are complicated functions of r, a, θ. We have evaluated these integrals using Maxima. See Appendix B. The result is as follows:

$$\begin{aligned}\tilde{F} &= \iint_\Sigma \tilde{f}_z r^2 \sin\theta\, d\theta\, d\phi = 2\pi r^2 \int_0^\pi \tilde{f}_z \sin\theta\, d\theta \\ &= 2\pi r^2[\mathcal{I}(M^2) + \mathcal{I}(Mm) + \mathcal{I}(m^2)],\end{aligned}$$

$$\begin{aligned}\text{where}\quad \mathcal{I}(M^2) &= \int_0^\pi \tilde{f}_z(M^2)\sin\theta\, d\theta = 0, \\ \mathcal{I}(Mm) &= \int_0^\pi \tilde{f}_z(Mm)\sin\theta\, d\theta = \frac{12Mm}{a^4r^2}, \\ \mathcal{I}(m^2) &= \int_0^\pi \tilde{f}_z(m^2)\sin\theta\, d\theta = 0.\end{aligned} \tag{6.36}$$

$$\text{Hence,}\quad \tilde{F} = \frac{24\pi Mm}{a^4}.$$

Because of the relation (6.30) the true force $\mathbf{F}_\mathrm{m}$ acting on the dipole **M** is $\frac{\mu_0}{16\pi^2}$ times the force $\tilde{F}$. Hence

$$\mathbf{F}_\mathrm{m} = \frac{3\mu_0 Mm}{2\pi a^4}\mathbf{e}_z. \tag{6.37}$$

We have thus verified that the stress tensor has given us the same force that we obtained in Eq. (6.33) using standard formulas of magnetostatics.

We have worked out three examples to bring out the meaning of Maxwell's stress tensor for electric and magnetic fields. The reader may wonder why we should go through such a tortuous road to get answers that can be easily obtained using simpler formulas of electrostatics and magnetostatics? Isn't it like demolishing a mud wall with a cannon?

Every cannon needs a mud wall to ensure its trust-worthiness before deployment in a true situation. Maxwell's stress tensor is destined to play a bigger role, in constructing the conservation equation for field momentum, and later under the watchful eye of Special Relativity, in building up the covariant expression for conservation of energy and momentum. The three examples we have worked out were intended to be an intellectual exercise to instill confidence in the mathematical expressions of $\widehat{\mathcal{T}}^{(\mathrm{e})}$ and $\widehat{\mathcal{T}}^{(\mathrm{m})}$ before crowning them for their majestic role.

Our next example is not a mud wall. It shows how Maxwell's stress tensor can solve a difficult problem directly.

6.4. Example 4: The Force Between Two Hemispheres of a Charged Sphere

Consider a uniformly charged sphere of radius R, and carrying a total charge Q. What is the (repulsive) force that the lower hemisphere exerts on the upper hemisphere.

Finding the force by a naive application of Coulomb's law can be difficult.

The solution of this problem can be found in Griffiths.[c] However, Griffiths employs Cartesian system. We shall use the spherical coordinate system to obtain the result compactly. We have illustrated the geometry in Fig. 6.6.

We divide the boundary surface into two parts: (1) the upper surface S_{top}, on which the normal vector is $\mathbf{e}_r$, (2) the lower surface S_{bottom} on which the normal vector is $-\mathbf{e}_z = \mathbf{e}_\theta$. The $\mathbf{E}$-field is radial on both. We shall find the stress vectors $\boldsymbol{\tau}^{(\mathrm{t})}, \boldsymbol{\tau}^{(\mathrm{b})}$, and their normal components $\tau_n^{(\mathrm{t})}, \tau_n^{(\mathrm{b})}$ on both surfaces, and by integrating, shall get the answer. Let us first get the stress tensors on the top and the bottom surfaces. The (reduced) electric fields are as follows:

$$\mathcal{E}^{(\mathrm{t})} = \frac{Q}{r^2}\mathbf{e}_r; \quad \mathcal{E}^{(\mathrm{b})} = \frac{Qr}{R^3}\mathbf{e}_r. \tag{6.38}$$

[c]See [14, p. 368].

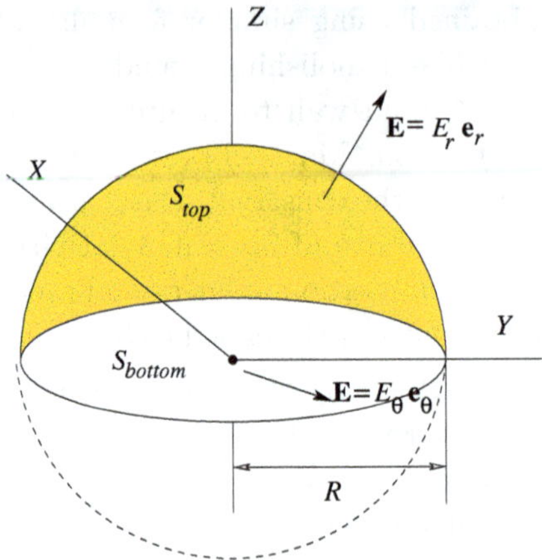

Fig. 6.6. Charged hemisphere.

Using Eq. (6.19), the (reduced) stress tensor takes the following forms:

$$\tilde{\mathcal{T}}^{(\mathrm{t})} = \frac{Q^2}{2R^4}\begin{pmatrix} 1 & 0 & 0 \\ 0 & -1 & 0 \\ 0 & 0 & -1 \end{pmatrix}, \quad \tilde{\mathcal{T}}^{(\mathrm{b})} = \frac{Q^2 r^2}{2R^6}\begin{pmatrix} 1 & 0 & 0 \\ 0 & -1 & 0 \\ 0 & 0 & -1 \end{pmatrix}. \tag{6.39}$$

Note that the normal to the top surface is $\mathbf{n} = \mathbf{e}_r$, and the normal to the bottom surface is $\mathbf{n} = -\mathbf{e}_z = \mathbf{e}_\theta$. Since the net force is in the z-direction, we shall consider only the (z, r)-component of the stress vector on the top surface, and the (z, θ)-component on the bottom surface

$$\begin{aligned} \tau_{zr}^{(\mathrm{t})} &= \mathbf{e}_z \cdot \tilde{\mathcal{T}}^{(\mathrm{t})} \cdot \mathbf{e}_r = (\cos\theta \mathbf{e}_r - \sin\theta \mathbf{e}_\theta) \cdot \tilde{\mathcal{T}}^{(\mathrm{t})} \cdot \mathbf{e}_r \\ &= \cos\theta \mathcal{T}_{rr}^{(\mathrm{t})} = \frac{Q^2}{2R^4}\cos\theta, \\ \tau_{z\theta}^{(\mathrm{b})} &= \mathbf{e}_z \cdot \tilde{\mathcal{T}}^{(\mathrm{b})} \cdot \mathbf{e}_\theta = -\mathbf{e}_\theta \cdot \tilde{\mathcal{T}}^{(\mathrm{b})} \cdot \mathbf{e}_\theta = -\tilde{\mathcal{T}}_{\theta\theta}^{(\mathrm{b})} = \frac{Q^2 r^2}{2R^6}. \end{aligned} \tag{6.40}$$

It is seen from the last equation that the stress vector is pointing *into* the volume above the surface, though the field $\mathcal{E}$ is parallel to the surface. This may appear strange on first sight, but conforms to Eq. (6.15), and the conclusions following them.

Integrating the stress vectors given in Eq. (6.40) over the respective surfaces, we get the force on the upper hemisphere.

$$\begin{aligned}\mathcal{F}^{(\mathrm{t})} &= \frac{\varepsilon_0 Q^2}{2R^4}\iint_S \cos\theta R^2 \sin\theta\, d\theta\, d\phi = \frac{Q^2\pi}{2R^2},\\ \mathcal{F}^{(\mathrm{b})} &= \frac{\varepsilon_0 Q^2}{2R^6}\iint_S r^2\, dr\, r\, d\phi = \frac{Q^2}{2R^6}\times\frac{2\pi R^4}{4} = \frac{Q^2\pi}{4R^2}.\end{aligned} \tag{6.41}$$

Adding the above two forces and multiplying with $\frac{1}{16\pi^2\varepsilon_0}$ (see Eq. (6.19)), we get the total force:

$$\mathbf{F} = \frac{1}{4\pi\epsilon_0}\frac{3Q^2}{16R^2}\mathbf{e}_z. \tag{6.42}$$

6.5. Maxwell's Stress Tensor for the Electromagnetic Field and Momentum Conservation

We had introduced Maxwell's stress tensor for static electric and static magnetic fields, with suitable applications, in Secs. 6.2 and 6.3. These applications demonstrated that the force acting on static distributions of electric charges and currents lying within a bounded volume $\mathcal{V}$ is equal to the stress vector integrated over the surface $\mathcal{S}$ bounding this volume. The attribute "static" implied that the objects considered in our discussion, e.g. isolated charges and isolated current carrying loops, were fixed with a kind of "glue" making them immobile in spite of the electric and magnetic forces acting on them. We shall now remove that glue and see what role can now be played by the same stress tensors.

At this point, we shall make a subtle distinction between force and stress. Force acts on material objects which may be discrete charged particles or a localized continuous material media, e.g. a plasma. The stress considered here acts on the field, which is a kind of ethereal medium, as conceived by Maxwell and his contemporary physicists. In the absence of any glue holding them, the charges (e.g. electrons, nuclei) and currents (e.g. current loops) will be free to move and gain momentum. However, the momentum need not be confined to material objects. It can be shared by the field as well. Therefore, we shall make the following conjecture.

Conjecture 6.1. *There exists a Maxwell's stress tensor* $\widehat{\mathcal{T}}^{(\mathrm{em})}$ *for the electromagnetic field, and it is given as*

$$\widehat{\mathcal{T}}^{(\mathrm{em})} \equiv \widehat{\mathcal{T}}^{(\mathrm{e})} + \widehat{\mathcal{T}}^{(\mathrm{m})} = \varepsilon_0[\mathbf{E}\mathbf{E} - \frac{1}{2}E^2\widehat{\mathbb{1}}] + \frac{1}{\mu_0}[\mathbf{B}\mathbf{B} - \frac{1}{2}B^2\widehat{\mathbb{1}}], \tag{6.43}$$

such that

$$\frac{d}{dt}\left(\iiint_{\mathcal{V}} \boldsymbol{g}\, d^3r\right) + \frac{d}{dt}\left(\iiint_{\mathcal{V}} \mathbf{P}\, d^3r\right) = \iint_{S} \widehat{\mathcal{T}}^{(\mathrm{em})} \cdot \mathbf{n}(\mathbf{r})\, da, \qquad (6.44)$$

where $\boldsymbol{g}$ *and* $\mathbf{P}$ *are, respectively, the field momentum density and the material momentum density, the latter being governed by Newton–Minkowski–Lorentz-force equation*

$$\frac{\partial \mathbf{P}}{\partial t} = \rho\mathbf{E} + \mathbf{J}\times\mathbf{B}. \qquad (6.45)$$

Moral: It is seen from (6.44) that the force transmitted by the Maxwell stress tensor $\widehat{\mathcal{T}}_{(\mathrm{em})}$ across a closed surface S contributes to the *total momentum* inside the volume V (bounded by the same surface), and has two parts, namely, the *mechanical part* and the *field part.*

In Appendix A.3, we have shown all the 3×3 components of $\widehat{\mathcal{T}}^{(\mathrm{em})}$ explicitly.

The right-hand side of Eq. (6.44) gives the stress transmitted across the boundary $\mathcal{S}$. The right-hand side of Eq. (6.45) gives the density of Lorentz force acting on all charged matter lying within the volume $\mathcal{V}$. We shall convert the surface integral on the right-hand side of (6.44) into a volume integral, using Gauss's theorem (see Sec. 5.2.4) so that each term in this equation is a volume integral, and then remove the integral sign reducing the same equation to an equality among three density functions:

$$\frac{\partial \boldsymbol{g}}{\partial t} + \frac{\partial \mathbf{P}}{\partial t} = \boldsymbol{\nabla}\cdot\widehat{\mathcal{T}}^{(\mathrm{em})}, \qquad (6.46\mathrm{a})$$

$$\text{or}\quad \frac{\partial \boldsymbol{g}}{\partial t} + \rho\mathbf{E} + \mathbf{J}\times\mathbf{B} = \boldsymbol{\nabla}\cdot\widehat{\mathcal{T}}^{(\mathrm{e})} + \boldsymbol{\nabla}\cdot\widehat{\mathcal{T}}^{(\mathrm{m})}. \qquad (6.46\mathrm{b})$$

We shall now show that the above conjecture is right, that starting from Maxwell's equations we are able to find an expression for the field momentum density such that the momentum conservation of matter and field together falls into the scheme suggested in Eq. (6.46). Our task is made simple by the identity (6.7) we had established in Sec. 6.2. We shall do the work in two stages: (1) set $\mathbf{E}$ for $\mathbf{A}$ in (6.7), and use Maxwell's equations: $\boldsymbol{\nabla}\cdot\mathbf{E} = \rho/\varepsilon_0$; $\boldsymbol{\nabla}\times\mathbf{E} = -\frac{\partial \mathbf{B}}{\partial t}$, (2) set $\mathbf{B}$ for $\mathbf{A}$ and use Maxwell's equations: $\boldsymbol{\nabla}\cdot\mathbf{B}=0$; $\boldsymbol{\nabla}\times\mathbf{B}=\mu_0(\mathbf{J}+\varepsilon_0\frac{\partial\mathbf{E}}{\partial t})$. Hence,

$$\boldsymbol{\nabla}\cdot\widehat{\mathcal{T}}^{(\mathrm{e})}=\boldsymbol{\nabla}\cdot\varepsilon_0\left[\mathbf{EE}-\frac{1}{2}E^2\widehat{\mathbf{1}}\right]=\varepsilon_0\left[(\boldsymbol{\nabla}\cdot\mathbf{E})\mathbf{E}-\mathbf{E}\times(\boldsymbol{\nabla}\times\mathbf{E})\right]$$

$$=\rho\mathbf{E}+\varepsilon_0\mathbf{E}\times\frac{\partial\mathbf{B}}{\partial t},\tag{6.47a}$$

$$\boldsymbol{\nabla}\cdot\widehat{\mathcal{T}}^{(\mathrm{m})}=\boldsymbol{\nabla}\cdot\frac{1}{\mu_0}\left[\mathbf{BB}-\frac{1}{2}B^2\widehat{\mathbf{1}}\right]=\frac{1}{\mu_0}\left[(\boldsymbol{\nabla}\cdot\mathbf{B})\mathbf{B}-\mathbf{B}\times(\boldsymbol{\nabla}\times\mathbf{B})\right]$$

$$=-\mathbf{B}\times(\mathbf{J}+\varepsilon_0\frac{\partial\mathbf{E}}{\partial t})=\mathbf{J}\times\mathbf{B}+\varepsilon_0\frac{\partial\mathbf{E}}{\partial t}\times\mathbf{B},\tag{6.47b}$$

$$\boldsymbol{\nabla}\cdot\widehat{\mathcal{T}}^{(\mathrm{em})}=\frac{\partial}{\partial t}(\varepsilon_0\mathbf{E}\times\mathbf{B})+(\rho\mathbf{E}+\mathbf{J}\times\mathbf{B}).\tag{6.47c}$$

Equation (6.47c) is obtained by adding Eqs. (6.47a) and (6.47b), and using definition of $\widehat{\mathcal{T}}^{(\mathrm{em})}$ as given in (6.43). It confirms validity of our conjecture and identifies the field momentum density as

$$\boxed{\boldsymbol{g}=\varepsilon_0(\mathbf{E}\times\mathbf{B}).}\tag{6.48}$$

We shall like to recast Eq. (6.46a) into the general format of the conservation equation

$$\frac{\partial}{\partial t}(\text{volume density})+\boldsymbol{\nabla}\cdot(\textbf{flux density})=0.\tag{6.49}$$

In this case, the momentum flux density $\widehat{\boldsymbol{\Phi}}^{(\mathrm{em})}$ is to be identified as

$$\widehat{\boldsymbol{\Phi}}^{(\mathrm{em})}=-\widehat{\mathcal{T}}^{(\mathrm{em})}.\tag{6.50}$$

Equation (6.46a) now reads like a true momentum conservation equation:

$$\frac{\partial}{\partial t}(\boldsymbol{g}+\mathbf{P})+\boldsymbol{\nabla}\cdot\widehat{\boldsymbol{\Phi}}^{(\mathrm{em})}=\mathbf{0}.\tag{6.51}$$

It may be easier to comprehend the meaning of the above conservation equation by writing its three Cartesian components. For example, the x-component of the above equation will be

$$\frac{\partial P_x}{\partial t}+\frac{\partial g_x}{\partial t}+\boldsymbol{\nabla}\cdot\boldsymbol{\Phi}_x=0,\tag{6.52a}$$

$$\text{where} \quad \boldsymbol{\Phi}_x = \widehat{\boldsymbol{\Phi}} \cdot \mathbf{e}_x = -\widehat{\boldsymbol{\mathcal{T}}}^{(\text{em})} \cdot \mathbf{e}_x \tag{6.52b}$$

$$= -\varepsilon_0 \left[\mathbf{e}_x \frac{1}{2}(E_x^2 - E_y^2 - E_z^2) + \mathbf{e}_y E_y E_x + \mathbf{e}_z E_z E_x \right]$$

$$- \frac{1}{\mu_0} \left[\mathbf{e}_x \frac{1}{2}(B_x^2 - B_y^2 - B_z^2) + \mathbf{e}_y B_y B_x + \mathbf{e}_z B_z B_x \right]. \tag{6.52c}$$

The first two terms in Eq. (6.52a) give the rate of increase of the x-component of total momentum (consisting of field momentum and material momentum) per unit volume, the third term gives the rate of outflux of the x-component of the field momentum per unit volume. Conservation of momentum implies that the sum of the two must be zero.

Before leaving this topic let us recall the expressions for the field energy density w and the field energy flux density $\mathbf{S}$ (i.e. the Poynting's vector, Sec. 12.3)

$$w \equiv \frac{\varepsilon_0}{2}[E^2 + c^2 B^2] \quad \textbf{Field Energy Density}, \tag{6.53a}$$

$$\mathbf{S} \equiv \varepsilon_0 c[\mathbf{E} \times c\mathbf{B}] \quad \textbf{Field Energy Flux Density}. \tag{6.53b}$$

It is immediately noticed that

$$\mathbf{S} = c^2 \boldsymbol{g}. \tag{6.54}$$

When the electromagnetic field is a radiation field, $E = cB$ and $\mathbf{E}\times c\mathbf{B} = E^2\mathbf{n}$ where $\mathbf{n}$ is the direction of the Poynting's vector, giving the direction of the flow of radiation energy. For such radiation fields,

$$w = \varepsilon_0 E^2, \quad \mathbf{S} = cw\mathbf{n}, \quad \boldsymbol{g} = \frac{w}{c}\mathbf{n}, \quad w = cg. \tag{6.55}$$

The last equality is a reminder of the relation $E = cp$ between the energy E and the momentum p of a photon.

We are still not too clear about the true meaning of the momentum flux density $\widehat{\boldsymbol{\Phi}}$. To get familiarity with it let us consider a plane electromagnetic wave propagating in the x-direction, polarized in the y-direction. For such a field $\mathbf{E} = E\mathbf{e}_y$, $c\mathbf{B} = E\mathbf{e}_z$. It is a simple exercise to evaluate $\widehat{\boldsymbol{\Phi}}$ by setting $E_x = 0$, $E_y = E$, $E_z = 0$; $cB_x = 0$, $cB_y = 0$, $cB_z = E$ in the expression for $\boldsymbol{\Phi}_x$ in Eq. (6.52c) and similar expressions for $\boldsymbol{\Phi}_y, \boldsymbol{\Phi}_z$ and obtain

$$\widehat{\boldsymbol{\Phi}} = \boldsymbol{\Phi}_x \mathbf{e}_x + \boldsymbol{\Phi}_y \mathbf{e}_y + \boldsymbol{\Phi}_z \mathbf{e}_z = (\varepsilon_0 E^2 \mathbf{e}_x)\mathbf{e}_x = c\boldsymbol{g}\mathbf{e}_x = \boldsymbol{g}\mathbf{c}. \tag{6.56}$$

Here $\mathbf{c} = c\mathbf{e}_x$ represents the "velocity" of light, being the speed c multiplied with a unit vector in the direction of propagation. If we now consider a plane perpendicular to the X-axis, so that $\mathbf{n} = \mathbf{e}_x$, then the outflux of field momentum per unit area across the plane will be $\widehat{\mathbf{\Phi}} \cdot \mathbf{n} = \widehat{\mathbf{\Phi}} \cdot \mathbf{e}_x = c\boldsymbol{g}$.

Generalization of Eq. (6.56) is obvious. If there is a source of radiation at the origin (say, an antenna, or an accelerating charged particle), then far away from the origin, the momentum flux density tensor $\widehat{\mathbf{\Phi}}$ has the form

$$\widehat{\mathbf{\Phi}} = cg\mathbf{e}_r\mathbf{e}_r = \boldsymbol{g}c\mathbf{e}_r = \boldsymbol{g}\mathbf{c}, \tag{6.57}$$

where $\mathbf{e}_r$ is the unit vector in the radial direction, also identified with the direction of propagation of the electromagnetic wave. The tensor $\widehat{\mathbf{\Phi}}$ gives the measure of how much momentum is crossing a spherical surface per unit area per unit time. The momentum density is $\boldsymbol{g} = \Pi\mathbf{e}_r$, and it is propagating in the radial direction with velocity $\mathbf{c} = c\mathbf{e}_r$.

Part III
Physics in Four Dimensions

Chapter 7

Space–Time and Its Inhabitants

7.1. World Line in Space–Time

It was noticed from the Lorentz transformation formulas derived and written in Secs. 3.1 and 3.2 that the space coordinates and time coordinates of any event θ in the frame S' is a linear combination of the space and time coordinates in S, and vice versa. Therefore, in a certain sense, the borderline between the space coordinates on the one hand and the time coordinate on the other, looks blurred in Relativity. We fancy therefore a four-dimensional *world* of *events* where the time coordinate takes an equal status along with the three space coordinates. This composite world, integrating time with space, is called *Space–Time*.[a]

Space–time needs four axes, namely, the time axis cT, and the space axes X, Y, Z, and four coordinates (ct, x, y, z). It will be convenient to label the coordinate axes as X^μ, and write the four coordinates of an event as (x^μ), with $\mu = 0$ for the time axis/coordinate ct, and $\mu = 1, 2, 3$ for the space axes/coordinates x, y, z, respectively. That is, $(x^0 = ct,\ x^1 = x,\ x^2 = y,\ x^3 = z)$. We shall follow this convention in this book. Also, we shall use Greek indices, e.g. μ, ν, α, β to mean *all* the four coordinates, and Roman indices, e.g. i, j, k to mean only the three space coordinates. For example, we may write $(x^\mu) = (x^0; x^k) = (ct, x, y, z)$.

Note that we are using the coordinate index μ as a *superscript*, i.e. as a *contravariant* index. The reader will understand the reason in Sec. 7.8,

[a]A detailed exposition of the original paper of Minkowski can be found in [41].

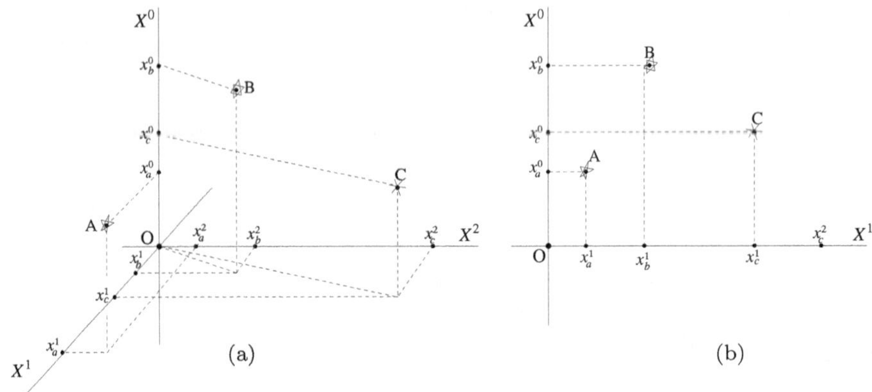

Fig. 7.1. Events A, B, C on space–time diagram: (a) Two space axes X, Y shown; (b) single space axis X shown.

where we shall make a distinction between a contravariant 4-vector and a covariant 4-vector.

In Fig. 7.1, we have presented a view of the space–time, which we shall refer to as a *space–time diagram* (or ST diagram), and displayed three events along with their coordinates: $\mathrm{A} = (x_a^0, x_a^1, x_a^2, x_a^3)$, $\mathrm{B} = (x_b^0, x_b^1, x_b^2, x_b^3)$, $\mathrm{C} = (x_c^0, x_c^1, x_c^2, x_c^3)$. In Fig. 7.1(a), we have suppressed the Z-axis, so that we can show events on paper, and marked the coordinates on the respective axes. In Fig. 7.1(b), we have made the ST diagram simpler by suppressing both Y- and Z-axes, exposing only the X-axis and the time axis.

The above mental construct of the four-dimensional world will not diminish the role of the familiar world of pure space dimensions X, Y, Z. It is in this space that we see objects like satellites, planets, locomotives. We shall call this space the *physical space*.

In Fig. 7.2(a), we have presented another view of space–time for describing the motion of a particle which is confined to move only on the XY-plane. $\mathcal{C}$ is its *physical trajectory*, and P is one point on it.

On the other hand, the trajectory of the particle in space–time, shown as Ω, is called the *world line* of the particle. The event that "the particle has reached P" is presented by the point Θ_P on the world line Ω, and is called a *world point* of the particle.

A particle that does not move at all in a given frame of reference S, still moves continuously along its world line, as depicted by the straight line Σ, directed upwards, because the time clock is continuously ticking.

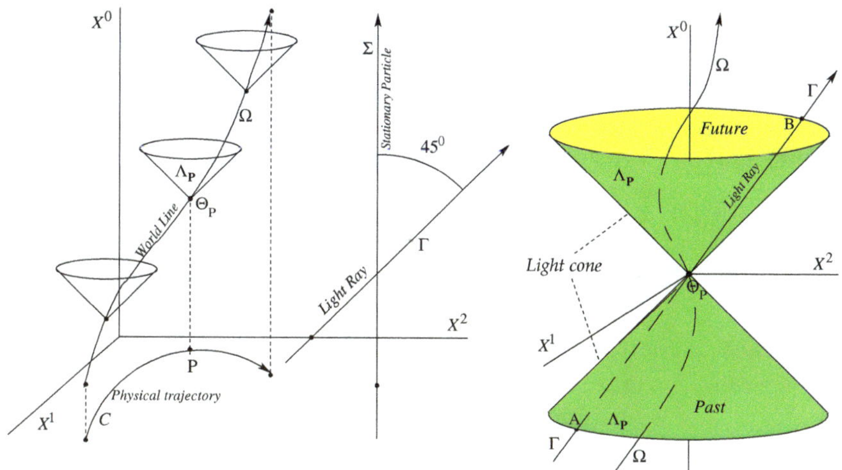

Fig. 7.2. World line and light cones.

The world line of a photon (i.e. a light quantum often represented by the symbol γ) propagating along the X^2-axis must make an angle of 45° with the X^0-axis, as illustrated by the straight line Γ. In fact, one can construct, at any event Θ_P, a *light cone* Λ_P whose surface will make the angle of 45° with the X^0-axis.

Figure 7.2(b) presents a better picture of the light cone and its significance. At any event point Θ_P the light cone carves out a region of space–time as *Future,* and another as *Past,* both of them confined within the light cone, the Future occupying the upper part, the Past occupying the lower one.

Standing at the event Θ_P I am entitled to have information of events that occurred only inside the "Past" segment of the light cone, everything outside remaining beyond my knowledge. In the same way, all future events which will originate from Θ_P, i.e. whose world lines will pass through Θ_P will lie within the "Future" segment of the light cone. The reason for this conclusion is that all information/knowledge is received/gathered through messengers that move with velocities that would never exceed the speed of light c. The fastest messenger is light, or radio signal, moving with the speed c.

The world line of a photon, marked Γ, having Θ_P as a world point, must be grazing the light cone, passing through two points A and B, lying on the Past segment and on the Future segment, respectively. Similarly, the world line of a material particle, marked Ω — progressing from the past to the future through the event Θ_P — must lie inside the light cone, as illustrated in the figure.

7.2. Hyperbolic World Line of a Particle Moving Under a Constant Force

A relativistic particle moving under a constant force undergoes a *constant acceleration* $\boldsymbol{a}$ *with respect to its instantaneous rest frame*, as we had found out on p. 97. If the force is $\mathbf{F}$, and the rest mass of the particle is m_o, then $\boldsymbol{a} = \mathbf{F}/m_o$.

Consider a constant force $\mathbf{F}$ acting on a particle in the X-direction. A good example can be a charged particle placed in a uniform electric field $\mathbf{E} = E_0\mathbf{e}_x$. Let us assume that at $t = 0$ this particle is instantaneously at rest, and located at the origin O, in a certain frame S. Then the (x, ct) coordinates of this particle are given in this frame, as functions of the proper time τ, as (see Eqs. (4.104) and (4.105))

$$x = \frac{c^2}{a}\left[\cosh\frac{a\tau}{c} - 1\right]; \quad ct = \frac{c^2}{a}\left[\sinh\frac{a\tau}{c}\right]. \tag{7.1}$$

The above parametric equation of the world line transforms into the familiar equation of a hyperbola, involving only the space and time coordinates:

$$(x+\rho)^2 - (ct)^2 = \rho^2, \quad \text{where } \rho = \frac{c^2}{a} = \text{unit length}. \tag{7.2}$$

We thus get a hyperbolic world line.

We have used Gnuplot to plot the hyperbolic world line, represented by Eq. (7.2) in Fig. 7.3, in which the X^0- and the X^1-axes are each graduated in the scale of $\rho = 1$ unit.

We have highlighted a few important features of the *world line* in Figs. 7.3(a) and 7.3(b). The *physical trajectory* of the particle is $+\infty \to$ AOB $\to +\infty$, i.e. a directed straight line merging with the X^1-axis, reversing its direction at O. The particle comes from infinity with velocity $\approx -c\mathbf{e}_x$, along the X^1-axis but in the negative direction, decelerates due to application of the force in the $+\mathbf{e}_x$-direction, stops momentarily at the origin O, then turns back and returns to infinity with velocity $\approx +c\mathbf{e}_x$ along the same X^1-axis, but now in the positive direction. A, O and B are three points on this axis reached by the particle at certain times during the inward and outward and journey. The corresponding events are Θ_A, Θ_O and Θ_B.

In Fig. 7.3(a), we have drawn a single light cone at the event Θ_O, and in Fig. 7.3(b) at each of the three events Θ_A, Θ_O and Θ_B. Note that as the

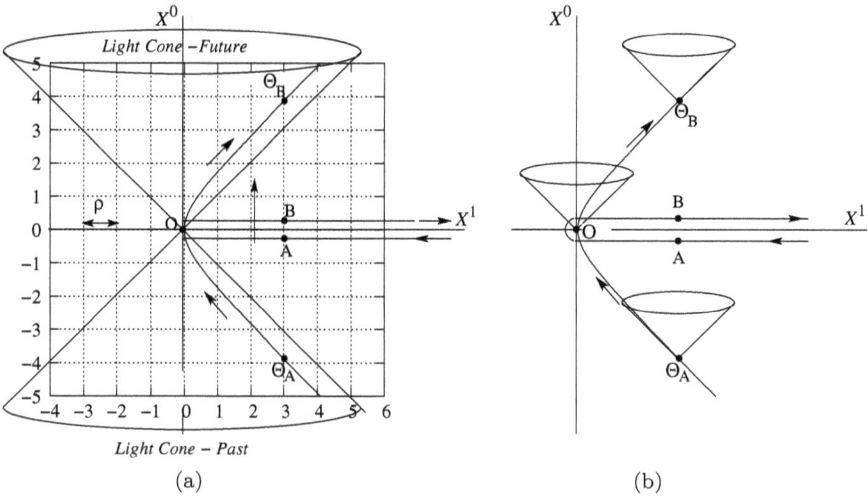

Fig. 7.3. World line of a charged particle under a uniform electric field.

velocity of the particle approaches c, its world line almost grazes the light cone, but never goes outside it.

We shall top up the above exercise with some realistic numerical estimates. Let us first define a characteristic time τ_o, such that the particle would reach the velocity c in this time, if non-relativistic mechanics had been applicable. That is,

$$a\tau_o = c, \quad \text{or } \tau_o = c/a. \tag{7.3a}$$

$$\text{Hence,} \quad \rho = \frac{c^2}{a} = c\tau_o \tag{7.3b}$$

We shall consider a charged particle, e.g. an electron, which is accelerated in a 30 m long linear accelerator (e.g. pelletron) to 30 MeV. The electric field through which the particle is accelerated is assumed to be uniform, and equal to $E = 10^6$ V/m. The charge and mass of the electron are $e = 1.6 \times 10^{-19}$ C, $m_0 = 9.11 \times \times 10^{-31}$ kg. The acceleration is then

$$a = \frac{eE}{m_0} = \frac{1.6 \times 10^{-19} \times 10^6}{9.11 \times \times 10^{-31}} = 0.17 \times 10^{18} \text{m/s}^2$$

$$\text{so that} \quad \tau_o = (3 \times 10^8)/(0.17 \times 10^{18}) = 17.6 \times 10^{-10}\,\text{s}. \tag{7.4}$$

$$\text{Hence,} \quad \rho = 17.6 \times 10^{-10} \times 3 \times 10^8 = 0.528\,\text{m}.$$

We go back to Sec. 4.4, copy formulas (4.58) and (4.59) which give the velocity and displacement the particle:

$$\begin{aligned} v &= c\beta = \frac{c}{\sqrt{1+(\frac{\tau_o}{t})^2}}, \\ x &= \rho\left[\sqrt{1+\left(\frac{t}{\tau_o}\right)^2} - 1\right], \end{aligned} \tag{7.5}$$

and make the following estimates:

$$\begin{aligned} &\text{at } t = \tau_o = 3.416\times 10^{-10}\,\text{s}, && v = \frac{1}{\sqrt{2}}\,c = 0.707\,c, \\ & && x = (\sqrt{2}-1)\rho = 0.414\,\rho; \\ &\text{at } t = 2\tau_o = 6.832\times 10^{-10}\,\text{s}, && v = \sqrt{\frac{4}{5}}\,c = 0.89\,c, \\ & && x = (\sqrt{5}-1)\rho = 1.236\,\rho; \\ &\text{at } t = 3\tau_o = 10.248\times 10^{-10}\,\text{s}, && v = \sqrt{\frac{9}{10}}\,c = 0.95\,c, \\ & && x = (\sqrt{10}-1)\rho = 2.162\,\rho. \end{aligned} \tag{7.6}$$

It is then seen that at $t = 3\tau_o = 10.248\times 10^{-10}\,\text{s}$. the particle has traversed 2.162 units of distance = 1.41 m from the origin, and has gained a speed of $0.95\,c$. We have marked this point on the X^1-axis with an upward arrow ↑.

7.3. Lorentz Transformation in Space–Time

7.3.1. *Graphical procedure*

How to represent Lorentz transformation in *space–time*? We need the answer for a better understanding of Special Relativity. In this section, we shall demonstrate Minkowski's graphical construction of Lorentz transformation, and use this construction to resolve the paradoxes of *length contraction* and *time dilation.*

We have explained the procedure in Fig. 7.4. In order to make the drawings less clumsy we have replaced the (X^1, X^0)-axes with (X, Y)-axes, and the (x^1, x^0) coordinates with (x, y) coordinates. We shall explain the procedure in two steps.

Step 1: SET UP THE COORDINATE AXES (X', Y'), WITH SCALES SHOWING TIC MARKS AT UNIT INTERVALS.

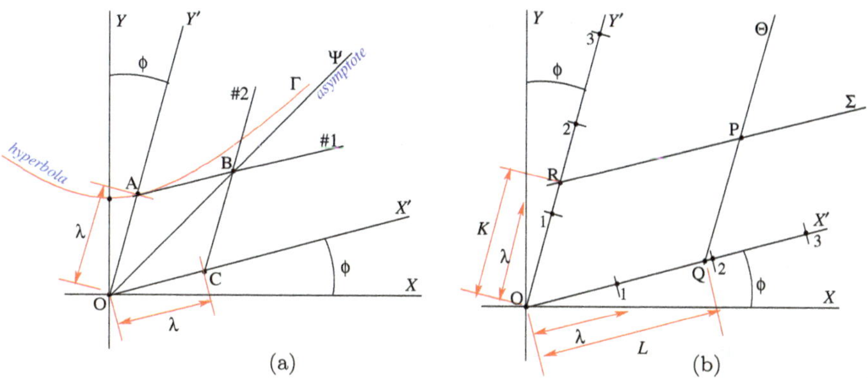

Fig. 7.4. Graphical construction of Lorentz transformation.

The first part is shown in Fig. 7.4(a). Draw the $X-Y$-axes (perpendicular to each other) with the origin at O. The straight line OY′ making an angle ϕ with the Y-axis ($\phi < 45°$) is the new Y'-axis of transformation. Γ is the hyperbola: $y^2 - x^2 = 1$ and Ψ is its asymptote: $y = x$. A is the point of intersection of Γ with the Y'-axis.

$$\begin{aligned}
&Y'\text{-axis:} && y = [\tan(\pi/2 - \phi)]\, x = x \cot\phi, \\
&\Gamma: && y^2 - x^2 = 1, \\
&\text{Intersection A: } (x_A, y_A) = \frac{1}{\sqrt{\cos^2\phi - \sin^2\phi}}(\sin\phi, \cos\phi).
\end{aligned} \tag{7.7}$$

Now scale the Y'-axis by defining one unit as the intercept OA, equal to

$$\lambda|_{y'} = \sqrt{x_A^2 + y_A^2} = \frac{1}{\sqrt{\cos^2\phi - \sin^2\phi}}, \tag{7.8}$$

to be called *scale factor*. Now draw the straight line #1, which is tangent to the hyperbola at A, interacting the asymptote Ψ at B.

$$\begin{aligned}
&\text{tangent:} && \left.\frac{dy}{dx}\right|_A = x_A/y_A = \tan\phi, \\
&\#1: && y - y_A = (x - x_A)\tan\phi, \\
&\Psi: && y = x, \\
&\text{Intersection B:} && (x_B, y_B) = \sqrt{\frac{\cos\phi + \sin\phi}{\cos\phi - \sin\phi}}(1, 1).
\end{aligned} \tag{7.9}$$

Now complete the *parallelogram* OABC. The line OC extended onward is the X'-axis. The point C lies at the intersection of the X'-axis and the line # 2.

$$\begin{aligned}
&\#2: && y - y_B = (x - x_B)\cot\phi, \\
&X'\text{-axis:} && y = x\tan\phi, \\
&\text{Intersection C:} && (x_C, y_C) = \frac{1}{\sqrt{\cos^2\phi - \sin^2\phi}}(\cos\phi, \sin\phi).
\end{aligned} \tag{7.10}$$

Now scale the X'-axis by defining one unit as the intercept OC, equal to

$$\lambda_{x'} = \sqrt{x_C^2 + y_C^2} = \frac{1}{\sqrt{\cos^2\phi - \sin^2\phi}}. \tag{7.11}$$

Note that $\lambda_{x'} = \lambda_{y'}$. Therefore, the two axes have a common scale factor

$$\begin{aligned}
\lambda \equiv \lambda_{x'} = \lambda_{y'} = \gamma\sqrt{1+\beta^2}, \\
\text{where} \quad \beta \equiv \tan\phi < 1, \ \gamma \equiv \tfrac{1}{\sqrt{1-\beta^2}}.
\end{aligned} \tag{7.12}$$

For the actual graphical construction of the LT, we have taken $\phi = 15°$. This gives $\beta = \tan\phi = 0.2679$; $\gamma = 1.038$; $\lambda = 1.074$.

Step 2: TRANSFORM THE (x, y) COORDINATES OF A POINT P TO (x', y'), USING THE SCALE FACTOR λ.

We have explained the steps in Fig. 7.4(b). Let us write the coordinates of the point P as (x_0, y_0). The straight line Θ drawn parallel to the Y'-axis and passing through P intersects X'-axis at Q, and the straight line Σ drawn parallel to the X'-axis and passing through P intersects Y'-axis at R. Then Q and R are the projections of P on the X'- and Y'-axes, respectively. Let us find the (x, y) coordinates of Q and R:

$$\begin{aligned}
&\Theta: && y - y_0 = (x - x_0)\cot\phi, \\
&X'\text{-axis:} && y = x\tan\phi, \\
&\text{Intersection Q:} && (x_Q, y_Q) = \lambda\gamma(x_0 - \beta y_0)(\cos\phi, \sin\phi),
\end{aligned} \tag{7.13}$$

where we have used the identity

$$\cos\phi(1 - \tan^2\phi) = \frac{1}{\lambda\gamma}. \tag{7.14}$$

Similarly,

$$\begin{aligned}
&\Sigma: && y - y_0 = (x - x_0)\tan\phi,\\
&Y'\text{-axis:} && y = x(\cot\phi), \qquad (7.15)\\
&\text{Intersection R:} && (x_R, y_R) = \lambda\gamma(y_0 - \beta x_0)(\sin\phi, \cos\phi).
\end{aligned}$$

The coordinates (x_0', y_0') of the point P, with respect to the X', Y'-axes are now calculated in the following way:

$$L \equiv \widehat{OQ} = \sqrt{x_Q^2 + y_Q^2} = \lambda\gamma(x_0 - \beta y_0), \qquad (7.16a)$$

$$K \equiv \widehat{OR} = \sqrt{x_R^2 + y_R^2} = \lambda\gamma(y_0 - \beta x_0), \qquad (7.16b)$$

$$x_0' \equiv L/\lambda = \gamma(x_0 - \beta y_0), \qquad (7.16c)$$

$$y_0' \equiv K/\lambda = \gamma(y_0 - \beta x_0). \qquad (7.16d)$$

Note that we have not used any principle of relativity in the above geometrical construction. It was a mathematical exercise in coordinate transformation $(x, y) \to (x', y')$ with the final result:

$$\left.\begin{aligned} y' &= \gamma(y - \beta x),\\ x' &= \gamma(x - \beta y). \end{aligned}\right\} \quad \beta < 1;\ \gamma = \frac{1}{\sqrt{1-\beta^2}}. \qquad (7.17)$$

This result is identical with the Lorentz transformation formulas derived in Eqs. (3.8). In this sense, the construction presented here can be called a Graphical Construction of Lorentz Transformation.

7.3.2. *Graphical construction of length contraction*

We shall resolve the paradox of length contraction graphically using space–time axes of the S and S' frames, as depicted in Fig. 7.5. It is a partial copy of Fig. 7.4(a) with the following important differences: (1) changed the (Y, X)-axes to the (X^0, X^1)-axes of the S frame, and similarly (Y', X') to $(X^{0'}, X^{1'})$ of the S' frame; (2) expanded the axes so that the tic marks are further apart.

However, since we shall depend on the construction shown in Fig. 7.4(a), we shall use the same coordinates used there, namely, (y, x); (y', x'), synonymously with the relativity coordinates (x^0, x^1); $(x^{0'}, x^{1'})$, respectively.

A meter stick, i.e. a rigid rod of *unit length*, is lying along the $X^{1'}$-axis of the inertial frame S' which is moving relative to S with the velocity βc in the X-direction. L and R represent its left and right ends. The world lines of these two ends are shown as the straight lines LM and RN. The “world view”

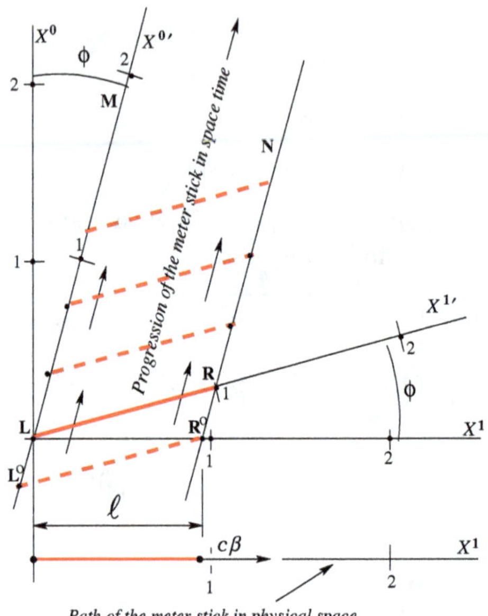

Fig. 7.5. Graphical construction of Lorentz contraction.

of the rod at $x^{0'} = 0$ is shown as a thick solid line of unit length, and at four other values of $x^{0'}$ as thick, but broken lines, also of unit length.

Let us first understand what we mean by the length of the rod in S, with respect to which it is moving. We had provided the answer in the paragraph following Eq. (2.26) on p. 41. It is the distance between the points L and R°, marking the left and the right end of the rod on the X^1-axis *simultaneously*, i.e. at same the instant $x^0 = 0$ (same as $t = 0$), as the rod was speeding away along this axis. We have shown the motion of the rod in the lower diagram.

Referring to Fig. 7.4(a), let R° be the intersection of the world line #2 (i.e. the straight line CB) on the X^1-axis. The intercept ℓ, in Fig. 7.5, is the x^1 coordinate of R°, and represents of length of the stick in S. We shall calculate ℓ using the (x_B, y_B) coordinates of B from Eq. (7.9).

$$\begin{aligned}
&\#2: && y - y_B = \cot\phi(x - x_B),\\
&X\text{-axis:} && y = 0,\\
&\text{At R}^\circ: && \ell = x_B - \tan\phi y_B = (1-\beta)\sqrt{\frac{\cos\phi + \sin\phi}{\cos\phi - \sin\phi}} = \tfrac{1}{\gamma}.
\end{aligned} \tag{7.18}$$

Conclusion: The length of a straight rod which is 1 m long in its rest frame is measured to be $\ell = 1/\gamma$ in the frame S with respect to which it is moving with velocity $c\beta$ parallel to its length. We have thus derived the length contraction formula by graphical construction.

7.3.3. *Graphical construction of time dilation*

We have explained the construction in Fig. 7.6. Again, the axes are the same as in Fig. 7.4(a), but the axes have been expanded even further so that the tic marks # 1 on the axes appear near the margins.

We have presented two events O and A, both occurring at the *same* spatial location in S', or, to be more precise, at the same space coordinate $x^{1'} = 0$, *but* at two different time instants, separated by a time interval of one unit in the frame S'. This means that $x^{0'} = 0$ for O, and $x^{0'} = 1$ for A (as defined while writing Eq. (7.8)). In this sense, the *proper time* between O and A is 1 unit.

We want to find out graphically the time interval between the same two events in the frame S. The procedure is very simple. Just find out

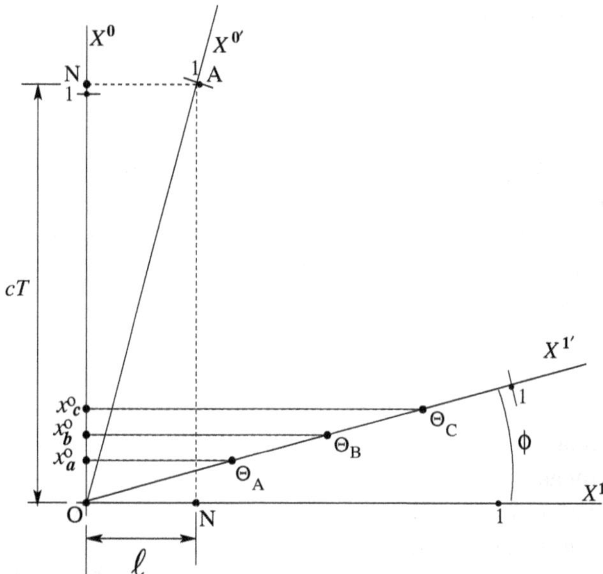

Fig. 7.6. Graphical construction of time dilation.

the y coordinate of A, from Eq. (7.7): $y_A = \frac{\cos\phi}{\sqrt{\cos^2\phi - \sin^2\phi}} = \gamma$. Hence the conclusion:

Conclusion: If the time interval between two events O and A occurring at the same spatial coordinates in a frame S' is 1 unit (so that the proper time between the events is one unit), then the time interval between the same two events as measured in the frame S which is moving uniformly relative to S' with velocity $c\boldsymbol{\beta}$, will be γ units.

Note from the figure that there is a certain spatial separation of ℓ between the two events in S.

7.3.4. *Simultaneity, or absence of it*

On the same diagram presented as Fig. 7.6 we have tried to resolve the paradox surrounding *simultaneity*. Three events $\Theta_{\text{A}}, \Theta_{\text{B}}, \Theta_{\text{C}}$, shown on the $X^{1'}$-axis are simultaneous in the frame S'. They occur at the same time $x^{0'} = 0$. Projecting these three events on the X^0-axis we find that they occur at different time coordinates x_a^0, x_b^0, x_c^0 in the frame S.

7.4. Minkowski Space–Time

What is the length of the segment of the world line between two event points Θ_A and Θ_B?

When I am sleeping, I am still walking a long way in space–time along the time axis. Can I say that the distance I have travelled is c times, say 6 hours of sleep?

During daytime I have commuted from Mysore to Bangalore, a distance of 140 km, in 3 hours. I have moved along some XY-plane, as well as along the time axis. Shall I apply the Pythagorean theorem to arrive at a distance of $\sqrt{(3c)^2 + (140)^2}$ km, covered in space–time?

We shall find an appropriate definition of "length" in space–time. The reader may ask, "What is the need for measuring length in an abstract four-dimensional world which we cannot even visualize before our eyes?" We need a measure of length because we are going to construct four-dimensional vectors in space–time. The most elementary such vector is a "directed straight line" from an event Θ to another event Φ. Length is the only invariant (i.e. something that does not change with a change of the coordinate system) associated with a straight line, whether is space–time or in physical space.

Let us review how length is defined in analytical geometry in terms of coordinates. Consider a straight rod whose endpoints A and B are at the coordinates (x_1, y_1, z_1) and (x_2, y_2, z_2) with reference to a Cartesian frame S. The length of the rod is given by the Pythagorean expression

$$\ell^2 = (x_2 - x_1)^2 + (y_2 - y_1)^2 + (z_2 - z_1)^2. \tag{7.19}$$

If we look at the rod from another angle, say, by shifting my telescope, or better still by rotating my frame of reference, I shall still measure the same length ℓ of the rod.

Suppose we perform a rotation of the axes by an angle θ about the Z-axis, to obtain a new frame S'. It is an elementary exercise[b] to show that the above rotation causes a transformation of coordinates from (x, y, z) to (x', y', z'), given by

$$\begin{aligned} x' &= x\cos\theta + y\sin\theta, \\ y' &= -x\sin\theta + y\cos\theta, \\ z' &= z. \end{aligned} \tag{7.20}$$

Therefore, if (x'_1, y'_1, z'_1) and (x'_2, y'_2, z'_2) be the Cartesian coordinates of A and B in S', the transformed length ℓ' of the rod in S' will be

$$\ell'^2 = (x'_2 - x'_1)^2 + (y'_2 - y'_1)^2 + (z'_2 - z'_1)^2. \tag{7.21}$$

Applying the transformation (7.20) we get

$$\begin{aligned} \ell'^2 &= \{(x_2 - x_1)\cos\theta + (y_2 - y_1)\sin\theta\}^2 \\ &\quad + \{-(x_2 - x_1)\sin\theta + (y_2 - y_1)\cos\theta\}^2 + (z_2 - z_1)^2 \\ &= (x_2 - x_1)^2 + (y_2 - y_1)^2 + (z_2 - z_1)^2 \\ &= \ell^2. \end{aligned} \tag{7.22}$$

In summary, the length expression as given in Eq. (7.19) remains invariant in our familiar physical space, under all transformations of coordinates due to rotation. Since Euclid's geometry is valid in this space, we shall often refer to this space as the *Euclidean space* and denote it by the symbol E^3. The square of the distance $d\ell$ between two neighbouring points (x, y, z) and $(x + dx, y + dy, z + dz)$ in E^3 is given by the expression

$$d\ell^2 = dx^2 + dy^2 + dz^2. \tag{7.23}$$

We shall call Equation (7.23) an expression for the *line element* in E^3. A formal name for the above expression is *metric*. Equation (7.23) expresses an *Euclidean metric*.

[b]The transformation of coordinates by rotation about the Z-axis [23].

We cannot expect invariance of the Euclidean metric (7.23) under a Lorentz transformation. This is because the length of a stick appears different to two different observers who are moving relative to each other, so that $\ell' \neq \ell$. On the other hand, if we consider two events Θ and Φ with coordinates (ct_1, x_1, y_1, x_1) and (ct_2, x_2, y_2, z_2) in some Lorentz frame S, then the value of the expression

$$\begin{aligned} s^2 &\equiv c^2(t_2 - t_1)^2 - \ell^2 \\ &= c^2(t_2 - t_1)^2 - [(x_2 - x_1)^2 + (y_2 - y_1)^2 + (z_2 - z_1)^2] \end{aligned} \tag{7.24}$$

remains invariant under a Lorentz transformation (cf. Eq. (3.11)). Therefore, if we are looking for a candidate to represent "length" in an invariant way in space–time, then the expression give in Eq. (7.24) should satisfy the requirement.

The letter s used above is the invariant "length" in relativity. However, since the word itself connotes a measure involving a meter-stick — as that of a reel of yarn or the width of a fabric — and since we cannot stretch a meter stick across space–time, it is better to suggest an alternative name, for which we choose "line interval".

The differential line interval between two infinitely close events having coordinates (ct, x, y, z) and $(ct + c\,dt, x + dx, y + dy, z + dz)$ has a greater relevance in the geometry of space–time. We shall write this as

$$ds^2 = c^2 dt^2 - dx^2 - dy^2 - dz^2. \tag{7.25}$$

Equation (7.25) will represent the line element, or the metric, better known as the *Minkowski metric*, for space–time. Any LT from a frame S to another frame S' will guarantee invariance of this metric:

$$ds^2 = c^2 dt^2 - dx^2 - dy^2 - dz^2 = c^2 dt'^2 - dx'^2 - dy'^2 - dz'^2. \tag{7.26}$$

This is a consequence of the definition of LT as written following Eq. (3.11).

When the reader will take up study of the General Theory of Relativity he/she will realize that it is not possible to obtain the simple metric of Eq. (7.25) in the presence of a true gravitational field. Massive gravitating objects, like pulsars, quasars, black holes, distort the geometry of space–time grossly thereby invalidating the Pythagorean expression (7.23) for the length of a "straight line". However, for our study of Special Relativity, from which gravity is excluded, we shall always assume the Minkowski metric of Eq. (7.25). The *space–time* whose geometry is described by the Mankowski metric is called *Minkowski space–time*, and will be denoted by the symbol M^4.

The reader may have already noticed that the right-hand side of Eqs. (7.24) and (7.25) are not necessarily positive. In fact, they can be positive, negative, even zero. When the events Θ, Φ for which we have written the line interval (7.24) are on the world line of a material particle, $s^2 > 0$, because all material particles move slower than light. In that case, we say that the line interval between these events, or the line element joining these events, is *time-like*. If $s^2 < 0$, the interval is *space-like*. On the other hand, if $s^2 = 0$, the interval is *light-like*. (We have already defined these terms in Sec. 3.4.)

Suppose a star, which is 10 light years away, has "exploded today" into a supernovae. Call this event Θ. We shall see this supernovae 10 years later in the form of a brilliant flash of light. Let this "seeing at a certain observatory" on earth be called the event Φ. Then the line interval between these two events is zero, according to Eq. (7.24), even though the intervening "distance" between the events is $10 \times 365 \times 24 \times 60 \times 60 \times 3 \times 10^5$ km!!

In Sec. 7.8, we shall characterize a 4-vector as time-like, space-like or light-like depending on whether the square of its length is greater than, less than, or equal to zero.

7.5. 3-Vectors, Contravariant and Covariant Families

We shall be using vectors and tensors in E^3 as well as M^4. The new species of vectors we shall be using in M^4 will be called *4-vectors*, a term coined by Minkowski. To distinguish our familiar vectors, used so far in non-relativistic physics, we may refer to them as 3-vectors.

It is easier to build the concept of vectors in E^3 and then extend the same to M^4. A vector in E^3 is a *directed straight line segment* with the additional property when two of them, **A** and **B** are added to obtain their sum $\mathbf{C} = \mathbf{A} + \mathbf{B}$, this operation is carried out by constructing a *vector triangle*.

The first one of these above two properties gives a geometrical character to a vector. If we have chosen a frame of reference, then the length and orientation of a vector **V** will determine its components V_x, V_y, V_z along the X-, Y-, Z-axes, which are obtained by projecting the vector along these axes.

Conversely, given the components V_x, V_y, V_z of a vector **V** with reference to a frame of reference, we can construct the vector **V** in space as a geometrical straight line. Therefore, it is often a practice to express a vector as an ordered triple of real numbers, which can be written either as a column, or as a row matrix.

$$\mathbf{V} = \begin{pmatrix} V_x \\ V_y \\ V_z \end{pmatrix}; \quad \text{alternatively,} \quad \mathbf{V} = (V_x, V_y, V_z). \tag{7.27}$$

In view of the above matrix representation, the triangle rule of vector addition is equivalent to a matrix addition:

$$\mathbf{A} + \mathbf{B} = \begin{pmatrix} A_x + B_x \\ A_y + B_y \\ A_z + B_z \end{pmatrix}; \qquad \text{alternatively,} \quad \mathbf{A} + \mathbf{B} = (A_x + B_x, A_y + B_y, A_z + B_z). \tag{7.28}$$

The most elementary vector, which also serves as the patriarch of one family of vectors known as *contravariant vectors* is the infinitesimal displacement from a point $\mathrm{P}(x, y, z)$ to a neighbouring point $\mathrm{Q}(x + dx, y + dy, z + dz)$, so that

$$d\mathbf{r} = (dx, dy, dz). \tag{7.29}$$

When we change over from a frame of reference S to another one S', which is rotated with respect to the first one, the coordinates get transformed according to Eq. (7.20). As a consequence, the vector $d\mathbf{r}$ changes its components from (dx, dy, dz) to (dx', dy', dz'), where,

$$\begin{aligned} dx' &= dx \cos\theta + dy \sin\theta, \\ dy' &= -dx \sin\theta + dy \cos\theta, \\ dz' &= dz. \end{aligned} \tag{7.30}$$

Starting from the primary vector $d\mathbf{r}$ one obtains, through multiplication with scalars (e.g. $1/dt$, m), and differentiation with respect to time t, other members of the contravariant family are obtained, e.g.

$$\begin{array}{lll} \text{velocity} & \mathbf{v} = \dfrac{1}{dt} \times d\mathbf{r} = \dfrac{d\mathbf{r}}{dt} & \text{(multiplication)}, \\ \text{acceleration} & \boldsymbol{a} = \dfrac{d\mathbf{v}}{dt} & \text{(differentiation)}, \\ \text{momentum} & \mathbf{p} = m \times \mathbf{v} = m\mathbf{v} & \text{(multiplication)}, \\ \text{force} & \mathbf{F} = \dfrac{d\mathbf{p}}{dt} & \text{(differentiation)}, \end{array} \tag{7.31}$$

and so on, all of which share the transformation of $d\mathbf{r}$. One can now *define* a set of three real numbers $\mathbf{V} = (V_x, V_y, V_z)$ to constitute a *contravariant vector*, if they transform under a change of coordinates exactly like the components of $d\mathbf{r}$. The rotation of the vector $\mathbf{V}$ will change its components from (V_x, V_y, V_z) to (V'_x, V'_y, V'_z), such that

$$\begin{aligned} V'_x &= V_x \cos\theta + V_y \sin\theta, \\ V'_y &= -V_x \sin\theta + V_y \cos\theta, \\ V'_z &= V_z. \end{aligned} \tag{7.32}$$

To be more precise and general, we replace the particularly simple transformation formula (7.20) to

$$x'^k = f^k(x^1, x^2, x^3) = f^k(x), \quad k = 1, 2, 3, \tag{7.33}$$

where we have written (x) to mean all the old coordinates (x^1, x^2, x^3). Note that we have used *superscripts* (1,2,3) on x to mean x, y, z, respectively. A superscript will be called a *contravariant* index. We shall follow this practice in the remaining part of this book.

Equation (7.33) represent a general change-over from the old coordinates (x) to (x') — most common examples being (1) Cartesian to Cartesian due to rotation of axes, just cited, (2) Cartesian (x, y, z) to the spherical coordinate system (r, θ, ϕ), or vice versa.[c]

The coordinate differentials (dx') in the old system, and (dx) in the new system, are connected by the *linear* expression

$$dx'^k = \sum_{j=1}^{3} \frac{\partial x'^k}{\partial x^j} dx^j = \sum_{j=1}^{3} R^k_{\cdot j} dx^j, \quad k = 1, 2, 3, \tag{7.34}$$

where the coefficients

$$R^k_{\cdot j} \equiv \frac{\partial x'^k}{\partial x^j} = \frac{\partial f^k}{\partial x^j}, \quad k, j = 1, 2, 3, \tag{7.35}$$

constitute a 3×3 matrix $\widehat{R}$. They are, in general, functions of the coordinates (x). For the special case of rotation, these coefficients are independent of the coordinates and constitute an *orthogonal matrix* $\widehat{R}$. An orthogonal

[c]One has to be careful before applying this transformation rule to curvilinear coordinates. A distinction between the "physical components" and tensor components has to be made, bringing in the scale factors before making any serious application, say, in mechanics.

matrix is one for which its inverse equals its transpose, as we have explained in Eq. (7.45) below.

Note the dot "." followed by a small space in the subscripts, to imply the second position for the subscript, i.e. its role as the column index. The superscript k and the subscript j in $R^k_{\cdot j}$ constitute the *row* index (first index) and the *column* index (second index), respectively.

For the case of the rotation with transformation represented in Eqs. (7.30) and (7.32), $R^1_{\cdot 1} = \cos\theta$; $R^1_{\cdot 2} = \sin\theta; \ldots$ We write the complete matrix:

$$\widehat{R} = \begin{pmatrix} \cos\theta & \sin\theta & 0 \\ -\sin\theta & \cos\theta & 0 \\ 0 & 0 & 1 \end{pmatrix}. \tag{7.36}$$

Though simple, the above matrix is used to construct the most complex rotation matrix of a rigid body, consisting of precession, spin and nutation.[d]

In order to avoid writing the summation symbol Σ repeatedly — for which occasions will arise copiously in the sequel — it has been a common practice among relativists to adopt the *Einstein Summation Convention.* For this purpose, we introduce a *contravariant index* as the one which appears either as a superscript in the numerator, or as a subscript in the denominator. Its opposite is the *covariant index*, which, therefore, appears either as a subscript in the numerator or as a superscript in the denominator. In the expression (7.34), for instance, k appears as a contravariant index on either side of the equation, whereas j appears first as a covariant index, and then as a contravariant index.

Einstein convention proposes that if in a certain term of an expression the same index appears twice, once in the contravariant form and another time in the covariant form, then that index is to be interpreted as a *summation index*, and the formal summation sign Σ is to be dropped.

Adopting this convention we rewrite Eq. (7.34) as (compare this with the convention adopted on page 147)

$$dx'^k = \frac{\partial x'^k}{\partial x^j} dx^j = R^k_{\cdot j} dx^j, \quad k = 1, 2, 3. \tag{7.37}$$

A summation index, e.g. j in Eq. (7.37), is a *dummy* index. A dummy index can be replaced by some other dummy index. For example

$$dx'^k = R^k_{\cdot j}\, dx^j = R^k_{\cdot m}\, dx^m. \tag{7.38}$$

[d]Goldstein, *op. cit.*, p. 147ff.

We shall complete the suggestion following Eq. (7.31) to define a *contravariant* vector **A** to be an ordered triple of real numbers (A^k) that transform under a coordinate change (x) to (x') of the form (7.33) to another triple (A'^j), such that

$$A'^k = R^k_{\cdot j}\, A^j, \quad \text{where } R^k_{\cdot j} \equiv \frac{\partial x'^k}{\partial x^j}, \tag{7.39}$$

with $k, j = 1, 2, 3$, are the components of a 3×3 matrix $\widehat{R}$.

Note that we have indicated the components of the contravariant vector with a *superscript.*

Equation (7.32) is a special case of (7.39), in which the components $R^k_{\cdot j}$ of $\widehat{R}$ are constants, i.e. independent of the coordinates. This happens in the case of Cartesian to Cartesian transformation, like rotation of the axes.

Another class of vectors extensively used in physics is the *covariant* family. Their prototype, the gradient of a scalar field $\psi(x, y, z)$ is denoted as $\nabla\psi$. Let us write

$$\begin{gathered}\mathbf{G} = \nabla\psi, \\ \text{having components} \quad G_k = \frac{\partial\psi}{\partial x^k}, \quad k = 1, 2, 3.\end{gathered} \tag{7.40}$$

Under the transformation (7.33), these components will change to $G_{k'} = \frac{\partial\psi}{\partial x'^k}$ in the system (x'), Using the chain rule of calculus, one then gets

$$\begin{gathered}G'_k = \frac{\partial\psi}{\partial x'^k} = \frac{\partial\psi}{\partial x^j}\frac{\partial x^j}{\partial x'^k} = G_j M^j_{\cdot k}, \\ \text{where} \quad M^j_{\cdot k} \equiv \frac{\partial x^j}{\partial x'^k}\end{gathered} \tag{7.41}$$

are the components of a 3×3 matrix $\hat{M}$.

Let $\hat{P} = \hat{M}\widehat{R}$ be the product of the two matrices $\hat{M}$ and $\widehat{R}$. Then

$$\begin{gathered}P^j_\ell = \frac{\partial x^j}{\partial x'^k}\frac{\partial x'^k}{\partial x^\ell} = \frac{\partial x^j}{\partial x^\ell} = \delta^j_\ell, \\ \text{or} \quad \hat{M}\widehat{R} = \widehat{1} = \text{identity matrix.}\end{gathered} \tag{7.42}$$

Equation (7.42) shows that the matrix $\hat{M}$ is identical with $\widehat{R}^{-1}$. We can therefore rewrite the transformation equation (7.41) for the gradient vector as

$$G'_k = G_j(\widehat{R}^{-1})^j_k. \tag{7.43}$$

Therefore, we define a covariant vector $\mathbf{V}$ as an ordered triple of real numbers (V_k) that transform like $\nabla\psi$ under a change of coordinates (x) to (x'). This means that the transformed components of the vector will be:

$$V'_k = V_j(\widehat{R}^{-1})^j_k. \tag{7.44}$$

Note that we have indicated the components of the covariant vector with a *subscript.*

It is then clear that the contravariant vectors and the covariant vectors have distinctively different characters. Isn't it then strange that in physics so far we have treated them alike and represented both by directed straight lines?

The reason is that the formal difference between Eqs. (7.39) and (7.44) disappears when we consider transformation from one Cartesian system S to another Cartesian system S' due to a rotation of the axes. The transformation matrix $\widehat{R}$ is then an *orthogonal* matrix, as stated following Eq. (7.35). By its definition, the inverse of an orthogonal matrix is its transpose. That is,

$$\begin{aligned} \widehat{\mathcal{O}}^{-1} &= \widehat{\mathcal{O}}^{\text{trans}}, \\ \text{or} \quad (\mathcal{O}^{-1})^j_k &= \mathcal{O}^k_j. \end{aligned} \tag{7.45}$$

Using this property, we can rewrite (7.44) as

$$V'_k = \sum_{k=1}^{3} R^k_{.j}\, V_j. \tag{7.46}$$

The above transformation rule is identical with (7.39). It is for this reason that the discriminatory labels "contravariant" and "covariant" are unnecessary in classical physics.

When the components of a 3-vector $\mathbf{V}$ change from $\{V_k\}$ to $\{V'_k\}$, through an orthogonal transformation, its length, by which we mean its magnitude, does not change, i.e.

$$\sum_{k=1}^{3}[V'_k]^2 = \sum_{j=1}^{3}[V_j]^2 \tag{7.47}$$

as exemplified in Eq. (7.22). We shall adopt Eq. (7.47) for defining a 3-vector.

7.6. 3-Tensors, Contravariant and Covariant Families

A vector is a geometrical object that does not change with a coordinate transformation, like a hexahedron which remains the same geometrical object no matter from which angle you look at it. This means that if under a coordinate transformation (x) to (x') the scalar components of a vector $\mathbf{V}$ change from (V_x, V_y, V_z) to (V'_x, V'_y, V'_z), then the unit vectors $(\mathbf{e}_x, \mathbf{e}_y, \mathbf{e}_z)$ will change to $(\mathbf{e}'_x, \mathbf{e}'_y, \mathbf{e}'_z)$ such that

$$V_x\mathbf{e}_x + V_y\mathbf{e}_y + V_z\mathbf{e}_z = V'_x\mathbf{e}'_x + V'_y\mathbf{e}'_y + V'_z\mathbf{e}'_z.$$

Same as, $V^j\mathbf{e}_j = V'^k\mathbf{e}'_k.$

However, $V'^k = R^k_{.j}V^j,$ from Eq. (7.39).

Hence, $\mathbf{e}_jV^j = \mathbf{e}'_kR^k_{.j}V^j,$

or $\mathbf{e}_j = \mathbf{e}'_kR^k_{.j},$

or $(R^{-1})^j_{.\ell}\mathbf{e}_j = \mathbf{e}'_kR^k_{.j}(R^{-1})^j_{.\ell} = \mathbf{e}'_k\delta^k_{.\ell} = \mathbf{e}'_l.$

Note that $(R^{-1})^j_{.\ell}$ represents the $j\ell$ component of the matrix $\widehat{R}^{-1}$, the inverse of $\widehat{R}$. Hence the theorem.

Theorem 7.1. *Let $\widehat{R}$ represent the transformation matrix corresponding to a change of coordinates $(x) \to (x')$, so that a contravariant vector $\mathbf{A}$ changes its components from (A^k) to (A'^k) by the rule given by Eq. (7.39). Then, the transformation of the unit vectors $(\mathbf{e}_j) \to (\mathbf{e}'_l)$ will be given by the rule:*

$$\mathbf{e}'_l = \mathbf{e}_j(R^{-1})^j_{.\ell}, \tag{7.48a}$$

and the converse,

$$\mathbf{e}_l = \mathbf{e}'_jR^j_{.\ell}. \tag{7.48b}$$

We shall obtain the transformation formula for the *contravariant 3-tensor* T^{ik} using Eq. (7.48)(b).

$$\text{Let}\quad \widehat{\mathbf{T}} = T^{ik}\mathbf{e}_i\mathbf{e}_k = T'^{j\ell}\mathbf{e}'_j\mathbf{e}'_l.$$

$$\text{Now,}\quad T^{ik}\mathbf{e}_i\mathbf{e}_k = T^{ik}\{\mathbf{e}'_jR^j_{.i}\}\{\mathbf{e}'_lR^\ell_{.k}\} = \{R^j_{.i}R^\ell_{.k}T^{ik}\}\mathbf{e}'_j\mathbf{e}'_l.$$

Hence,

$$\boxed{T'^{j\ell} = R^j_{.i}R^\ell_{.k}T^{ik}.} \tag{7.49}$$

Going back to Eq. (7.39) we draw the following conclusions:

- A scalar is a real number that does not change under a coordinate transformation.
- A vector is a row or column matrix of three real numbers that transforms through a single application of $\widehat{R}$ as shown in (7.39).
- A tensor is a 3×3 square matrix that transforms through a double application of $\widehat{R}$ as shown in (7.49).

We shall illustrate the concepts outlined above by citing the example of rotation about the Z-axis for which we had written the transformation matrix $\widehat{R}$ in Eq. (7.36). Its inverse $\widehat{R}^{-1}$ is its transpose. We shall write $\widehat{R}$ and $\widehat{R}^{-1}$ side by side for contrast:

$$\widehat{R} = \begin{pmatrix} \cos\theta & \sin\theta & 0 \\ -\sin\theta & \cos\theta & 0 \\ 0 & 0 & 1 \end{pmatrix}. \quad \widehat{R}^{-1} = \begin{pmatrix} \cos\theta & -\sin\theta & 0 \\ \sin\theta & \cos\theta & 0 \\ 0 & 0 & 1 \end{pmatrix}. \tag{7.50}$$

The base vectors should now transform as

$$\begin{aligned} &\mathbf{e}'_l = \mathbf{e}_k (R^{-1})^k_{.\ell}, \\ \text{so that,}\quad &\mathbf{e}'_1 = \mathbf{e}_1 (R^{-1})^1_{.1} + \mathbf{e}_2 (R^{-1})^2_{.1} = \cos\theta \mathbf{e}_1 + \sin\theta \mathbf{e}_2, \\ &\mathbf{e}'_2 = \mathbf{e}_1 (R^{-1})^1_{.2} + \mathbf{e}_2 (R^{-1})^2_{.2} = -\sin\theta \mathbf{e}_1 + \cos\theta \mathbf{e}_2, \\ &\mathbf{e}'_3 = \mathbf{e}_3. \end{aligned} \tag{7.51}$$

Comparing Eq. (7.51) with Eq. (7.32), we find that the vector components (V'_x, V'_y, V'_z) are the same linear combinations of (V_x, V_y, V_z), as the unit vectors $(\mathbf{e}'_1, \mathbf{e}'_2, \mathbf{e}'_3)$ are of $(\mathbf{e}_1, \mathbf{e}_2, \mathbf{e}_3)$, even though they follow different rules of transformation. This is because in the first case the matrix $\widehat{R}$ appears before, and in the second case the matrix $\widehat{R}^{-1}$ appears after, the quantities being transformed. Let us highlight this.

$$\begin{aligned} \begin{pmatrix} V'^1 \\ V'^2 \\ V'^3 \end{pmatrix} &= \begin{pmatrix} \cos\theta & \sin\theta & 0 \\ -\sin\theta & \cos\theta & 0 \\ 0 & 0 & 1 \end{pmatrix} \begin{pmatrix} V^1 \\ V^2 \\ V^3 \end{pmatrix} \\ &= \begin{pmatrix} V^1\cos\theta + V^2\sin\theta \\ -V^1\sin\theta + V^2\cos\theta \\ V^3 \end{pmatrix}, \end{aligned} \tag{7.52}$$

$$(\mathbf{e}'_1\,\mathbf{e}'_2\,\mathbf{e}'_3) = (\mathbf{e}_1\,\mathbf{e}_2\,\mathbf{e}_3)\begin{pmatrix} \cos\theta & -\sin\theta & 0 \\ \sin\theta & \cos\theta & 0 \\ 0 & 0 & 1 \end{pmatrix}$$

$$= (\cos\theta\mathbf{e}_1 + \sin\theta\mathbf{e}_2, \quad -\sin\theta\mathbf{e}_1 + \cos\theta\mathbf{e}_2, \quad \mathbf{e}_3). \tag{7.53}$$

We shall now obtain the transformed components of a contravariant tensor $\widehat{\mathbf{T}}$ under the same rotation of the axes (about the Z-axis) for which the transformation matrix $\widehat{R}$ is given in (7.50). We shall illustrate the procedure for only one component, namely, T'^{12}.

$$\begin{aligned} T'^{12} &= R^{1}_{.\,k}R^{2}_{.\,\ell}T^{k\ell} \\ &= R^{1}_{.\,1}R^{2}_{.\,1}T^{11} + R^{1}_{.\,1}R^{2}_{.\,2}T^{12} + R^{1}_{.\,1}R^{2}_{.\,3}T^{13} + R^{1}_{.\,2}R^{2}_{.\,1}T^{21} \\ &\quad + R^{1}_{.\,2}R^{2}_{.\,2}T^{22} + R^{1}_{.\,2}R^{2}_{.\,3}T^{23} + R^{1}_{.\,3}R^{2}_{.\,1}T^{31} + R^{1}_{.\,3}R^{2}_{.\,2}T^{32} \\ &\quad + R^{1}_{.\,3}R^{2}_{.\,3}T^{33} \\ &= (\cos\theta)(-\sin\theta)\,T^{11} + (\cos\theta)(\cos\theta)\,T^{12} \\ &\quad + (\sin\theta)(-\sin\theta)\,T^{21} + (\sin\theta)(\cos\theta)\,T^{22}. \end{aligned} \tag{7.54}$$

The reader should complete the work by working out the transformation of the other eight components. Here are all the transformed components:

$$\begin{aligned} T'^{11} &= T^{11}\cos^2\theta + (T^{12} + T^{21})\cos\theta\sin\theta + T^{22}\sin^2\theta, \\ T'^{12} &= T^{12}\cos^2\theta + (T^{22} - T^{11})\cos\theta\sin\theta - T^{21}\sin^2\theta, \\ T'^{13} &= T^{13}\cos\theta + T^{23}\sin\theta, \\ T'^{21} &= T^{21}\cos^2\theta + (T^{22} - T^{11})\cos\theta\sin\theta - T^{12}\sin^2\theta, \\ T'^{22} &= T^{22}\cos^2\theta - (T^{12} + T^{21})\cos\theta\sin\theta + T^{11}\sin^2\theta, \\ T'^{23} &= T^{23}\cos\theta - T^{13}\sin\theta, \\ T'^{31} &= T^{31}\cos\theta + T^{32}\sin\theta, \\ T'^{32} &= T^{32}\cos\theta - T^{31}\sin\theta, \\ T'^{33} &= T^{33} \end{aligned} \tag{7.55}$$

A contravariant 3-tensor like $\hat{\mathbf{T}}$ can find its companion in the *covariant 3-tensor* $\check{\mathbf{P}}$ such that $\check{\mathbf{P}} \cdot \hat{\mathbf{T}}$ is a 3-scalar. The transformation rules (7.71), (7.75) and (7.49) would suggest that the components of $\check{\mathbf{P}}$ should transform $\{P_{mn} \to P'_{mn}\}$ as

$$\boxed{P'_{mn} = P_{j\ell}\,(R^{-1})^{j}_{.\,m}(R^{-1})^{\ell}_{.\,n}.} \tag{7.56}$$

Equations (7.49) and (7.56) constitute the transformation rules for a contravariant and a covariant 3-tensor.

We shall show that the above transformation rule will ensure invariance of $\check{\mathbf{P}} \cdot \hat{\mathbf{T}}$ under the transformation $\widehat{R}$.

Proof.

$$\begin{aligned} P'_{mn}T'^{mn} &= \{P_{j\ell}(R^{-1})^{j}_{.\,m}(R^{-1})^{\ell}_{.\,n}\}\{R^{m}_{.\,i}R^{n}_{.\,k}T^{ik}\} \\ &= \{(R^{-1})^{j}_{.\,m}R^{m}_{.\,i}\}\{(R^{-1})^{\ell}_{.\,n}R^{n}_{.\,k}\}P_{j\ell}T^{ik} \\ &= \delta^{j}_{.\,i}\delta^{\ell}_{.\,k}P_{j\ell}T^{ik} = P_{j\ell}T^{j\ell}. \end{aligned} \tag{7.57}$$

(QED)

There is one tensor which is not related to any physical quantity in physics. It is the *metric tensor*, which we shall denote as $\check{\boldsymbol{g}}$, and its 3×3 components as $\{g_{ij};\ i, j = 1, 2, 3\}$. In Relativity, particularly in the General Theory of Relativity, $\check{\boldsymbol{g}}$ plays the most dominant role, being synonymous with the gravitational field itself.[e] The metric tensor defines the geometry of space, or space–time, through an expression of the line element. For a Cartesian coordinate system in E^3, the expression of the line element is written in Eq. (7.23). For a spherical coordinate system (r, θ, ϕ), $\{x^1 = r;\ x^2 = \theta,\ x^3 = \phi\}$ and the same line element is written as

$$d\ell^2 = dr^2 + r^2 d\theta^2 + r^2 \sin^2\theta. \tag{7.58}$$

The metric tensor's role in E^3 is to write all line elements as

$$d\ell^2 = g_{ij}dx^i dx^j. \tag{7.59}$$

It is then seen from (7.23) and (7.58) that

$$\begin{aligned} \widehat{g} = \widehat{g}^{(\text{cart})} = \delta_{ij} &= \begin{pmatrix} 1 & 0 & 0 \\ 0 & 1 & 0 \\ 0 & 0 & 1 \end{pmatrix}; \text{ Cartesian coordinate system,} \\ \widehat{g} = \widehat{g}^{(\text{sph})} &= \begin{pmatrix} 1 & 0 & 0 \\ 0 & (x^1)^2 & 0 \\ 0 & 0 & (x^1 \sin x^2)^2 \end{pmatrix}; \text{ spherical coordinate system} \end{aligned} \tag{7.60}$$

[e] "Metric as the foundation of all"— p. 304 of Ref. [5]

Why do we call $\breve{\boldsymbol{g}}$, whose components are defined by Eq. (7.59), and exemplified by (7.60) a tensor? The reason lies in the *Quotient Theorem* stated in Sec. 7.9.4. The left-hand side of Eq. (7.59) is a scalar, whereas $dx^i dx^j$ is a contravariant tensor of rank 2 (having two superscripts). Therefore, g_{ij} should stand for a covariant tensor of rank 2 (i.e. with two subscripts).

We shall demonstrate the above tensor character of $\breve{\boldsymbol{g}}$ by showing that under an orthogonal transformation (e.g. rotation of coordinate axes), the transformation rule (7.56) will convert $\widehat{g}^{(\text{cart})}$ (same as δ_{ij}) to itself.

Proof.

$$\begin{aligned} g'_{mn} &= \delta_{ij}(R^{-1})^{i}_{\cdot m}(R^{-1})^{j}_{\cdot n} = (R^{-1})^{j}_{\cdot m}(R^{-1})^{j}_{\cdot n} \\ &= R^{m}_{\cdot j}(R^{-1})^{j}_{\cdot n} = \delta_{mn}. \end{aligned} \tag{7.61}$$

(QED)

We shall further demonstrate the tensor character of $\breve{\boldsymbol{g}}$ by showing that when the coordinate system changes from Cartesian to spherical, the same transformation rule will convert $\widehat{g}^{(\text{cart})}$ to $\widehat{g}^{(\text{sph})}$.

7.7. Transformation of the Metric 3-Tensor from Cartesian to Spherical

As in Eq. (7.61),

$$g'_{mn} \equiv \widehat{g}^{(\text{sph})}_{mn} = \delta_{ij}(R^{-1})^{i}_{\cdot m}(R^{-1})^{j}_{\cdot n} = (R^{-1})^{j}_{\cdot m}(R^{-1})^{j}_{\cdot n}. \tag{7.62}$$

The transformation from (x) to (x') is given in Eq. (5.35). $(R^{-1})^{i}_{\cdot m} = \frac{\partial x^i}{\partial x'^m}$. Therefore, $(R^{-1})^{1}_{\cdot 1} = \frac{\partial x}{\partial r} = \sin\theta\cos\phi$; $(R^{-1})^{1}_{\cdot 2} = \frac{\partial x}{\partial \theta} = r\cos\theta\cos\phi$, etc. We therefore have the following (inverse) transformation matrix:

$$R^{-1} = \begin{pmatrix} \sin\theta\cos\phi & r\cos\theta\cos\phi & -r\sin\theta\sin\phi \\ \sin\theta\sin\phi & r\cos\theta\sin\phi & r\sin\theta\cos\phi \\ \cos\theta & -r\sin\theta & 0 \end{pmatrix}. \tag{7.63}$$

It then follows from (7.62) that

$$\widehat{g}^{(\text{sph})}_{11} = [(R^{-1})^{1}_{\cdot 1}]^2 + [(R^{-1})^{2}_{\cdot 1}]^2 + [(R^{-1})^{3}_{\cdot 1}]^2 = 1.$$

In this way, the other components of $\widehat{g}^{(\text{sph})}$ as given in (7.60) can be obtained.

In writing equations of physics that respect relativity, $\breve{g}$ remains in the background, changing indices — contravariant to covariant and vice versa, and in the process helps us write the laws and principles of physics in the relativistic, covariant language.

7.8. 4-Vectors in Relativity

The simplest 4-vector, by which we shall mean a vector in M^4, that comes to one's mind is a directed straight line segment $\delta\overrightarrow{\mathbf{r}}$ stretching from some event Θ (which could have occurred yesterday) to some other event Ψ which might have occurred in the past, or may occur tomorrow. Since such a vector spans time as well, we cannot represent it with an earthly rod. 4-vectors cannot be "pictured". We can only write them on a piece of paper as an ordered quadruple of four real numbers, e.g.

$$\delta\overrightarrow{\mathbf{r}} = (\delta x^0, \delta x^1, \delta x^2, \delta x^3), \tag{7.64}$$

or draw a space–time diagram using 4 axes cT, X, Y, Z with one of the axes, e.g. the Z-axis suppressed, and show it as a geometrical straight line, with an arrowhead showing its direction from Θ to Ψ.

Note that, the last three numbers, i.e. the ordered triple $(\delta x^1, \delta x^2, \delta x^3)$, constitute a 3-vector $\delta\mathbf{r}$, stretching from the spatial location of Θ to the spatial location of Ψ. We shall often find it convenient to write a 4-vector as a 1+3-component object, e.g. $(\delta x^0, \delta\mathbf{r})$.

The primordial 4-vector, or, the *most elementary 4-vector*, the ancestor of all 4-vectors of the contravariant 4-vector family, is the infinitesimal displacement 4-vector $d\overrightarrow{\mathbf{r}}$ in M^4, stretching from an event Θ to an infinitesimally close-by-event Ψ, the time–space separation between them being (dx^0, dx^1, dx^2, dx^3). Depending on our convenience, we shall write this vector, and its progenies, in four different ways (cf. Sec. 7.1):

$$d\overrightarrow{\mathbf{r}} = (dx^\mu) = (dx^0, dx^1, dx^2, dx^3) = (dx^0, d\mathbf{r}). \tag{7.65}$$

Note that in the last equation we have clubbed the *three spatial components* of the elementary 4-vector $\delta\overrightarrow{\mathbf{r}}$ as $d\mathbf{r} = (dx^1, dx^2, dx^3) = (dx, dy, dz)$. These spatial components, isolated out under the banner $d\mathbf{r}$, constitute a 3-vector, the object of our discussion in Sec. 7.5. We shall give a better definition of 3-vector below Eq. (7.71).

Under a general coordinate transformation (x^μ) to (x'^μ), $\mu = 0, 1, 2, 3$,

$$x'^\mu = f^\mu(x), \tag{7.66}$$

the coordinate differentials (dx^μ) will transform to (dx'^μ), according to the rule

$$dx'^\mu = \frac{\partial x'^\mu}{\partial x^\nu} dx^\nu = \Omega^\mu_{\cdot\,\nu} dx^\nu, \quad \mu = 0, 1, 2, 3, \tag{7.67}$$

where the coefficients

$$\Omega^\mu_{\cdot\,\nu} \equiv \frac{\partial x'^\mu}{\partial x^\nu}, \quad \mu, \nu = 0, 1, 2, 3, \tag{7.68}$$

constitute a 4×4 matrix $\hat{\Omega}$. In the case of Lorentz transformation, the transformation coefficients are constants, as in the case of rotation in E^3, cf. Eq. (7.36). It is the same LT matrix written as Eq. (3.14) in Sec. 3.2, and copied into Eq. (7.69) below, in which we have also spelt out the time index 0, and space indices 1, 2, 3 for the four rows (arranged vertically on the left) and the four columns (arranged horizontally at the top):

$$\hat{\Omega} = \begin{matrix} & \begin{matrix} 0 & 1 & 2 & 3 \end{matrix} \\ \begin{matrix} 0 \\ 1 \\ 2 \\ 3 \end{matrix} & \begin{pmatrix} \gamma & -\gamma\beta & 0 & 0 \\ -\gamma\beta & \gamma & 0 & 0 \\ 0 & 0 & 1 & 0 \\ 0 & 0 & 0 & 1 \end{pmatrix} \end{matrix}. \tag{7.69}$$

By analogy with Eq. (7.39), we now define a *contravariant 4-vector* to be a vector

$$\vec{\mathbf{A}} = (A^\mu) = (A^0, \mathbf{A}), \tag{7.70}$$

whose components transform from (x) to (x') according to the rule:

$$A'^\mu = \Omega^\mu_{\cdot\,\nu} A^\nu, \quad \mu = 0, 1, 2, 3. \tag{7.71}$$

Note that A^0 is the *time component*, and $\mathbf{A}$ is the *space component* (or, rather the three spatial components taken together) of the 4-vector $\vec{\mathbf{A}}$. We shall treat $\mathbf{A}$ as a 3-*vector*, as mentioned below Eq. (7.65). Confining ourselves to Cartesian systems (so that we do not have to invoke metric tensor) a 3-vector $\mathbf{A}$ possesses the property of invariance under an orthogonal transformation (e.g. rotation of the coordinate axes, but *not* boost), as per the definition given below Eq. (7.47).

We shall now come to the four-dimensional *covariant vector*. Imagine the gradient vector $\boldsymbol{\nabla}\psi$ upgraded from E^3 to M^4. We get the 4-gradient of a 4-scalar field $\overleftarrow{\boldsymbol{\nabla}}\Psi$.

A 4-*scalar field* $\Psi(x)$ as a single component field, i.e. represented by a single function of the four coordinate (x^μ), such that the value of the function does not change at any event point "Θ" under a change of the coordinates from (x) to (x') due to a transformation of the type (7.66). In other words, if $\Psi(x) \to \Psi'(x')$ under the above coordinate transformation, then $\Psi'(x') = \Psi(x)$.

Now we define the *covariant* 4-*gradient* of such a scalar field Ψ as the 4-component field:

$$\overleftarrow{\mathbf{G}} = \overleftarrow{\boldsymbol{\nabla}}\Psi,$$

$$\text{having components} \quad G_\mu = \nabla_\mu \Psi \equiv \frac{\partial \Psi}{\partial x^\mu}, \quad \mu = 0, 1, 2, 3, \tag{7.72}$$

$$\text{or} \quad \overleftarrow{\mathbf{G}} = (\nabla_\mu \Psi) = \left(\frac{1}{c}\frac{\partial \Psi}{\partial t}, \frac{\partial \Psi}{\partial x}, \frac{\partial \Psi}{\partial y}, \frac{\partial \Psi}{\partial z}\right).$$

Note that we have used a *leftward arrow* to imply a covariant 4-vector, and a *subscript* to indicate its components, in contrast with the contravariant 4-vector (in Eq. (7.70) for which we have used a *rightward arrow*, and *superscript* to indicate its components.

Under the coordinate transformation (7.66), the 4-vector $\overleftarrow{\mathbf{G}}$ changes to $\overleftarrow{\mathbf{G}}'$, and their components change:

$$\nabla_\mu \Psi \to \nabla'_\mu \Psi' = \frac{\partial \Psi'}{\partial x'^\mu} = \frac{\partial \Psi}{\partial x^\nu}\frac{\partial x^\nu}{\partial x'^\mu}, \quad \mu = 0, 1, 2, 3. \tag{7.73}$$

Analogous to Eq. (7.42), we have here

$$\left(\frac{\partial x^\nu}{\partial x'^\mu}\right)\left(\frac{\partial x'^\mu}{\partial x^\eta}\right) = \frac{\partial x^\nu}{\partial x^\eta} = \delta^\nu_\eta. \tag{7.74}$$

Going back to Eq. (7.68) we now identify $\left(\frac{\partial x'^\mu}{\partial x^\eta}\right)$ as the $\Omega^\mu_{\cdot\eta}$ component of the matrix $\overleftrightarrow{\Omega}$, and now using (7.74), $\left(\frac{\partial x^\nu}{\partial x'^\mu}\right)$ as the $(\Omega^{-1})^\nu_{\cdot\mu}$ component of the inverse matrix $(\overleftrightarrow{\Omega})^{-1}$.

We therefore *define a covariant 4-vector* $\overleftarrow{\mathbf{B}} = B_\mu = (B_0, B_1, B_2, B_3)$ to be an ordered quadruple of real numbers that transform under the coordinate transformation (7.66) as

$$B'_\mu = B_\alpha(\Omega^{-1})^{\alpha}_{\cdot\,\mu}. \tag{7.75}$$

The covariant gradient $\overleftarrow{\mathbf{G}} = \overleftarrow{\nabla}\Psi$ obviously satisfies this requirement.

As a concrete example we shall write down the Lorentz transformation equations for a contravariant 4-vector $\overrightarrow{\mathbf{A}}$ and a covariant 4-vector $\overleftarrow{\mathbf{B}}$ corresponding to a boost $c\beta\mathbf{e}_1$ in the X^1-direction. The relevant matrix has already been given in Eq. (7.69). Its inverse can be determined either rigorously using matrix algebra, or by a short-cut, replacing β with $-\beta$.

$$\hat{\Omega}^{-1} = \begin{array}{c} \\ 0 \\ 1 \\ 2 \\ 3 \end{array} \begin{array}{c} \begin{array}{cccc} 0 & 1 & 2 & 3 \end{array} \\ \begin{pmatrix} \gamma & \gamma\beta & 0 & 0 \\ \gamma\beta & \gamma & 0 & 0 \\ 0 & 0 & 1 & 0 \\ 0 & 0 & 0 & 1 \end{pmatrix} \end{array}. \tag{7.76}$$

Using the matrices (7.69) and (7.76) in the transformation formulas (7.71) and (7.75), respectively, we get the transformed components of these two 4-vectors:

$$\begin{aligned} A'^0 &= \gamma(A^0 - \beta A^1), \\ A'^1 &= \gamma(A^1 - \beta A^0), \\ A'^2 &= A^2, \\ A'^3 &= A^3. \end{aligned} \tag{7.77a}$$

$$\begin{aligned} B'_0 &= \gamma(B_0 + \beta B_1), \\ B'_1 &= \gamma(B_1 + \beta B_0), \\ B'_2 &= B_2, \\ B'_3 &= B_3. \end{aligned} \tag{7.77b}$$

For future use we shall also write the LT formula for the contravariant vector $\overrightarrow{\mathbf{A}}$ corresponding to a general boost: $[\,S(c\boldsymbol{\beta})S'\,]$ along any arbitrary direction $\mathbf{n}$. That is $\boldsymbol{\beta} = \beta\mathbf{n}$. For this purpose, we adopt the formula (3.18) given in Sec. 3.2, and replace ct with A^0 and $\mathbf{r}$ with $\mathbf{A}$. The result is [24]

$$A'^0 = \gamma(A^0 - \boldsymbol{\beta}\cdot\mathbf{A}) \tag{7.78a}$$

$$= A^0 + [\,(\gamma-1)A^0 - \gamma\boldsymbol{\beta}\cdot\mathbf{A}], \tag{7.78b}$$

$$\mathbf{A}' = (\gamma\mathbf{A}_{\parallel} + \mathbf{A}_{\perp}) - \gamma\boldsymbol{\beta}A^0 \tag{7.78c}$$

$$= \mathbf{A} + \left[\frac{\gamma-1}{\beta^2}(\boldsymbol{\beta}\cdot\mathbf{A})\boldsymbol{\beta} - \gamma\boldsymbol{\beta}A^0\right] \tag{7.78d}$$

$$= \mathbf{A} + \mathbf{n}[(\gamma-1)(\mathbf{n}\cdot\mathbf{A}) - \gamma\beta A^0]. \tag{7.78e}$$

The corresponding Lorentz transformation matrix is the same as Eq. (3.20). The inverse of the above transformation, corresponding to the boost: $[S'(-\boldsymbol{\beta})S]$, is obtained by replacing $\boldsymbol{\beta}$ with $-\boldsymbol{\beta}$ (same as replacing $\mathbf{n}$ with $-\mathbf{n}$) and $(A^0, \mathbf{A})$ with $(A'^0, \mathbf{A}')$:

$$A^0 = \gamma(A'^0 + \boldsymbol{\beta}\cdot\mathbf{A}') \tag{7.79a}$$

$$= A'^0 + [(\gamma-1)A'^0 + \gamma\boldsymbol{\beta}\cdot\mathbf{A}'], \tag{7.79b}$$

$$\mathbf{A} = (\gamma\mathbf{A}'_{\parallel} + \mathbf{A}'_{\perp}) + \gamma\boldsymbol{\beta}A'^0 \tag{7.79c}$$

$$= \mathbf{A}' + \left[\frac{\gamma-1}{\beta^2}(\boldsymbol{\beta}\cdot\mathbf{A}')\boldsymbol{\beta} + \gamma\boldsymbol{\beta}A'^0\right] \tag{7.79d}$$

$$= \mathbf{A}' + \mathbf{n}[(\gamma-1)(\mathbf{n}\cdot\mathbf{A}') + \gamma\beta A'^0]. \tag{7.79e}$$

As for a graphical representation of vectors in M^4, we shall continue to rely on the straight line picture for the contravariant vectors. The covariant vectors can be represented by a succession of parallel planes. The orientation of the planes (i.e. the normal drawn on them) will indicate the "direction" and the spacing the magnitude. The closer the spacing, the larger the magnitude.

7.9. 4-Tensors in Relativity

7.9.1. *Contravariant, covariant and mixed tensors*

Tensors of relativity will be defined as matrices subject to specific transformation rules. A 4-scalar is a tensor of rank 0. A contravariant vector and a covariant vector are both tensors of rank 1 and can be represented as a 4×1 or 1×4 (i.e. column or row) matrices. A tensor of rank 2 is a 4×4 matrix.

By an extension of the defining equation (7.49), we shall call the elements of a 4×4 matrix $(T^{\mu\nu})$ as constituting a *Contravariant Tensor* of rank 2, if under a coordinate transformation of the form (7.33), they change into $(T'^{\mu\nu})$ according to the following rule:

$$T^{\mu\nu} \to T'^{\mu\nu} = \Omega^{\mu}_{\cdot\,\alpha}\Omega^{\nu}_{\cdot\,\beta}T^{\alpha\beta}. \tag{7.80}$$

Similarly, by an extension of the defining equation (7.56), a 4×4 matrix $(K_{\mu\nu})$ which transforms into $(K'_{\mu\nu})$ according to the rule:

$$K_{\mu\nu} \to K'_{\mu\nu} = K_{\alpha\beta}(\Omega^{-1})^{\alpha}_{\cdot\,\mu}(\Omega^{-1})^{\beta}_{\cdot\,\nu} \tag{7.81}$$

will constitute a *Covariant Tensor* of rank 2.

On the other hand, the matrix $(Q^{\mu}_{\cdot\,\nu})$ which transforms into $(Q'^{\mu}_{\cdot\,\nu})$ according to the rule:

$$Q^{\mu}_{\cdot\,\nu} \to Q'^{\mu}_{\cdot\,\nu} = \Omega^{\mu}_{\cdot\,\alpha}Q^{\alpha}_{\cdot\,\beta}(\Omega^{-1})^{\beta}_{\cdot\,\nu} \tag{7.82}$$

will constitute a *Mixed Tensor* of rank 2, having contravariant character with respect to the first index, and covariant character with respect to the second one. Note that $Q^{\mu}_{\cdot\,\nu}$ and $Q_{\nu}^{\,;\mu}$ are not the same tensor.

By obvious generalization, one can extend the definition to any $4^{[1]} \times 4^{[2]} \times \cdots \times 4^{[n]}$ matrix and call it a tensor of rank n by ascribing to it the necessary transformation properties. For example, the $4 \times 4 \times 4$ matrix $(A^{\mu}_{\cdot\,\nu\sigma})$ will represent a tensor of rank 3, in which a contravariant index is followed by two covariant indices, if

$$A^{\mu}_{\cdot\,\nu\sigma} \to A'^{\mu}_{\cdot\,\nu\sigma} = \Omega^{\mu}_{\cdot\,\alpha}A^{\alpha}_{\cdot\,\beta\gamma}(\Omega^{-1})^{\beta}_{\cdot\,\nu}(\Omega^{-1})^{\gamma}_{\cdot\,\sigma}. \tag{7.83}$$

As an illustration, we shall obtain the transformation equations for a contravariant and a covariant tensor of rank 2, corresponding to the boost: $S(\beta, 0, 0)S'$.

First the contravariant tensor. We shall illustrate the procedure by obtaining the transformed component T'^{00}. The relevant LT matrix $\hat{\Omega}$ is shown in Eq. (7.69):

$$\begin{aligned} T'^{00} &= \Omega^{0}_{\cdot\,\alpha}\Omega^{0}_{\cdot\,\beta}T^{\alpha\beta} \\ &= \Omega^{0}_{\cdot\,0}\Omega^{0}_{\cdot\,0}T^{00} + \Omega^{0}_{\cdot\,0}\Omega^{0}_{\cdot\,1}T^{01} + \Omega^{0}_{\cdot\,1}\Omega^{0}_{\cdot\,0}T^{10} + \Omega^{0}_{\cdot\,1}\Omega^{0}_{\cdot\,1}T^{11} \\ &= \gamma^2[T^{00} - \beta(T^{01} + T^{10}) + \beta^2 T^{11}]. \end{aligned} \tag{7.84}$$

Note that we avoided writing zero terms, like $\Omega^{0}_{\cdot\,1}\Omega^{0}_{\cdot\,2}T^{12}$, etc.

Next, the covariant tensor. We shall demonstrate the procedure for the transformed component K'_{00}. The relevant inverse LT matrix is shown in Eq. (7.76). Hence,

$$K'_{00} = K_{\alpha\beta}(\Omega^{-1})^{\alpha}_{.0}(\Omega^{-1})^{\beta}_{.0} = \gamma^2[K_{00} + \beta(K_{01} + K_{10}) + \beta^2 K_{11}]. \quad (7.85)$$

We shall now write all the transformed components corresponding to the boost: $S(\beta, 0, 0)S'$.

$$\begin{aligned}
T'^{00} &= \gamma^2[T^{00} - \beta(T^{01} + T^{10}) + \beta^2 T^{11}], & T'^{02} &= \gamma(T^{02} - \beta T^{12}),\\
T'^{01} &= \gamma^2[T^{01} - \beta(T^{00} + T^{11}) + \beta^2 T^{10}], & T'^{20} &= \gamma(T^{20} - \beta T^{21}),\\
T'^{10} &= \gamma^2[T^{10} - \beta(T^{00} + T^{11}) + \beta^2 T^{01}], & T'^{03} &= \gamma(T^{03} - \beta T^{13}),\\
T'^{11} &= \gamma^2[T^{11} - \beta(T^{01} + T^{10}) + \beta^2 T^{00}], & T'^{30} &= \gamma(T^{30} - \beta T^{31}),\\
T'^{12} &= \gamma(T^{12} - \beta T^{02}), & T'^{21} &= \gamma(T^{21} - \beta T^{20}),\\
T'^{13} &= \gamma(T^{13} - \beta T^{03}), & T'^{31} &= \gamma(T^{31} - \beta T^{30}),\\
T'^{22} &= T^{22}, \quad T'^{23} = T^{23}, \quad T'^{32} = T^{32}; & T'^{33} &= T^{33};
\end{aligned} \quad (7.86)$$

$$\begin{aligned}
K'_{00} &= \gamma^2[K_{00} + \beta(K_{01} + K_{10}) + \beta^2 K_{11}], & K'_{02} &= \gamma(K_{02} + \beta K_{12}),\\
K'_{01} &= \gamma^2[K_{01} + \beta(K_{00} + K_{11}) + \beta^2 K_{10}], & K'_{20} &= \gamma(K_{20} + \beta K_{21}),\\
K'_{10} &= \gamma^2[K_{10} + \beta(K_{00} + K_{11}) + \beta^2 K_{01}], & K'_{03} &= \gamma(K_{03} + \beta K_{13}),\\
K'_{11} &= \gamma^2[K_{11} + \beta(K_{01} + K_{10}) + \beta^2 K_{00}], & K'_{30} &= \gamma(K_{30} + \beta K_{31}),\\
K'_{12} &= \gamma(K_{12} + \beta K_{02}), & K'_{21} &= \gamma(K_{21} + \beta K_{20}),\\
K'_{13} &= \gamma(K_{13} + \beta K_{03}), & K'_{31} &= \gamma(K_{31} + \beta K_{30}),\\
K'_{22} &= K_{22}, \quad K'_{23} = K_{23}, \quad K'_{32} = K_{32}, & K'_{33} &= K_{33}.
\end{aligned} \quad (7.87)$$

7.9.2. *Equality of two tensors*

If the components of two tensors $\widehat{\mathbf{A}}$ and $\widehat{\mathbf{B}}$ are equal in a preferred coordinate system, say S', then their components are equal in any general coordinate system S, and we say that $\widehat{\mathbf{A}} = \widehat{\mathbf{B}}$.

We shall prove this theorem assuming that $\widehat{\mathbf{A}}, \widehat{\mathbf{B}}$ are contravariant tensors of rank two. The two tensors being equal in S', the components of the tensor $\widehat{\mathbf{N}} \equiv \widehat{\mathbf{A}} - \widehat{\mathbf{B}}$ must the zero in S'. That is $N'^{\mu\nu} = 0$. To get its components $N^{\mu\nu}$ in S, we apply the inverse of the transformation given in (7.80):

$$N^{\alpha\beta} = N'^{\mu\nu}(\Omega^{-1})^{\alpha}_{.\mu}(\Omega^{-1})^{\beta}_{.\nu} = 0. \quad (7.88)$$

(QED)

7.9.3. *The metric tensor of the Minkowski space–time*

We had introduced the concept of metric, and metric tensor, in Sec. 7.4, through Eqs. (7.23), (7.25), and again in Sec. 7.6, through Eq. (7.59), and casually mentioned the pivotal role it plays in relativity, especially General Relativity. The metric tensor $g_{\mu\nu}$ of Minkowski space–time is defined as

$$ds^2 = g_{\mu\nu}\, dx^\mu\, dx^\nu. \tag{7.89}$$

The Minkowski metric was written in (7.25), which we rewrite in our new index format as

$$ds^2 = (dx^0)^2 - (dx^1)^2 - (dx^2)^2 - (dx^3)^2. \tag{7.90}$$

Comparing the two equations, we get the Minkowski metric as the following 4×4 matrix:

$$g_{\mu\nu} = \begin{array}{c} \\ 0 \\ 1 \\ 2 \\ 3 \end{array} \begin{array}{c} \begin{array}{cccc} 0 & 1 & 2 & 3 \end{array} \\ \begin{pmatrix} 1 & 0 & 0 & 0 \\ 0 & -1 & 0 & 0 \\ 0 & 0 & -1 & 0 \\ 0 & 0 & 0 & -1 \end{pmatrix} \end{array}. \tag{7.91}$$

We shall now establish the tensor character of $g_{\mu\nu}$. We cannot follow the path of (7.61), because in this case the transformation is not an orthogonal one. We shall demand that under the transformation $(x) \to (x')$ the metric must remain invariant. That is,

$$ds^2 = g_{\mu\nu}\, dx^\mu\, dx^\nu = g'_{\alpha\beta}\, dx'^\alpha\, dx'^\beta. \tag{7.92}$$

The right-hand side is

$$g'_{\alpha\beta}(\Omega^\mu_{\cdot\,\alpha}\, dx^\mu)(\Omega^\nu_{\cdot\,\beta}\, dx^\nu),$$

leading to the transformation rule

$$g_{\mu\nu} = g'_{\alpha\beta}\Omega^\mu_{\cdot\,\alpha}\Omega^\nu_{\cdot\,\beta}; \tag{7.93a}$$

$$\text{and its inverse:} \quad g'_{\alpha\beta} = g_{\mu\nu}(\Omega^{-1})^\mu_{\cdot\,\alpha}(\Omega^{-1})^\nu_{\cdot\,\beta}. \tag{7.93b}$$

Equation (7.93b) conforms to the transformation rule for a covariant tensor, as defined in Eq. (7.81). Hence, $g_{\mu\nu}$ is a covariant tensor of rank 2.

(QED)

If the transformation $(x) \to (x')$ is an LT, so that the changeover is from one Cartesian system to another, then as per (7.92)

$$ds^2 = (dx'^0)^2 - (dx'^1)^2 - (dx'^2)^2 - (dx'^3)^2, \tag{7.94}$$

so that

$$g'_{\alpha\beta} = \begin{pmatrix} 1 & 0 & 0 & 0 \\ 0 & -1 & 0 & 0 \\ 0 & 0 & -1 & 0 \\ 0 & 0 & 0 & -1 \end{pmatrix} = g_{\alpha\beta}. \tag{7.95}$$

The same structure of the metric tensor is preserved under LT. This can also be seen by working out the components of $g'_{\alpha\beta}$ directly from (7.93b). It now follows from Eq. (7.93b) that

$$g_{\mu\nu} = g_{\alpha\beta}\Omega^{\mu}_{\cdot\,\alpha}\Omega^{\nu}_{\cdot\,\beta}. \tag{7.96}$$

We shall now define a *Lorentz transformation* to be a *linear transformation* — from one Cartesian system (x) to another (x') — such that the transformation matrix $\hat{\Omega}$ satisfies Eq. (7.96). The simplest example of such a matrix $\hat{\Omega}$ is (7.69), and a more general one is (3.20). Also note that the above definition is a restatement of the property of the LT given in Eq. (3.44).

7.9.4. *New tensors from old ones, index gymnastics*

A special significance of the metric tensor is that it is instrumental in generating a class of "converts" through two operations, namely, *lowering* and *raising* an index.

Lowering an index

In Eq. (7.64) we had written the primordial contravariant 4-vector $d\vec{\mathbf{r}}$ as $dx^\mu = (dx^0, dx^1, dx^2, dx^3) = (c\,dt, dx, dy, dz)$. One can now define four infinitesimals dx_μ by the operation:

$$dx_\mu = g_{\mu\nu}dx^\nu. \tag{7.97}$$

It follows from the $g_{\mu\nu}$ matrix given in (7.91) that

$$dx_\mu = (dx^0, -dx^1, -dx^2, -dx^3) = (c\,dt, -dx, -dy, -dz). \tag{7.98}$$

In general, any *contravariant* vector V^μ can find its *covariant* counterpart V_μ through the operation:

$$V_\mu = g_{\mu\nu} V^\nu. \tag{7.99}$$

This V_μ, as much as dx_μ, is a *covariant vector*. We shall prove.

Proof. Consequent to Lorentz transformation V_μ should change to V'_μ, as per the rule:

$$\begin{aligned} V'_\mu &= g'_{\mu\tau} V'^\tau = g_{\alpha\beta} (\Omega^{-1})^{\alpha}_{\cdot\,\mu} (\Omega^{-1})^{\beta}_{\cdot\,\tau} \Omega^{\tau}_{\cdot\,\sigma} V^\alpha \\ &= g_{\alpha\beta} V^\alpha (\Omega^{-1})^{\alpha}_{\cdot\,\mu} \delta^\beta_\sigma = g_{\alpha\beta} V^\beta (\Omega^{-1})^{\alpha}_{\cdot\,\mu} \\ &= V_\alpha (\Omega^{-1})^{\alpha}_{\cdot\,\mu}. \end{aligned} \tag{7.100}$$

This confirms the covariant character of V_μ. (QED)

The operation (7.99) is called *lowering an index*. It consists of multiplication with $g_{\mu\nu}$, followed by a summation over the contravariant index that is to be lowered. We can geneleralize this lowering operation to cover any tensor that has at least one contravariant index. For example, consider the tensor $T^{\mu}_{\cdot\,\nu}{}^{\sigma}$. We shall show two operations, the first one will lower the first contravariant index μ, and the second one will lower the second contravariant index σ.

$$T^{\mu}_{\cdot\,\nu}{}^{\sigma} \to \overset{\mu}{\downarrow} \to T_{\mu\nu}^{\cdot\,\cdot\,\sigma} \equiv g_{\mu\alpha} T^{\alpha}_{\cdot\,\nu}{}^{\sigma}; \tag{7.101a}$$

$$T^{\mu}_{\cdot\,\nu}{}^{\sigma} \to \overset{\sigma}{\downarrow} \to T^{\mu}_{\cdot\,\nu\sigma} \equiv T^{\mu}_{\cdot\,\nu}{}^{\alpha} g_{\alpha\sigma}. \tag{7.101b}$$

The reader should prove that the new matrices (a) $T_{\mu\nu}^{\cdot\,\cdot\,\sigma}$, (b) $T^{\mu}_{\cdot\,\nu\sigma}$ — obtained by the lowering operations are tensors.

Raising an index

The converse of the lowering operation is the *raising of an index*, for which one employs the *contravariant metric tensor* $g^{\mu\nu}$, defined by the relation:

$$g^{\mu\nu} g_{\mu\nu} \stackrel{\text{def}}{=} \delta^\mu_\nu. \tag{7.102}$$

The above equation means that in any coordinate system, $(g^{\mu\nu})$ is the inverse of $(g_{\mu\nu})$. In particular, if the coordinate system is Cartesian, then

(reader verify)

$$g^{\mu\nu} = g_{\mu\nu} = \begin{pmatrix} 1 & 0 & 0 & 0 \\ 0 & -1 & 0 & 0 \\ 0 & 0 & -1 & 0 \\ 0 & 0 & 0 & -1 \end{pmatrix}. \tag{7.103}$$

As in the case of the lowering operation, we define the *raising operation* as multiplication with $g^{\mu\nu}$, followed by a summation over the covariant index (the tensor being operated upon must have one or more) which is being raised. We shall show two examples:

$$F_\mu \rightarrow \underset{\mu}{\uparrow} \rightarrow F^\mu \equiv g^{\mu\alpha} F_\alpha; \tag{7.104a}$$

$$T^\mu_{.\nu\sigma} \rightarrow \underset{\sigma}{\uparrow} \rightarrow T^{\mu\ \sigma}_{.\nu} \equiv T^\mu_{.\nu\alpha}\, g^{\alpha\sigma} \tag{7.104b}$$

The reader should prove that if the operated matrix (in this case $F_\mu, T^\mu_{.\nu\sigma}$) is a tensor, the end matrix (in this case $F^\mu, T^{\mu\ \sigma}_{.\nu}$) is also a tensor with a reshuffle of one of its indices.

The reader may have found out by now that in either the lowering or in the raising operation the *time component does not change, whereas the space components change their signs only.* For example,

$$\begin{array}{lll} \text{if} & A^\mu = (A^0, \mathbf{A}), & K_\mu = (K_0, \mathbf{K}), \\ \text{then} & A_\mu = (A^0, -\mathbf{A}), & K^\mu = (K_0, -\mathbf{K}). \end{array} \tag{7.105}$$

We have provided a few examples of raising and lowering in Appendix A.2

There are three other ways of creating new tensors, that do not involve the metric tensor directly. We shall briefly explain them.

Multiplication, division, contraction, scalar product

The "tensor product" of two tensors A and B of ranks m and n, respectively, is a tensor $C = AB$ of rank $m + n$. For example, if $A^{\mu\nu}$ and $B_{\sigma\ \beta}^{.\alpha}$ are two tensors, then

$$C^{\mu\nu}{}_{..\sigma.\beta}^{\ \ \alpha} \equiv A^{\mu\nu} B_{\sigma}^{\cdot\alpha}{}_{\beta} \tag{7.106}$$

is a tensor of rank 5, having three contravariant and two covariant indices. This is the *Product Theorem.*

The inverse of multiplication is quotient. If $C = AB$ and C and B are tensors of ranks m and n, and $m > n$, the quotient $A = C/B$ is a tensor of rank $(m-n)$. For example, if the relationship (7.106) is known to hold, and if $C^{\mu\nu}{}_{..\sigma.\beta}^{\ \ \alpha}$ and $B_{\sigma}^{\cdot\alpha}{}_{\beta}$ have been confirmed to possess the desired tensor properties, then $A^{\mu\nu}$ must be a tensor of rank 2. This is the *Quotient Theorem.*

If F is a tensor of rank $n \geq 2$, and if it has atleast one contravariant and at least one covariant index, then by equating one contravariant index to one covariant index and summing over it, one gets a contracted tensor of rank $n-2$. For example, if $F^{\alpha\beta}_{..\mu\tau}$ is a tensor, then

$$F^{\alpha}_{.\tau} \equiv F^{\alpha\beta}_{..\beta\tau} \tag{7.107}$$

is a tensor of rank 2. This is the *Contraction Theorem.* The reader must prove the above three theorems.

Using a combination of lowering (or raising), tensor product and contraction operations, we shall obtain the *scalar product* of two contravariant (or two covariant) vectors. Let $\vec{\mathbf{A}} = (A^{\mu}), \vec{\mathbf{B}} = (B^{\mu})$ be two contravariant vectors. Using the lowering operation, product operation and contraction, we get

$$T^{\mu\nu} \equiv A^{\mu}B^{\nu}; \rightarrow T^{\mu}_{.\nu} = A^{\mu}B_{\nu} \ \rightarrow T^{\mu}_{.\mu} = A^{\mu}B_{\mu}. \tag{7.108}$$

We shall call $A^{\mu}B_{\mu}$ the *scalar product* of the two contravariant vectors $\vec{\mathbf{A}}$ and $\vec{\mathbf{B}}$, and write this as

$$\vec{\mathbf{A}} \cdot \vec{\mathbf{B}} = A^{\mu}B_{\mu} = A^0B^0 - A^1B^1 - A^2B^2 - A^3B^3 = A^0B^0 - \mathbf{A}\cdot\mathbf{B}. \tag{7.109}$$

The object $\vec{\mathbf{A}} \cdot \vec{\mathbf{B}}$ is a tensor of rank zero, having been formed by contraction of a tensor of rank 2. Therefore, it is a 4-*scalar.* Its value will be the same in all inertial frames. Since two inertial frames are connected by a Lorentz transformation, a 4-scalar is also called *Lorentz invariant.*

In particular, if we take the scalar product of a 4-vector, contravariant or covariant, with itself, we get the *square of the magnitude of the 4-vector*, often called the *norm* of the vector. For example, the norm of the contravariant vector $\overrightarrow{\mathbf{A}}$ is

$$(\overrightarrow{\mathbf{A}})^2 \equiv A^\mu A_\mu = (A^0)^2 - (A^1)^2 - (A^2)^2 - (A^3)^2 = (A^0)^2 - \mathbf{A}^2. \quad (7.110)$$

The norm of a 4-vector is Lorentz invariant.

The gradient and the d'Alembartian operator

We had introduced the 4-gradient operator ∇_μ in Eq. (7.72), as a vehicle which itself is not a tensor, but changes a 4-scalar field into a covariant 4-vector field. Using "index gymnastics", we shall obtain the contravariant form ∇^μ and the contracted form $\nabla_\mu \nabla^\mu$ of the same gradient operator, as all of these three forms will have important applications in the remaining chapters of this book.

It follows from (7.72) that

$$\nabla_\mu = \left(\frac{\partial}{\partial x^0}, \frac{\partial}{\partial x^1}, \frac{\partial}{\partial x^2}, \frac{\partial}{\partial x^3}\right) = \left(\frac{1}{c}\frac{\partial}{\partial t}, \frac{\partial}{\partial x}, \frac{\partial}{\partial y}, \frac{\partial}{\partial z}\right),$$

$$\nabla^\mu = g^{\mu\alpha}\nabla_\alpha = \left(\frac{\partial}{\partial x^0}, -\frac{\partial}{\partial x^1}, -\frac{\partial}{\partial x^2}, -\frac{\partial}{\partial x^3}\right) = \left(\frac{1}{c}\frac{\partial}{\partial t}, -\frac{\partial}{\partial x}, -\frac{\partial}{\partial y}, -\frac{\partial}{\partial z}\right),$$

$$\Box^2 \equiv \nabla_\mu \nabla^\mu = \left(\frac{1}{c^2}\frac{\partial^2}{\partial t^2} - \frac{\partial^2}{\partial x^2} - \frac{\partial^2}{\partial y^2} - \frac{\partial^2}{\partial z^2}\right) = \left(\frac{1}{c^2}\frac{\partial^2}{\partial t^2} - \nabla^2\right), \quad (7.111)$$

where $\nabla^2 = \frac{\partial^2}{\partial x^2} + \frac{\partial^2}{\partial y^2} + \frac{\partial^2}{\partial z^2}$ is the familiar *Laplacian* operator, and the operator $\Box^2$ is called *d'Alembertian operator.* It is then obvious that if $\Phi(x)$ is a 4-scalar field, then (i) $\nabla_\mu \Phi(x)$ is a covariant 4-vector field, (ii) $\nabla^\mu \Phi(x)$ is a contravariant 4-vector field, and (iii) $\Box^2 \Phi(x)$ is a 4-scalar field.

For future convenience, we shall rewrite Eqs. (7.111a) and (7.111b) as

$$\begin{aligned}
\overleftarrow{\nabla} &= (\nabla_\mu) = (\nabla_0, \nabla_1, \nabla_2, \nabla_3) \equiv \left(\frac{1}{c}\frac{\partial}{\partial t}, \frac{\partial}{\partial x}, \frac{\partial}{\partial y}, \frac{\partial}{\partial z}\right) \\
&= \left(\frac{1}{c}\frac{\partial}{\partial t}, \boldsymbol{\nabla}\right), \\
\overrightarrow{\nabla} &= (\nabla^\mu) = (\nabla^0, \nabla^1, \nabla^2, \nabla^3) \equiv \left(\frac{1}{c}\frac{\partial}{\partial t}, -\frac{\partial}{\partial x}, -\frac{\partial}{\partial y}, -\frac{\partial}{\partial z}\right) \\
&= \left(\frac{1}{c}\frac{\partial}{\partial t}, -\boldsymbol{\nabla}\right).
\end{aligned} \quad (7.112)$$

Chapter 8

Four Vectors of Relativistic Mechanics

Chapter 4 outlined the salient concepts and formulas of Relativistic Mechanics. However, the style and format were non-relativistic, in the sense that the kinematic and dynamical quantities — like velocity, momentum and force — were written as 3-vectors, whereas true spirit of relativity would require their expressions as 4-vectors. This requirement will be justified in Chapter 11 in the context of the Principle of Covariance. Our immediate concern is now to find 4-vectors which will satisfactorily represent the above quantities. We shall sometimes (but not always) follow the convention set at the beginning of Sec. 7.8, in particular in Eq. (7.65), i.e. first write it as a geometrical object with no reference to its components, followed by a specific reference to it as a 4-component object.

Let us be specific that the 4-vectors of relativistic mechanics that we are going to construct and deploy for our purpose are contravariant vectors, e.g. 4-displacement $d\vec{\mathbf{r}}$, 4-velocity $\vec{\mathbf{V}}$, 4-acceleration $\vec{\mathcal{A}}$, 4-momentum $\vec{\mathcal{P}}$, 4-force $\vec{\mathcal{F}}$. Our immediate purpose is to build up the equation of motion in the Minkowski space–time M^4, and then use them in the larger objective of constructing the energy–momentum stress tensor in Chapter 12. The above 4-vectors will play a key role in achieving this primary objective.

8.1. 4-Displacement

Consider once again the motion of a point particle in Fig. 8.1. In Fig. 8.1(a), we have depicted its *physical trajectory* C in E^3, the Euclidean 3-space.

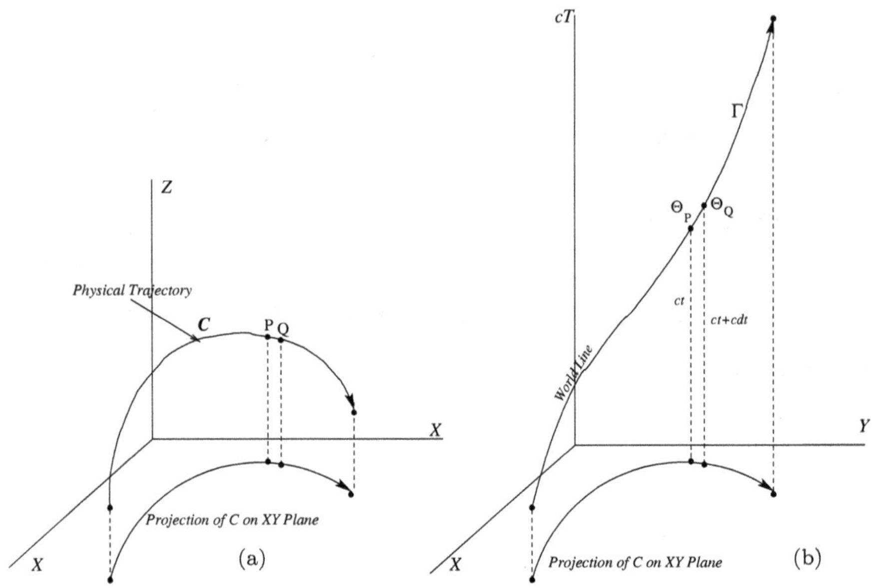

Fig. 8.1. (a) Trajectory of a particle in E^3; (b) world line in M^4.

In Fig. 8.1(b), we have shown its *world line* Γ in the four-dimensional Minkowski space M^4, suppressing the Z-axis.

P (x, y, z) and Q $(x + dx, y + dy, z + dz)$ are two infinitesimally close points on the trajectory C, reached by the particle at times t and $t + dt$, respectively. The coordinates of these events are

$$\Theta_P = (x^\mu) = (ct, \mathbf{r}) = (ct, x, y, z),$$

$$\Theta_Q = (x^\mu + dx^\mu) = (ct + c\,dt, \mathbf{r} + d\mathbf{r}) = (ct + c\,dt, x + dx, y + dy, z + dz).$$

The infinitesimal 4-displacement from Θ_P to Θ_Q

$$d\overrightarrow{\mathbf{r}} = (dx^\mu) = (dx^0, dx^1, dx^2, dx^3) = (c\,dt, dx^1, dx^2, dx^3) \tag{8.1}$$

is the "primordial" contravariant 4-vector from which all other "truly" contravariant 4-vectors (i.e. those which have not been converted by the "raising" operation) by multiplication with scalars and differentiation (cf. Sec. 7.5).

We shall define the basis vectors along the $\{cT, X, Y, Z\}$ axes as $\overrightarrow{\mathbf{e}_\mu} = (\overrightarrow{\mathbf{e}_t}, \overrightarrow{\mathbf{e}_x}, \overrightarrow{\mathbf{e}_y}, \overrightarrow{\mathbf{e}_z})$, in terms of which we shall write the above displacement

4-vector as

$$d\overrightarrow{\mathbf{r}} = \overrightarrow{\mathbf{e}_\mu}\, dx^\mu = (c\, dt,\, d\mathbf{r})$$
$$\text{where}\quad d\mathbf{r} = \text{displacement 3-vector from P to Q.} \tag{8.2}$$

The time interval dt, which was treated as a scalar while obtaining the $\mathbf{v}, \boldsymbol{a}, \mathbf{p}$ and $\mathbf{F}$ vectors in Eq. (7.31), is no longer a scalar in the M^4 context, because the "time interval" transforms under an LT. The quantity which should now replace dt is $d\tau$, the "proper time" interval between the events Θ_P and Θ_Q. It is then imperative that we should establish the 4-scalar nature of $d\tau$.

The norm of the 4-displacement dx^μ is, according to Eqs. (7.110) and (7.90):

$$ds^2 = c^2\, dt^2 - d\mathbf{r}^2, \tag{8.3}$$

and is therefore a 4-scalar. In the instantaneous rest frame of the particle, $dt = d\tau$, and $d\mathbf{r} = \mathbf{0}$. Therefore,

$$ds^2 = c^2\, d\tau^2, \quad \text{so that} \quad d\tau = ds/c. \tag{8.4}$$

Since ds is a 4-scalar, and c is a universal constant, $d\tau$ is a 4-scalar. (QED)

8.2. 4-Velocity

We shall now define the 4-velocity $\overrightarrow{\mathbf{V}} = (V^\mu)$ of the particle at the event (x^μ) as the following contravariant vector:

$$\overrightarrow{\mathbf{V}} \equiv \frac{1}{d\tau} \times d\overrightarrow{\mathbf{r}} = \frac{d\overrightarrow{\mathbf{r}}}{d\tau} = \left(\frac{c\, dt}{d\tau}, \frac{d\mathbf{r}}{d\tau}\right) = \left(\frac{c\, dt}{d\tau}, \frac{dx}{d\tau}, \frac{dy}{d\tau}, \frac{dz}{d\tau}\right). \tag{8.5}$$

Also, as we explained in Eq. (4.37),

$$\frac{dt}{d\tau} = \Gamma = \frac{1}{\sqrt{1 - u^2/c^2}}, \quad \text{so that} \quad \frac{d}{d\tau} = \Gamma \frac{d}{dt}. \tag{8.6}$$

Therefore,

$$\overrightarrow{\mathbf{V}} = \overrightarrow{\mathbf{e}_\mu}\, V^\mu = \Gamma \left(\frac{c\, dt}{dt}, \frac{d\mathbf{r}}{dt}\right) \tag{8.7a}$$

$$= (\Gamma c, \Gamma \mathbf{v}) \tag{8.7b}$$

$$= \Gamma(c, v_x, v_y, v_z). \tag{8.7c}$$

To avoid confusion in future, we shall denote the *Cartesian components of the velocity 3-vectors in lower case, and with subscripts* 1,2,3 to mean

x, y, z. We shall rewrite (8.7) as

$$\vec{\mathbf{V}} = \vec{\mathbf{e}_\mu} V^\mu = (\Gamma c, \Gamma \mathbf{v}) = \Gamma(c, v_1, v_2, v_3). \tag{8.8}$$

What is the magnitude of the 4-velocity $\vec{\mathbf{V}}$? As suggested through Eq. (7.110), the quantity that comes closest to "magnitude" is "norm" which in this case is

$$\vec{\mathbf{V}} \cdot \vec{\mathbf{V}} = (\vec{\mathbf{V}})^2 = V^\alpha V_\alpha = \Gamma^2(c^2 - v^2) = \Gamma^2 c^2(1 - v^2/c^2) = c^2. \tag{8.9}$$

Therefore, the magnitude of any velocity is

$$\sqrt{(\vec{\mathbf{V}})^2} = c. \tag{8.10}$$

All 4-velocities have the *same* "magnitude", equal to the speed of light c. Mount Everest and a 1 GeV proton move with the same absolute magnitude of speed. The reason is not difficult to find. You may be the fastest globe trotter. Yet when you view the world from your jet plane, you find yourself at rest, and the rest of the world, inducing Mount Everest, moving at jet speed. Since motion is relative, it is not possible to make an absolute judgement of which is slow and which is fast. Therefore, all velocities, "big", "small" even "zero," have the same absolute magnitude. Relativity is a great equalizer.

8.3. 4-Acceleration

The 4-acceleration $\vec{\mathcal{A}}$ is a contravariant vector defined as $\frac{d\vec{\mathbf{V}}}{d\tau}$. We shall write it more explicitly as follows:

$$\begin{aligned} &\vec{\mathcal{A}} \stackrel{\text{def}}{=} \frac{d\vec{\mathbf{V}}}{d\tau}; \\ &\text{Or } \vec{\mathcal{A}} = \vec{\mathbf{e}_\mu} \mathcal{A}^\mu = \vec{\mathbf{e}_\mu}\left(\frac{dV^\mu}{d\tau}\right) = \left(c\frac{d^2t}{d\tau^2}, \frac{d^2x}{d\tau^2}, \frac{d^2y}{d\tau^2}, \frac{d^2z}{d\tau^2}\right). \end{aligned} \tag{8.11}$$

It is easy to see that the 4-vector $\vec{\mathcal{A}}$ is "orthogonal" to the 4-vector $\vec{\mathbf{V}}$:

$$\vec{\mathcal{A}} \cdot \vec{\mathbf{V}} = 0. \tag{8.12}$$

To see this take the proper-time-derivative of both sides of (8.9). The right side reduces to 0. The left-hand side becomes $\frac{d}{d\tau}(\vec{\mathbf{V}} \cdot \vec{\mathbf{V}}) = 2\vec{\mathbf{V}} \cdot \vec{\mathcal{A}}$.

(QED)

We shall go a little bit deeper and examine in what way the 4-acceleration $\vec{\mathcal{A}}$ is connected with the 3-acceleration

$$\boldsymbol{a} \stackrel{\text{def}}{=} \frac{d\mathbf{v}}{dt} \equiv \dot{\mathbf{v}}, \tag{8.13}$$

as in N.R. mechanics. In the equations below, as above, "dot" ($\dot{}$) would mean $\frac{d}{dt}$. Referring to Eq. (8.7b) for differentiation,

$$\begin{aligned} \vec{\mathcal{A}} &= \tfrac{d\vec{\mathbf{V}}}{d\tau} = \Gamma \tfrac{d\vec{\mathbf{V}}}{dt} = \Gamma \tfrac{d}{dt}[\Gamma(c,\, \mathbf{v})] = \Gamma\dot{\Gamma}(c,\, \mathbf{v}) + \Gamma^2(0,\, \boldsymbol{a}), \\ \text{or } \vec{\mathcal{A}} &= \Gamma(\dot{\Gamma}c,\, \dot{\Gamma}\mathbf{v} + \Gamma\boldsymbol{a}). \end{aligned} \tag{8.14}$$

We shall now use (8.12) to eliminate $\dot{\Gamma}$.

$$\begin{aligned} &\Gamma(\dot{\Gamma}c,\, \dot{\Gamma}\mathbf{v} + \Gamma\boldsymbol{a}) \cdot \Gamma(c,\, \mathbf{v}) = 0, \\ &\text{or } \Gamma^2[\dot{\Gamma}c^2 - (\dot{\Gamma}\mathbf{v}\cdot\mathbf{v} + \Gamma\boldsymbol{a}\cdot\mathbf{v})] = 0, \\ &\text{or } \dot{\Gamma}(c^2 - v^2) - \Gamma\boldsymbol{a}\cdot\mathbf{v} = 0. \end{aligned} \tag{8.15}$$

$$\text{Hence, using (4.15)} \quad \dot{\Gamma} = \frac{\Gamma^3}{c^2}(\boldsymbol{a}\cdot\mathbf{v}).$$

Using the above result, we can now rewrite (8.14) in the more useful form:

$$\vec{\mathcal{A}} = \left(\frac{\Gamma^4}{c}(\boldsymbol{a}\cdot\mathbf{v}),\; \frac{\Gamma^4}{c^2}(\boldsymbol{a}\cdot\mathbf{v})\mathbf{v} + \Gamma^2\boldsymbol{a}\right). \tag{8.16}$$

8.4. 4-Momentum, or En-Mentum

Multiplying 4-velocity $\vec{\mathbf{V}}$ with the rest mass m_o of the particle we get the *4-momentum* $\vec{\mathcal{P}}$ of the particle, which is an important member of the contravariant family.

$$\vec{\mathcal{P}} \equiv m_o\vec{\mathbf{V}} = m_o\Gamma(c, \mathbf{v}) = m_o\Gamma(c, v_x, v_y, v_z). \tag{8.17}$$

According to Eqs. (4.45), (4.46), (4.79)

$$m = \text{relativistic mass} = \Gamma m_o. \tag{8.18a}$$

$$\mathbf{p} = \text{relativistic momentum} = \Gamma m_o\mathbf{v} = m\mathbf{v}. \tag{8.18b}$$

$$\mathcal{E} = \text{total energy} = \Gamma m_o c^2 = mc^2. \tag{8.18c}$$

Note that we have used two different symbols for energy, namely E in Chapter 4, and now $\mathcal{E}$ in the present chapter. The reason is that we shall prefer to reserve "E" for the magnitude of the electric field, as we are going to take up the covariant equations of electrodynamics in Chapter 11.

It follows from (8.17) and (8.18) that the time component of the 4-momentum is *energy* (divided by c). Therefore,

$$\vec{\mathcal{P}} = \vec{\mathbf{e}_\mu}\, p^\mu = \left(\frac{\mathcal{E}}{c},\, \mathbf{p}\right) \equiv \overrightarrow{\textbf{En-Mentum}}. \tag{8.19}$$

Energy is thus integrated into momentum into a single 4-vector, the 4-momentum, in which energy is the zeroth component, i.e. the time component, followed by three space components. It will be easier to remember the constituents of this 4-vector, and their ordering, if we give the 4-momentum an alternative and generic name *En-Mentum.*

There is a more important justification. The name 4-vector, or 4-momentum, makes us expect four components in the 4-vector involved. Most often this is not true. There can be 2-components, or 3-components when the motion is one-dimensional, or two-dimensional. The name En-Mentum will steer clear of this confusion.

This fusion of two distinct quantities, energy and momentum, into a single entity En-Mentum was expected, if not overdue, following our awakening to the fact that the energy and momentum conservation laws could not be separated into a conservation of one of them without the other, as we saw in Sec. 4.6. Either we get the conservation of the united En-Mentum or no conservation at all. Read our observation following Eq. (4.92). We can therefore interpret the energy–momentum transformation equations (4.89) and (4.90) as the transformation of En-Mentum, which we rewrite here as the transformation of its time–space components:

$$p'^0 = \gamma(p^0 - \beta p^1) \Rightarrow \frac{\mathcal{E}'}{c} = \gamma\left(\frac{\mathcal{E}}{c} - \beta p_x\right), \tag{8.20a}$$

$$p'^1 = \gamma(p^1 - \beta p^0) \Rightarrow p'_x = \gamma\left(p_x - \beta\frac{\mathcal{E}}{c}\right), \tag{8.20b}$$

$$p'^2 = p^2;\ p'^3 = p^3 \Rightarrow p'_y = p_y;\ p'_z = p_z. \tag{8.20c}$$

Suppose in the frame S we have pure energy $\mathcal{E}$, but no momentum (we can think of a ball of radiation, propagating isotropically in all directions). Thanks to $\mathcal{E} = mc^2$, this radiation will have a rest mass $m_o = \mathcal{E}/c^2$. According to (8.20b), this "ball" will have a momentum $p'_x = -\gamma c\beta m_o$, same as that of a particle of rest mass m_o, moving with the velocity $-\beta c$, in agreement with (8.18b). Pure energy in one frame transforms into a combination of energy and momentum in another.

It is seen from (8.18c) that the energy $\mathcal{E}$ of a particle is proportional to the Lorentz factor Γ, which can be used as a representative of the energy. From (8.20a), we find its transformation from S to S' under the boost: $S(\beta, 0, 0)S'$:

$$\Gamma' = \gamma\Gamma(1 - \beta v_x/c). \tag{8.21}$$

The magnitude of 4-momentum (its "norm") has a special significance. By Eqs. (8.9) and (8.17)

$$(\vec{\mathcal{P}})^2 = m_o{}^2(V)^2 = m_o{}^2c^2, \tag{8.22}$$

so that the magnitude of 4-momentum is $m_o c$. However, from Eq. (8.19)

$$(\vec{\mathcal{P}})^2 = (p^0)^2 - \mathbf{p}^2 = \frac{\mathcal{E}^2}{c^2} - \mathbf{p}^2. \tag{8.23}$$

Equating the right sides of the above two equations, we rediscover the energy–momentum relation given earlier in Eq. (4.83):

$$\mathcal{E}^2 = (m_o c^2)^2 + \mathbf{p}^2c^2. \tag{8.24}$$

For a massless particle, e.g. photon, the above equation yields

$$\mathcal{E} = |\mathbf{p}|\, c, \tag{8.25}$$

as in (4.84). We shall therefore specialize the definition of 4-momentum given in (8.19) for a photon:

$$\vec{\mathcal{P}} = \vec{\mathbf{e}}_\mu\, p^\mu = \left(\frac{\mathcal{E}}{c}, \frac{\mathcal{E}}{c}\mathbf{n}\right) \quad \text{for a photon,} \tag{8.26}$$

where $\mathbf{n}$ is the direction of propagation of the photon. For a photon,

$$(\vec{\mathcal{P}})^2 = 0 \tag{8.27}$$

so that the En-Mentum of a photon is a null vector.

8.5. 4-Force, or Pow-Force and Minkowski's Equation of Motion of a Point Particle

The structure of 4-force $\vec{\mathcal{F}}$ should be such that it will recast Newton's equation of motion $\mathbf{F} = \frac{d\mathbf{p}}{dt}$ in E^3 to *Minkowski's equation of motion* (to be referred to as EoM frequently) in M^4 in the form

$$\boxed{\frac{d\vec{\mathcal{P}}}{d\tau} = \vec{\mathcal{F}}.} \tag{8.28}$$

The 4-force $\vec{\boldsymbol{\mathcal{F}}}$ that appears on the right is known as *Minkowski force.* However, we may like to call it *Pow-Force*, for reasons that will follow. Let us write

$$\vec{\boldsymbol{\mathcal{F}}} = \vec{\mathbf{e}_\mu}\,\mathcal{F}^\mu = (\mathcal{F}^o, \boldsymbol{\mathcal{F}}). \tag{8.29}$$

The EoM (8.28) combined with the definition of $\vec{\boldsymbol{\mathcal{P}}}$ given in (8.18), (8.19) means that

$$\frac{1}{c}\frac{d\mathcal{E}}{d\tau} = \frac{1}{c}\Gamma\frac{d\mathcal{E}}{dt} = \Gamma\frac{\Pi}{c} = \mathcal{F}^0, \tag{8.30a}$$

$$\frac{d\mathbf{p}}{d\tau} = \Gamma\frac{d\mathbf{p}}{dt} = \boldsymbol{\mathcal{F}}, \tag{8.30b}$$

where Π (capital pi) $= \frac{d\mathcal{E}}{dt}$. It stands for the *power* received by the particle (same as energy received by the particle per unit time), due to (i) work done on it by external forces, and/or (ii) by absorption of radiation or heat (thereby changing its rest mass). In the case of a particle whose rest mass does not change, Π is the same as the power delivered by the force $\mathbf{F}$; cf. Eq. (4.81).

According to Newton's second law of motion, Eq. (4.47), $\frac{d\mathbf{p}}{dt} = \mathbf{F}$. Hence the space component of $\vec{\boldsymbol{\mathcal{F}}}$ is identified as $\Gamma\mathbf{F}$, and the time component as $\mathcal{F}^0 = \Gamma\Pi/c$. Then

$$\vec{\boldsymbol{\mathcal{F}}} = \Gamma\left(\frac{\Pi}{c}, \mathbf{F}\right) \equiv \overrightarrow{\textbf{Pow-Force}}. \tag{8.31}$$

Since the time component of $\vec{\boldsymbol{\mathcal{F}}}$ is *Power* (multiplied by Γ/c), it may be easier to remember the constituents of this 4-vector, and their ordering, by the alternative and generic name *Pow-Force.*

Since $\frac{d}{d\tau} = \Gamma\frac{d}{dt}$, Minkowski's EoM (8.28) gets resolved into the following time and space components:

Time component:	$\dfrac{d\mathcal{E}}{dt} = \Pi.$ (8.32a)
Space component:	$\dfrac{d\mathbf{p}}{dt} = \mathbf{F}.$ (8.32b)

The *space component* of Minkowski's equation of motion is a restatement of Newton's second law of motion. The corresponding *time component* expresses conservation of energy. It simply states that the rate of change of the kinetic energy of a particle equals the power delivered by the external agents (e.g. forces) acting on it.

Let us consider an important application of the EoM (8.28) to a point particle with a constant rest mass m_0. Due to (8.17), then the above EoM takes the alternative form

$$m_o \vec{\mathcal{A}} = \vec{\mathcal{F}}. \tag{8.33}$$

The above equation shows that this 4-force is orthogonal to 4-velocity, a fact that follows from (8.12):

$$\vec{\mathcal{F}} \cdot \vec{\mathbf{V}} = 0. \tag{8.34}$$

Therefore, the four components of $\vec{\mathcal{F}}$ are not independent. We shall fix the time component using (8.7), (8.31) and (8.34).

$$\left(\Gamma \frac{\Pi}{c}\right)(\Gamma c) - (\Gamma \mathbf{F}) \cdot (\Gamma \mathbf{v}) = 0 \Rightarrow \Pi = \mathbf{F} \cdot \mathbf{v}. \tag{8.35}$$

Hence, for a *particle with constant rest mass* the 4-force must have the form:

$$\vec{\mathcal{F}} = \Gamma \left(\frac{\mathbf{F} \cdot \mathbf{v}}{c}, \mathbf{F} \right). \tag{8.36}$$

Now the equation of motion (8.28) in M^4 will have the time and space components:

Time component:	$\dfrac{d\mathcal{E}}{dt} = \mathbf{F} \cdot \mathbf{v}.$	(8.37a)
Space component:	$\dfrac{d\mathbf{p}}{dt} = \mathbf{F}.$	(8.37b)

8.6. Force on a Particle with a Variable Rest Mass

Let us now be more general and consider a 4-force that can change the rest mass of the particle. We shall then write

$$\vec{\mathcal{F}}_{\text{tot}} = \vec{\mathcal{F}} + \vec{\mathcal{K}}, \tag{8.38}$$

where $\vec{\mathcal{F}}$ is the force that does not alter the mass of the particle, and is orthogonal to the 4-velocity. It has the same properties as in (8.33) – (8.37).

$\vec{\mathcal{K}}$ can be called a *convective force*, as it can alter the mass of the particle. The time–space components of $\vec{\mathcal{K}}$ can be written as

$$\vec{\mathcal{K}} = \Gamma\left(\frac{\Pi}{c}, \boldsymbol{\mathcal{K}}\right) \text{ in } S; \quad \vec{\mathcal{K}} = \left(\frac{\Pi_o}{c}, \boldsymbol{\mathcal{K}}_o\right) \text{ in } S_o \qquad (8.39a)$$

$$\vec{\mathcal{K}}_{\text{tot}} \cdot \vec{\mathbf{V}} = \vec{\mathcal{K}} \cdot \vec{\mathbf{V}} = \Gamma^2(\Pi - \boldsymbol{\mathcal{K}} \cdot \mathbf{v}) = \Pi_o, \qquad (8.39b)$$

$$\text{where} \quad \Pi_o = \frac{dm_o}{d\tau}c^2 \qquad (8.39c)$$

is the convective power absorbed in S_o and m_o is the proper mass of the particle. We shall illustrate this case with the example of a Relativistic Rocket in Sec. 9.

8.7. Lorentz Transformation of the 4-Vectors of Relativistic Mechanics

We shall apply Lorentz transformation to the 4-vectors of Relativistic Mechanics and from the resulting relations extract the transformation formulas of the corresponding 3-vectors, and make sure that they agree with the transformations obtained in Sec. 4.1 and 4.2.

8.7.1. *LT of 4-velocity*

We shall apply the general transformation given in Eq. (7.78), to the velocity 4-vector $\vec{\mathbf{V}}$ shown in (8.7b). We shall rewrite the same as

$$\vec{\mathbf{V}} = \Gamma c(1, \boldsymbol{\nu}), \qquad (8.40)$$

where $\boldsymbol{\nu} = \mathbf{v}/c$ as in (4.22), so that the quantities inside the brackets are dimensionless:

$$(7.78a) \Rightarrow \quad \Gamma' = \gamma\Gamma(1 - \boldsymbol{\beta} \cdot \boldsymbol{\nu}), \qquad (8.41a)$$

$$(7.78c) \Rightarrow \quad \Gamma'\boldsymbol{\nu}' = \Gamma[(\gamma\boldsymbol{\nu}_\parallel + \boldsymbol{\nu}_\perp) - \gamma\boldsymbol{\beta}], \qquad (8.41b)$$

$$\text{or,} \quad \boldsymbol{\nu}' = \frac{\Gamma[(\gamma\boldsymbol{\nu}_\parallel + \boldsymbol{\nu}_\perp) - \gamma\boldsymbol{\beta}]}{\Gamma'}, \qquad (8.41c)$$

$$\text{using (8.41a)}, \quad \boldsymbol{\nu}'_\parallel + \boldsymbol{\nu}'_\perp = \boldsymbol{\nu}' = \frac{(\gamma\boldsymbol{\nu}_\parallel + \boldsymbol{\nu}_\perp) - \gamma\boldsymbol{\beta}}{\gamma(1 - \boldsymbol{\beta} \cdot \boldsymbol{\nu})} \qquad (8.41d)$$

$$\text{from which} \quad \boldsymbol{\nu}'_{\parallel} = \frac{\boldsymbol{\nu}_{\parallel} - \boldsymbol{\beta}}{(1 - \boldsymbol{\beta} \cdot \boldsymbol{\nu})}, \tag{8.41e}$$

$$\text{and} \quad \boldsymbol{\nu}'_{\perp} = \frac{\boldsymbol{\nu}_{\perp}}{\gamma(1 - \boldsymbol{\beta} \cdot \boldsymbol{\nu})}. \tag{8.41f}$$

We get back (4.23a) for the special case of x-orientation.

Equation (8.41a) will be found useful in many applications. Hence, we shall write it, and its inverse, as two separate equations:

$$\Gamma' = \gamma\Gamma(1 - \boldsymbol{\beta} \cdot \boldsymbol{\nu}), \tag{8.42a}$$

$$\Gamma = \gamma\Gamma'(1 + \boldsymbol{\beta} \cdot \boldsymbol{\nu}'). \tag{8.42b}$$

8.7.2. *LT of 4-acceleration*

The components of $\vec{\mathcal{A}}$ are given in (8.16). To make the formulas less complicated we shall specialize the expression for $\vec{\mathcal{A}}$ to rectilinear motion, along the X-axis, and get the following expression for $\vec{\mathcal{A}}$, having only t and x components:

$$\mathcal{A}^0 = \frac{\Gamma^4 av}{c}, \quad \mathcal{A}^1 = \frac{\Gamma^4 av^2}{c^2} + \Gamma^2 a = \Gamma^2 a\left\{\frac{\Gamma^2 v^2}{c^2} + 1\right\} = \Gamma^4 a. \tag{8.43}$$

In short,

$$\vec{\mathcal{A}} = \Gamma^4 a\left(\frac{v}{c}, 1\right), \tag{8.44}$$

where a is the acceleration along the X-axis. We shall now let $\vec{\mathbf{A}} \to \vec{\mathcal{A}}$, and apply the transformation (7.77a).

$$A'^0 = \gamma(A^0 - \beta A^1) \Rightarrow \Gamma'^4 a'\frac{v'}{c} = \gamma\Gamma^4 a\left(\frac{v}{c} - \beta\right), \tag{8.45a}$$

$$A'^1 = \gamma(A^1 - \beta A^0) \Rightarrow \Gamma'^4 a' = \gamma\Gamma^4 a\left(1 - \beta\frac{v}{c}\right). \tag{8.45b}$$

Let us now take the S' frame to be the IRF, so that

$$v' = 0,\ \Gamma' = 1,\ v = \beta c,\ \gamma = \Gamma.$$

Now go back to (8.45). Equation (8.45a) yields $0 = 0$. Equation (8.45b) gives the acceleration in the IRF, also called *proper acceleration*, as

$$a' = \Gamma^5 a(1 - \beta^2) = \Gamma^3 a, \tag{8.46}$$

same result as in (4.29).

8.7.3. *LT of 4-momentum*

We shall take the same steps as for 4-velocity. Recall the time and space components of 4-momentum as given in (8.19):

$$\vec{\mathcal{P}} = \left(\frac{\mathcal{E}}{c}, \mathbf{p}\right), \quad \mathcal{E} = \Gamma m_o c^2, \quad \mathbf{p} = \Gamma m_o \mathbf{v}. \tag{8.47}$$

Apply the general transformation given in Eq. (7.78) to these components. First the time component. From (7.78a)

$$\begin{aligned} \frac{\mathcal{E}'}{c} &= \gamma\left(\frac{\mathcal{E}}{c} - \boldsymbol{\beta}\cdot\mathbf{p}\right), \\ \text{or } \Gamma' m_o c &= \gamma(\Gamma m_o c - \boldsymbol{\beta}\cdot\Gamma m_o \mathbf{v}), \\ \text{or } \Gamma' &= \gamma\Gamma(1 - \boldsymbol{\beta}\cdot\boldsymbol{\nu}). \end{aligned} \tag{8.48}$$

For one-dimensional motion,

$$\begin{aligned} \Gamma' &= \gamma\Gamma(1 - \beta\,\nu), \\ \Gamma &= \gamma\Gamma'(1 + \beta\,\nu'). \end{aligned} \tag{8.49}$$

We get back Eq. (8.42). Now the space component. From (7.78e)

$$\mathbf{p}' = \mathbf{p} + \mathbf{n}\left[(\gamma - 1)(\mathbf{n}\cdot\mathbf{p}) - \gamma\beta\,\frac{\mathcal{E}}{c}\right]. \tag{8.50}$$

We shall get back Eq. (4.23) if we use the definitions of $\mathcal{E}$ and $\mathbf{p}$ as given in (8.47).

The inverse transformation, corresponding to the boost: $\{S'(\boldsymbol{\beta})S\}$ is obtained by replacing $\mathbf{n}$ with $-\mathbf{n}$:

$$\mathbf{p} = \mathbf{p}' + \mathbf{n}\left[(\gamma - 1)(\mathbf{n}\cdot\mathbf{p}') + \gamma\beta\,\frac{\mathcal{E}'}{c}\right]. \tag{8.51}$$

8.7.4. *LT of 4-force*

The components of the 4-force are given in (8.31) which we rewrite here

$$\vec{\mathcal{F}} = \Gamma\left(\frac{\Pi}{c}, \mathbf{F}\right). \tag{8.52}$$

Here $\mathbf{F}$ is the 3-force as defined by Newton's second law of motion $\frac{d\mathbf{p}}{dt} = \mathbf{F}$. Apply the general transformation given in Eq. (7.78) to these components.

First the time component. From (7.78a)

$$\frac{\Gamma'\Pi'}{c} = \gamma\left(\frac{\Gamma\Pi}{c} - \boldsymbol{\beta}\cdot\Gamma\,\mathbf{F}\right),$$

or

$$\Pi' = \frac{\gamma\Gamma}{\Gamma'}(\Pi - \mathbf{v}\cdot\mathbf{F}), \tag{8.53}$$

$$\text{using (8.42), } \Pi' = \frac{\Pi - \mathbf{v}\cdot\mathbf{F}}{1 - \boldsymbol{\beta}\cdot\boldsymbol{\nu}}.$$

Now the space component. From (7.78e)

$$\begin{aligned} \Gamma'\mathbf{F}' &= \Gamma\mathbf{F} + \mathbf{n}\left[(\gamma-1)(\mathbf{n}\cdot\Gamma\mathbf{F}) - \gamma\beta\frac{\Gamma\Pi}{c}\right], \\ \text{using (8.42), } \mathbf{F}' &= \frac{\mathbf{F} + \mathbf{n}\left[(\gamma-1)(\mathbf{n}\cdot\mathbf{F}) - \gamma\beta\frac{\Pi}{c}\right]}{\gamma(1 - \boldsymbol{\beta}\cdot\boldsymbol{\nu}).} \end{aligned} \tag{8.54}$$

For a particle moving with constant rest mass, $\Pi = \mathbf{F}\cdot\mathbf{v} = \mathbf{F}\cdot c\boldsymbol{\nu}$. In this case

$$\mathbf{F}' = \frac{\mathbf{F} + \mathbf{n}\left[(\gamma-1)(\mathbf{n}\cdot\mathbf{F}) - \gamma\beta\,\mathbf{F}\cdot\boldsymbol{\nu}\right]}{\gamma(1 - \boldsymbol{\beta}\cdot\boldsymbol{\nu})}. \tag{8.55}$$

The inverse transformation, corresponding to the boost: $\{S'(\boldsymbol{\beta})S\}$, is obtained by replacing $\mathbf{n}$ with $-\mathbf{n}$:

$$\mathbf{F} = \frac{\mathbf{F}' + \mathbf{n}\left[(\gamma-1)(\mathbf{n}\cdot\mathbf{F}') + \gamma\beta\,\mathbf{F}'\cdot\boldsymbol{\nu}'\right]}{\gamma(1 + \boldsymbol{\beta}\cdot\boldsymbol{\nu}')}. \tag{8.56}$$

Corollaries of the force transformation formulas:

(1) A longitudinal force $\mathbf{F}_{\|}$ (i.e. a force in the direction of the boost velocity $c\boldsymbol{\beta}$) is invariant under LT.
Proof: Let S' be the rest frame of the particle. Hence, $\boldsymbol{\nu}' = \mathbf{0}$. By assumption, $\mathbf{n}(\mathbf{n}\cdot\mathbf{F}') = \mathbf{F}'$. Hence, from (8.55)

$$\mathbf{F}_{\|} = \mathbf{F}'_{\|}. \tag{8.57}$$

(QED)

(2) A transverse force $\mathbf{F}'_{\perp}$ in the rest frame S' of the particle transforms to the $\mathbf{F}_{\perp}/\gamma$ in the Lab frame S.
Proof: By assumption $\mathbf{n}\cdot\mathbf{F}' = 0$. By a similar argument

$$\mathbf{F}_{\perp} = \frac{\mathbf{F}'_{\perp}}{\gamma}. \tag{8.58}$$

(QED)

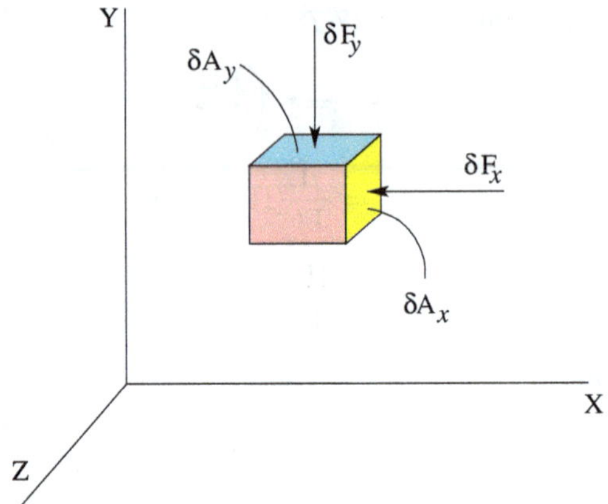

Fig. 8.2. Longitudinal and transverse force and area cross-section in the rest frame.

(3) Pressure p in a perfect fluid is Lorentz invariant.

Proof: By perfect fluid we mean a fluid in which there is no shear stress. If we consider an infinitesimal rectangular box inside the fluid (Fig. 8.2), the stress on all the bounding surfaces will be the same normal compressive stress p, called pressure.

Let the pressure inside the fluid at a certain event Θ, in the corresponding instantaneous rest frame S', be p'. Referring to Fig. 8.2, the force on the right face is $\delta F'_x$ and the area is $\delta A'_x$. Similarly, the force on the top face is $\delta F'_y$ and the area is $\delta A'_y$. Hence, $p' = \frac{\delta F'_x}{\delta A'_x} = \frac{\delta F'_y}{\delta A'_y}$.

Apply Rules 3, 4 of Sec. 2.7 to the above surface areas. The longitudinal dimension contracts, but the transverse dimensions do not. Hence, $\delta A'_y \to \delta A_y = \delta A'_y/\gamma$; $\delta A'_x \to \delta A_x = \delta A'_x$. If we now apply (8.57) and (8.58), we get

$$p_x \equiv \frac{\delta F_x}{\delta A_x} = \frac{\delta F'_x}{\delta A'_x} = p'.$$

$$p_y \equiv \frac{\delta F_y}{\delta A_y} = \frac{\delta F'_y/\gamma}{\delta A'_y/\gamma} = p'.$$

$$\text{Hence,} \quad p(x) = p_x(x) = p_y(x) = p'(x). \tag{8.59}$$

(QED)

8.8. Conservation of 4-Momentum of a System of Particles

8.8.1. *Zero momentum frame, equivalence of $\mathcal{E}$ and mass*

Consider a system $\mathfrak{S}$ of N *non-interacting* particles, their individual rest masses, velocities and En-Menta being $\{m_0^i, \mathbf{v}^i, \overrightarrow{\mathbf{p}}^i \equiv \Gamma^i m_0^i(c, \mathbf{v}^i);\ i = 1, 2, \ldots, N\}$, with respect to some inertial frame S. The total energy, total Momentum and total En-Mentum of this system are as follows:

$$\mathcal{E} = \sum_{i=1}^{n} \Gamma^i m_0^i c^2, \quad \mathbf{P} = \sum_{i=1}^{n} \Gamma^i m_0^i \mathbf{v}^i, \quad \overrightarrow{\mathcal{P}} = (\mathcal{E}/c, \mathbf{P}). \tag{8.60}$$

In performing the summation in the above equation, we count each of the N particles at the same instant of time, i.e. *simultaneously.* On the space–time diagram the counting events lie on one XYZ-hyperplane.

Let us consider the boost: $\{S' \to (\mathbf{n}\beta) \to S\}$, and apply the Lorentz transformation formula (7.79a) and (7.79e) to the En-Mentum $\overrightarrow{\mathcal{P}}$:

$$\mathcal{E}/c = \gamma[\mathcal{E}'/c + \boldsymbol{\beta}\cdot\mathbf{P}'], \tag{8.61a}$$

$$\mathbf{P} = \mathbf{P}' + \mathbf{n}[(\gamma - 1)(\mathbf{n}\cdot\mathbf{P}') + \gamma\beta(\mathcal{E}'/c)]. \tag{8.61b}$$

The above transformation equations are valid, a priori, for the En-Menta $(\mathcal{E}^i/c, \mathbf{p}^i)$ of individual particles referred to in (8.60). However, due to the linear character of the transformation equations, they are also valid for the total En-Mentum. The 4-vectors $(\mathcal{E}/c, \mathbf{p})$, $(\mathcal{E}'/c, \mathbf{p}')$, in Eq. (8.61) now stand for the total En-Mentum 4-vector in S and S', respectively.

For obtaining $\mathcal{E}', \mathbf{P}'$ in the above equation, we perform the summation by the same procedure as for Eq. (8.60). However, in this case the counting events are *simultaneous* in S', but not so in S. This should not cause any problem, because the particles being non-interactive, and under no external forces, keep their individual En-Menta constant as they progress along their respective world lines.

What happens when two of them collide. The collision of two particles is one event, a single point on the space–time diagram — same point whether in S or in S', although their coordinates will differ.

In Sec. 4.9 we have presented an example of elastic collision between a photon and an electron, known as Compton scattering. The En-Mentum is conserved, and the CM of the system keeps moving on as a single particle along its world line.

Suppose the collision is perfectly *inelastic*, i.e. they coalesce and form a new particle C. Even then the En-Mentum of C will be the sum of the En-Menta of A+B, just after collision. That is, $\mathcal{E}_C = \mathcal{E}_A + \mathcal{E}_B$; $\mathbf{p}_C = \mathbf{p}_A + \mathbf{p}_B$. However, in this case part of the energy $\mathcal{E}_C$ goes into the inner energy of the coalesced particle C, say in the form of "excitation energy". Problem 4.6 gives a good example of this process. In this case also the CM keeps moving on as a single particle, just after the reaction.

Let us go back to (8.61). Of all frames S' there will be one in which $\mathbf{P}' = \mathbf{0}$. We shall identify this frame as $\mathcal{S}_{\text{rest}}$, and call it the *rest frame* of the system $\mathfrak{S}$, or better still the *zero momentum frame* (ZMF or ZM frame) of the system. This is the analogue of centre of mass frame of non-relativistic classical mechanics.

We go back to Eq. (8.61), set $\mathbf{P}' = \mathbf{0}$, denote the energy in $\mathcal{S}_{\text{rest}}$ as $\mathcal{E}^0$ and obtain:

$$\mathcal{E} = \gamma\mathcal{E}^0, \tag{8.62a}$$

$$\mathbf{P} = \mathbf{n}\gamma\beta(\mathcal{E}^0/c) = \gamma\mathbf{v}(\mathcal{E}^0/c^2) = \gamma\mathbf{v}M_0 = \Gamma\mathbf{v}M_0, \tag{8.62b}$$

$$\text{where,} \quad M_0 = \mathcal{E}^0/c^2 = \text{rest mass of the system } \mathfrak{S} \text{ in the frame } S. \tag{8.62c}$$

This brings us to the following conclusion:

Conclusion: The system of particles $\mathfrak{S}$ can be considered to be a single particle at rest in $\mathcal{S}_{\text{rest}}$, and moving with the boost velocity $\mathbf{v}$ with respect to S. It has a rest energy $\mathcal{E}_0$, rest mass M_0, and the two are related by the relation $\mathcal{E}_0 = M_0c^2$. Its momentum $\mathbf{P}$ as shown in (8.62b), is in accordance with the definition of the momentum of a single particle as given in (4.44).

We have replaced the Lorentz factor γ associated with the LT, with the dynamic Lorentz factor Γ associated with the motion of a moving particle (see Eq. (4.38)). In this case, the particle velocity is identical with the boost velocity.

Thus the *inner kinetic energy* T_0 *of the system contributes to the inertial mass* of the system by an amount equal to T_0/c^2.

We shall illustrate the concepts with a few examples.

8.9. Illustrative Numerical Examples III

We shall now specialize our discussion to a system in which there are two particles before a collision, and two or more particles after the collision.

The collision may result in the production of a new system of particles which may or may not bear any resemblance with the original ones. One such example is the production of a pi-meson like $\pi^0, \pi^\pm$ from a nucleon–nucleon collision ($p+p$, $p+n$), another can be the example of the creation of an electron–positron pair when a γ-ray hits an electron. In such cases, the En-Mentum of the system may not have the simplest expression as given in (8.17) in which the rest mass m_o and the dynamic Lorentz factor Γ have been delineated clearly.

Even when the En-Mentum looks different, we can still filter out the rest mass and the dynamic Lorentz factor by recasting the expression suitably, as in the following example:

$$\vec{\mathcal{P}} = (A, B, C, D) = m_o \left(\frac{A}{m_o c}\right)\left(c, \frac{B\,c}{A}, \frac{C\,c}{A}, \frac{D\,c}{A}\right). \tag{8.63}$$

This is in the form of Eq. (8.17), if we identify

$$\frac{A}{m_o\, c} = \Gamma; \ \frac{B\,c}{A} = v_x; \text{ etc.} \tag{8.64}$$

8.9.1. *Example 1: Relativistic billiard balls*

The example we shall now discuss, and work out the details, will clarify the notion of the ZM frame, in particular how to find this frame, and illustrate how this intermediary will help us find a solution to the problem.

Figure 8.3(a) shows two identical billiard balls A and B of equal rest mass m_o. B is at rest, and A is approaching B along the X-axis with relativistic velocity $c\boldsymbol{\nu}_{\text{A-in}} = c\,\nu_{\text{A-in}}\,\mathbf{e}_x$. The dynamic Lorentz factor corresponding this velocity is $\Gamma_{\text{A-in}} = 1/\sqrt{1-\nu^2_{\text{A-in}}}$. After the collision, A and B bounce out with velocities $c\boldsymbol{\nu}_{\text{A-out}}$ and $c\boldsymbol{\nu}_{\text{B-out}}$, respectively. Our task is to calculate these velocities and the angles θ and ϕ they make with the X-axis.

Figure 8.3(b) shows the billiard ball experiment from the ZM frame. Due to symmetry both balls approach each other and then fly apart with the *same* speed $c\nu'$, making angle θ' with the X-axis. The dynamic Lorentz factor for *both* particles is $\Gamma' = 1/\sqrt{1-\nu'^2}$.

We shall proceed towards the answers in several steps.

Step 1: Find the velocity of the ZM frame with respect to the Lab frame.

For a while we shall suppress the subscript "A-in" in $\nu_{\text{A-in}}, \Gamma_{\text{A-in}}$, so that the equations look less cumbersome. Γ and ν in the equations below will mean $\Gamma_{\text{A-in}}$ and $\nu_{\text{A-in}}$.

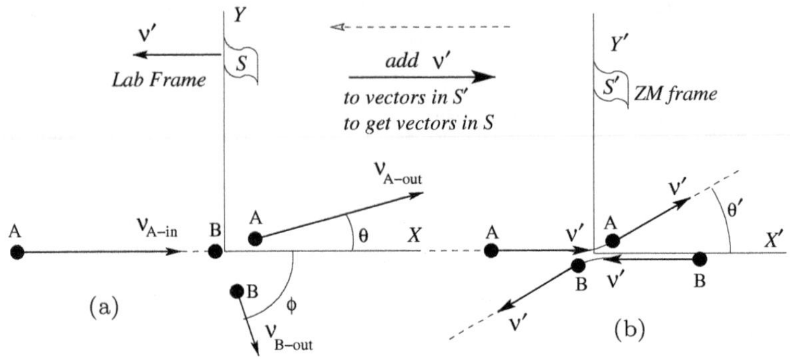

Fig. 8.3. Elastic collision between two equal balls seen from S and S′.

Let $c\beta$ be the boost velocity, $\gamma = 1/\sqrt{1-\beta^2}$ the boost Lorentz factor. We shall write the individual and total En-Menta before the collision. Since the encounter takes place on the XY-plane, we shall write only three components of En-Mentum in the format $\overrightarrow{\mathcal{P}} = \overrightarrow{\mathbf{e}}_\mu p^\mu = (p^0, p^1, p^2)$ (as in Eq. (8.17), but without p^3). In the *Lab frame*, represented by S, the relevant 4-vectors are as follows:

$$\overrightarrow{\mathcal{P}}_{\text{A-in}} = \overrightarrow{\mathbf{e}}_\mu\, p^\mu_{\text{A-in}} = \Gamma m_\circ c\,(1, \nu, 0), \tag{8.65a}$$

$$\overrightarrow{\mathcal{P}}_{\text{B-in}} = \overrightarrow{\mathbf{e}}_\mu\, p^\mu_{\text{B-in}} = m_\circ c\,(1, 0, 0), \tag{8.65b}$$

$$\overrightarrow{\mathcal{P}}_{\text{in}} = \overrightarrow{\mathcal{P}}_{\text{A-in}} + \overrightarrow{\mathcal{P}}_{\text{B-in}} = m_\circ c(\Gamma + 1, \Gamma\nu, 0). \tag{8.65c}$$

LT will convert $\overrightarrow{\mathcal{P}}_{\text{in}} \to \overrightarrow{\mathcal{P}}'_{\text{in}}$ with components $(P'^0_{\text{in}}, P'^1_{\text{in}}, P'^2_{\text{in}})$ of which the first two are non-zero:

$$P'^0_{\text{in}} = \gamma\,((\Gamma + 1) - \beta\,\Gamma\nu)\, m_\circ c, \tag{8.66a}$$

$$P'^1_{\text{in}} = \gamma\,(\Gamma\nu - \beta\,(\Gamma + 1))\, m_\circ c. \tag{8.66b}$$

To identify the ZM frame we must set $P'^1_{\text{in}} = 0$. This leads to

$$\beta = \frac{\Gamma\,\nu}{\Gamma + 1}. \tag{8.67}$$

Going back to (8.66a), and using (8.67),

$$\begin{aligned} P_{\text{in}}^{\prime 0} &= \gamma \left\{ (\Gamma + 1) - \frac{(\Gamma\nu)^2}{\Gamma + 1} \right\} m_o c \\ &= \frac{\gamma\, m_o c}{\Gamma + 1} \left\{ \Gamma^2 + 2\Gamma + 1 - \Gamma^2 \nu^2 \right\} \\ &= 2\gamma\, m_o c. \end{aligned} \tag{8.68}$$

We have used the fourth identity from Eq. (2.22). The total En-Mentum of the 2-particle system in the ZM frame can now be written as

$$\vec{\mathcal{P}}'_{\text{in}} = 2\gamma m_o c\,(1, 0, 0). \tag{8.69}$$

Particles A and B, together representing total energy $(\Gamma + 1)m_o c^2$ in the Lab frame S (see Eq. (8.65c)), transforms into a *stationary single particle* with rest mass energy $2\gamma m_o c^2$ in the ZM frame S'.

We can find a relation between Γ and γ by transforming $\vec{\mathcal{P}}' \to \vec{\mathcal{P}}$, from the ZM frame S' to the Lab frame S (using Eq. (7.77a) and setting $\beta \to -\beta$).

$$P_{\text{in}}^{0} = \gamma(P_{\text{in}}^{\prime 0} + \beta P_{\text{in}}^{\prime 1}) = \gamma(2\gamma + 0) m_o c = 2\gamma^2 m_o c.$$

$$\text{But, } P_{\text{in}}^{0} = (\Gamma + 1) m_o c, \quad \text{from Eq. (8.65c).} \tag{8.70}$$

Hence, $\Gamma + 1 = 2\gamma^2$.

We can transform this into another useful relation, using the identity $\gamma^2 - 1 = \gamma^2\beta^2$:

$$\Gamma = 2\gamma^2 - 1 = \gamma^2(1 + \beta^2). \tag{8.71}$$

Step 2: *Find the En-Menta of the balls before and after the collision.*

In the *ZM frame*, represented by S', the relevant En-Menta are as follows:

$$\begin{aligned} &\text{(i)}\ \ \vec{\mathcal{P}}'_{\text{A-in}} = \vec{\mathbf{e}}_\mu\, p'^{\mu}_{\text{A-in}} = \Gamma' m_o c\,(1, \nu', 0); \\ &\text{(ii)}\ \ \vec{\mathcal{P}}'_{\text{B-in}} = \vec{\mathbf{e}}_\mu\, p'^{\mu}_{\text{B-in}} = \Gamma' m_o c\,(1, -\nu', 0); \\ &\text{(iii)}\ \ \vec{\mathcal{P}}'_{\text{A-out}} = \vec{\mathbf{e}}_\mu\, p'^{\mu}_{\text{A-out}} = \Gamma' m_o c\,(1, \nu' \cos\theta', \nu' \sin\theta'); \\ &\text{(iv)}\ \ \vec{\mathcal{P}}'_{\text{B-out}} = \vec{\mathbf{e}}_\mu\, p'^{\mu}_{\text{B-out}} = \Gamma' m_o c\,(1, -\nu' \cos\theta', -\nu' \sin\theta'). \end{aligned} \tag{8.72}$$

We shall now find the components of the same 4-vectors in the *Lab frame* S which is the rest frame of B. Hence, the boost velocity $-\beta$ is the same as the velocity of B in S', and the boost Lorentz factor γ is the same as the dynamic Lorentz factor $\Gamma\,'$:

$$\beta = \nu\,'; \;\; \gamma = \Gamma\,' = \frac{1}{1-\nu\,'^2} = \frac{1}{\sqrt{1-\beta^2}}. \tag{8.73}$$

First, the *in-coming* components, by LT formulas (7.77a), with boost velocity $-\beta$, applied to the components shown in lines (i), (ii) of Eq. (8.72).

$$\begin{aligned} p^0_{\text{A-in}} &= \gamma(p'^0_{\text{A-in}} + \beta\, p'^1_{\text{A-in}}) = \gamma\Gamma' m_o c\,(1+\beta\times\nu\,') \\ &= \gamma^2(1+\beta^2)\, m_o c, \\ p^1_{\text{A-in}} &= \gamma(p'^1_{\text{A-in}} + \beta\, p'^0_{\text{A-in}}) = \gamma\Gamma' m_o c(\nu' + \beta\times 1) \\ &= (2\gamma^2\beta)\, m_o c, \\ p^2_{\text{A-in}} &= p'^2_{\text{A-in}} = 0; \end{aligned} \tag{8.72 a}$$

$$\begin{aligned} p^0_{\text{B-in}} &= \gamma(p'^0_{\text{B-in}} + \beta\, p'^1_{\text{B-in}}) = \gamma\Gamma' m_o c\,(1+\beta\,(-\nu')) \\ &= \gamma^2(1-\beta^2) m_o c = m_o c, \\ p^1_{\text{B-in}} &= \gamma(p'^1_{\text{B-in}} + \beta\, p'^0_{\text{B-in}}) = \gamma\Gamma' m_o c(-\nu' + \beta\times 1) = 0, \\ p^2_{\text{B-in}} &= p'^2_{\text{B-in}} = 0. \end{aligned} \tag{8.72 b}$$

Next, the *out-going* components, by LT of the components shown in lines (iii), (iv) of Eq. (8.72):

$$\begin{aligned} p^0_{\text{A-out}} &= \gamma(p'^0_{\text{A-out}} + \beta\, p'^1_{\text{A-out}}) = \gamma\Gamma' m_o c\,(1+\beta\nu\,'\cos\theta') \\ &= \gamma^2\,(1+\beta^2\cos\theta')\, m_o c, \\ p^1_{\text{A-out}} &= \gamma(p'^1_{\text{A-out}} + \beta\, p'^0_{\text{A-out}}) = \gamma\Gamma' m_o c\,(\nu\,'\cos\theta' + \beta\times 1) \\ &= \gamma^2\beta(1+\cos\theta')\, m_o c, \\ p^2_{\text{A-out}} &= p'^2_{\text{A-out}} = \Gamma'\nu\,'\sin\theta'\, m_o c, \\ &= \gamma\beta\sin\theta'\, m_o c. \end{aligned} \tag{8.72 c}$$

$$
\begin{aligned}
p^0_{\text{B-out}} &= \gamma(p'^0_{\text{B-out}} + \beta\, p'^1_{\text{B-out}}) = \gamma\Gamma' m_o c\,(1+\beta(-\nu')\cos\theta') \\
&= \gamma^2(1-\beta^2\cos\theta')\, m_o c. \\
p^1_{\text{B-out}} &= \gamma(p'^1_{\text{B-out}} + \beta\, p'^0_{\text{B-out}}) = \gamma\Gamma' m_o c\,(-\nu'\cos\theta' + \beta\times 1) \\
&= \gamma^2\,\beta(1-\cos\theta')\, m_o c. \\
p^2_{\text{B-out}} &= p'^2_{\text{B-out}} \\
&= -\gamma\beta\,\sin\theta' m_o c.
\end{aligned}
\tag{8.72 d}
$$

Compactly,

$$
\begin{aligned}
\vec{\mathcal{P}}_{\text{A-in}} &= m_o c\,(\gamma^2(1+\beta^2),\, 2\gamma^2\beta,\, 0) = m_o c\gamma^2(1+\beta^2)\left(1,\, \frac{2\beta}{1+\beta^2},\, 0\right), \\
\vec{\mathcal{P}}_{\text{B-in}} &= m_o c\,(1,\, 0,\, 0),
\end{aligned}
\tag{8.74}
$$

$$
\begin{aligned}
\vec{\mathcal{P}}_{\text{A-out}} &= m_o c\,(\gamma^2(1+\beta^2\cos\theta'),\, \gamma^2\beta(1+\cos\theta'),\, \gamma\beta\sin\theta') \\
&= m_o c\gamma^2(1+\beta^2\cos\theta')\left(1,\, \frac{\gamma\beta(1+\cos\theta')}{\gamma(1+\beta^2\cos\theta')},\, \frac{\beta\sin\theta'}{\gamma(1+\beta^2\cos\theta')}\right), \\
\vec{\mathcal{P}}_{\text{B-out}} &= m_o c\,(\gamma^2(1-\beta^2\cos\theta'),\, \gamma^2\beta(1-\cos\theta'),\, -\gamma\beta\sin\theta') \\
&= m_o c\gamma^2(1-\beta^2\cos\theta')\left(1,\, \frac{\gamma\beta(1-\cos\theta')}{\gamma(1-\beta^2\cos\theta')},\, -\frac{\beta\sin\theta'}{\gamma(1+\beta^2\cos\theta')}\right).
\end{aligned}
\tag{8.75}
$$

Going back to Eqs. (8.74) and (8.75),

$$
\begin{aligned}
\vec{\mathcal{P}}_{\text{in}} &= \vec{\mathcal{P}}_{\text{A-in}} + \vec{\mathcal{P}}_{\text{B-in}} = 2\gamma^2 m_o c(1,\, \beta,\, 0), \\
\vec{\mathcal{P}}_{\text{out}} &= \vec{\mathcal{P}}_{\text{A-out}} + \vec{\mathcal{P}}_{\text{B-out}} = 2\gamma^2 m_o c(1,\, \beta\,, 0),
\end{aligned}
\tag{8.76}
$$

confirming that $\vec{\mathcal{P}}_{\text{in}} = \vec{\mathcal{P}}_{\text{out}}$ in the frame S.

Step 3: *Find the velocities of the balls before and after the collision.*

Going back to Eqs. (8.74) and (8.75),

$$\Gamma_{\text{A-in}} = \gamma^2(1+\beta^2); \qquad \nu_{\text{A-in}} = \frac{2\beta}{1+\beta^2}; \tag{8.77a}$$

$$\Gamma_{\text{A-out}} = \gamma^2(1+\beta^2\cos\theta'); \quad \nu_{\text{A-out-x}} = \frac{\gamma\beta(1+\cos\theta')}{\gamma(1+\beta^2\cos\theta')}; \tag{8.77b}$$

$$\nu_{\text{A-out-y}} = \frac{\beta\sin\theta'}{\gamma(1+\beta^2\cos\theta')}; \tag{8.77c}$$

$$\Gamma_{\text{B-out}} = \gamma^2(1-\beta^2\cos\theta'); \quad \nu_{\text{B-out-x}} = \frac{\gamma\beta(1-\cos\theta')}{\gamma(1-\beta^2\cos\theta'}); \tag{8.77d}$$

$$\nu_{\text{B-out-y}} = \frac{-\beta\sin\theta'}{\gamma(1-\beta^2\cos\theta')}. \tag{8.77e}$$

We can now find the angles θ and ϕ:

$$\tan\theta = \frac{\nu_{\text{A-out-y}}}{\nu_{\text{A-out-x}}} = \frac{\sin\theta'}{\gamma(1+\cos\theta')} = \frac{1}{\gamma}\tan\frac{\theta'}{2}, \tag{8.78a}$$

$$\tan\phi = \frac{\nu_{\text{B-out-y}}}{\nu_{\text{B-out-x}}} = \frac{\sin\theta'}{\gamma(1-\cos\theta')} = \frac{1}{\gamma}\cot\frac{\theta'}{2}, \tag{8.78b}$$

$$\tan\theta\tan\phi = \frac{1}{\gamma^2}. \tag{8.78c}$$

Note that the above relationship between θ and θ' on the one hand, and ϕ and θ' on the other is consistent with the angle transformation formula given in Eq. (4.133). The exercise to confirm this is left to the reader.

We can now use Eq. (8.77a) to express γ^2 in terms of $\Gamma_{\text{A-in}}$ using the identity $\gamma^2 - 1 = \beta^2\gamma^2$:

$$\begin{aligned} \Gamma_{\text{A-in}} &= 2\gamma^2 - 1 \\ \therefore \quad \gamma^2 &= \frac{\Gamma_{\text{A-in}}+1}{2}, \\ \text{where } \Gamma_{\text{A-in}} &= \frac{1}{\sqrt{1-\nu_{\text{A-in}}^2}}. \end{aligned} \tag{8.79}$$

We have now a more convenient equation for the angles in terms of the velocity of the impinging particle A.

$$\tan\theta\tan\phi = \frac{2}{\Gamma_{\text{A-in}}+1}. \tag{8.80}$$

The reader should work out the exercise **R1** in Sec. 8.1 to get a full appreciation of the equations presented in this section.

8.9.2. *Example 2: Threshold energy for a $p+p$ collision resulting in the production of a π^0 particle*

We shall consider the following example[a]:

$$p_1 + p_2 \rightarrow p_1' + p_2' + \pi^0.$$

The left side represents colliding protons # 1 and # 2, before collision, and the right side the same protons after collision. The proton p_2 represents a stationary target (hydrogen atom), and the proton p_1 the bombarding particle. There is a certain minimum kinetic energy K_0, called the *threshold energy*, above which the above reaction can take place. We shall determine this energy, using the ZM frame as a stepping stone.

We have shown the reaction in Fig. 8.4. In the Lab frame, the target proton p_2 is stationary before the collision. In the ZM frame, the target proton p_2 and the bombarding proton p_1 are moving with opposite velocities βc and $-\beta c$ before the collision and settle down to rest along with the new born π^0 after the collision.

We shall adopt the following *rest mass* values: $m_\pi = 135$ MeV, $m_\mathrm{p} = 938.26$ MeV. Now apply energy conservation in the ZM frame:

$$\begin{aligned} 2\Gamma m_\mathrm{p} &= 2m_\mathrm{p} + m_\pi, \\ \text{or,} \quad \Gamma &= 1 + \frac{m_\pi}{2m_\mathrm{p}} = 1 + \frac{135.0}{2 \times 938.26} = 1.0719. \\ \text{Now,} \quad \beta^2 &= 1 - \frac{1}{\Gamma^2} = 0.13. \\ \text{Hence,} \quad \beta &= 0.36. \end{aligned} \tag{8.81}$$

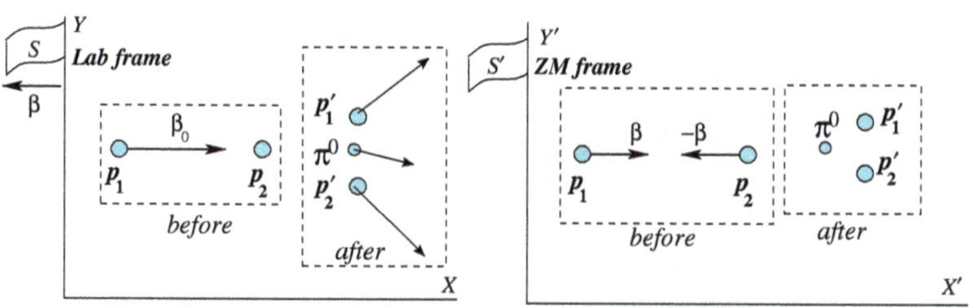

Fig. 8.4. Nuclear reaction $p+p \rightarrow p+p+\pi^0$ seen from two frames.

[a]See [25, p. 186]. For the actual reaction, see [37, p. 689].

To get the velocity β_0 of the bombarding particle p_1 in the Lab frame, we shall apply the velocity transformation formula (4.7), to go from the S' (ZM) frame to the S (Lab) frame.

$$\beta_0 = \frac{\beta + \beta}{1 + \beta \times \beta} = \frac{2 \times 0.36}{1 + 0.36 \times 0.36} = 0.637. \tag{8.82}$$

The corresponding dynamic Lorentz factor is

$$\Gamma_0 = \frac{1}{\sqrt{1 - {\beta_0}^2}} = \frac{1}{\sqrt{1 - 0.637^2}} = 1.3. \tag{8.83}$$

The kinetic energy of p_1 should be

$$K_0 = (\Gamma_0 - 1)m_p c^2 = 0.3 \times 936.26 = 281.478\,\text{MeV}. \tag{8.84}$$

This much is the minimum kinetic energy to be given to the bombarding proton to produce a mass which is less than half this value. The remaining half is shared by the three particles emerging from the reaction.

8.9.3. *Example 3: Threshold energy for a photon hitting an electron to produce an electron–positron pair*

We shall consider an example of $(e_+ e_-)$ pair production by photons[b]:

$$\gamma + e^- \to e^- + e^+ + e^-.$$

This example is similar to Example 2, except that a gamma ray, which is a particle of zero rest mass, now plays the role of the bombarding particle. We have made a picture of this reaction in Fig. 8.5.

Let $(\mathcal{E}_0, \mathcal{E})$ represent the photon energy in the ZM frame and the Lab frame, respectively, and similarly (Γ_0, Γ) the dynamic Lorentz factor of the

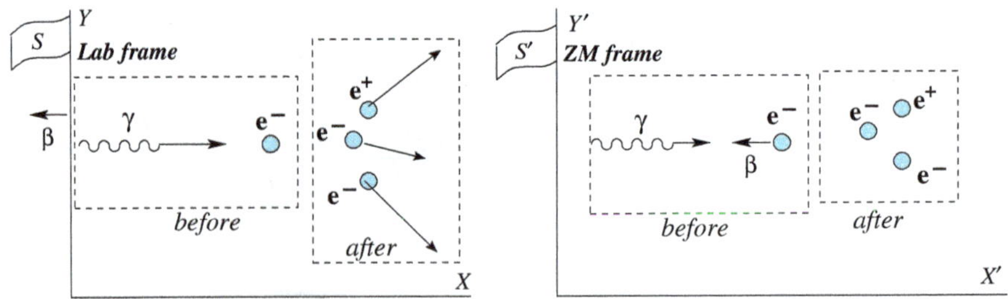

Fig. 8.5. Nuclear reaction $\gamma + e^- \to e^- + e^+ + e^-$ seen from two frames.

[b]See Ref. [26] and Problem 7-2 on p. 225 in Ref. [25].

electron in the same two frames, and let m be the rest mass of the electron. Let $-p_0$ be the electron momentum in ZM. Now apply the conservation equations in the ZM frame:

$$\text{Momentum}: \quad \mathcal{E}_0/c = p_0 = \Gamma_0 \beta mc.$$

$$\text{Energy}: \mathcal{E}_0 + \Gamma_0 mc^2 = 3mc^2.$$

$$\text{Hence,} \quad 3 - \Gamma_0 = \Gamma_0 \beta,$$

$$\text{or} \quad \Gamma_0(1+\beta) = \frac{1+\beta}{\sqrt{1-\beta^2}} = 3 \tag{8.85}$$

$$\text{from which} \quad \beta = \frac{4}{5}; \Gamma_0 = \frac{3}{1+\frac{4}{5}} = \frac{5}{3}.$$

$$\mathcal{E}_0 = p_0 c = \Gamma_0 \beta mc^2 = \frac{5}{3} \times \frac{4}{5} mc^2 = \frac{4}{3} mc^2.$$

Now consider the boost:$S'(-\beta)S$ from the ZM frame to the Lab frame. Apply the 4-momentum transformation (8.48) to the energy of the photon. In this case, the boost velocity β is the same as the electron velocity β in the ZM frame, and the boost Lorentz factor γ_{boost} is the same as Γ_0. Note that we have added a subscript to distinguish the boost Lorentz factor from the gamma ray:

$$\frac{\mathcal{E}}{c} = \gamma_{\text{boost}} \left[\frac{\mathcal{E}_0}{c} + \beta p_0\right] = \Gamma_0 \left[\frac{\mathcal{E}_0}{c} + \beta p_0\right],$$

$$\text{or } \mathcal{E} = \frac{5}{3} \times \left(\frac{4}{3} + \frac{4}{5} \times \frac{4}{3}\right) mc^2 = \left(\frac{5}{3} \times \frac{4}{3} \times \frac{9}{5}\right) mc^2 = 4\,mc^2 \tag{8.86}$$

$$= 4 \times 0.511 = 2.044\,\text{MeV}.$$

As in the previous example, we need a photon of minimum energy $4mc^2$ to produce an electron–positron pair of rest mass $2mc^2$. The balance $2mc^2$ goes to the kinetic energies of the pair (e^-, e^+) and the bombarding particle e^- emerging after the impact.

8.9.4. *Example 4: Compton scattering and inverse compton scattering*

1. (a) Consider a collision between two particles A and B, moving with velocities $\mathbf{u}_a$ and $\mathbf{u}_b$, their respective *rest* masses being m_{oa} and m_{ob}.

Their 4-velocities are $\overrightarrow{U_a}$ and $\overrightarrow{U_b}$. Show that

$$\overrightarrow{U_a} \cdot \overrightarrow{U_b} = c^2\Gamma(\beta), \tag{8.87}$$

where $\Gamma(\beta)$ is the Lorentz factor corresponding to relative velocity $c\beta$ between A and B.
Hint: Since the quantity is Lorentz invariant, evaluate the product in the frame of reference in which one of the particles, say B, is at rest. Now write the 4-velocities in this frame.
(b) Let their 4-momenta be $\overrightarrow{P_a}$ and $\overrightarrow{P_b}$. Show that

$$\overrightarrow{P_a} \cdot \overrightarrow{P_b} = c^2\Gamma(\beta) m_{oa} m_{ob}. \tag{8.88}$$

(c) Let the particle A represent a photon propagating with frequency ν_b in the rest frame of B. Show that

$$\overrightarrow{P_a} \cdot \overrightarrow{P_b} = h\nu_b m_{ob}. \tag{8.89}$$

(d) Suppose both the particles are photons. In the Lab frame, their frequencies are ν_a and ν_b, and the angle between their paths of propagation is θ. Show that

$$\overrightarrow{P_a} \cdot \overrightarrow{P_b} = \left(\frac{h}{c}\right)^2 \nu_a \nu_b (1 - \cos\theta). \tag{8.90}$$

2. Consider the elastic collision $\mathrm{A} + \mathrm{B} \to \mathrm{A}' + \mathrm{B}'$. Let A, A' represent a photon and B, B' a subatomic charged particle (e.g. an electron, a proton) before and after the scattering. The photon is propagating in the directions $\mathbf{n}$ and $\mathbf{n}'$ before and after collision. We write their 4-momenta, and the "magnitudes" of the corresponding 4-momenta, as defined in Sec. 7.9.4, and written in Eqs. (7.110), (8.19), (8.22), (8.26), (8.27):

$$\begin{aligned} &\overrightarrow{\mathcal{P}}_\gamma = (h\nu/c)(1, \mathbf{n}),\ \overrightarrow{\mathcal{P}}'_\gamma = (h\nu'/c)(1, \mathbf{n}'),\ (\overrightarrow{\mathcal{P}}_\gamma)^2 = (\overrightarrow{\mathcal{P}}'_\gamma)^2 = 0, \\ &\overrightarrow{\mathcal{P}}_e = m_o\Gamma(c, \mathbf{u}), \quad \overrightarrow{\mathcal{P}}'_e = m_o\Gamma'(c, \mathbf{u}'), \quad (\overrightarrow{\mathcal{P}}_e)^2 = (\overrightarrow{\mathcal{P}}'_e)^2 = m_o^2 c^2. \end{aligned} \tag{8.91}$$

From conservation of 4-momenta:

$$\begin{aligned} \overrightarrow{\mathcal{P}}_\gamma + \overrightarrow{\mathcal{P}}_e &= \overrightarrow{\mathcal{P}}'_\gamma + \overrightarrow{\mathcal{P}}'_e, \\ \text{or } \overrightarrow{\mathcal{P}}_\gamma + \overrightarrow{\mathcal{P}}_e - \overrightarrow{\mathcal{P}}'_\gamma &= \overrightarrow{\mathcal{P}}'_e \end{aligned} \tag{8.92}$$

Square either side and use Eq. (8.91)

$$\overrightarrow{\mathcal{P}}_\gamma \cdot \overrightarrow{\mathcal{P}}'_\gamma = \overrightarrow{\mathcal{P}}_e \cdot (\overrightarrow{\mathcal{P}}_\gamma - \overrightarrow{\mathcal{P}}'_\gamma), \tag{8.93a}$$

$$\text{or,} \quad \frac{h^2\nu\nu'}{c^2}(1 - \mathbf{n}\cdot\mathbf{n}') = \frac{m_o\Gamma h}{c}[(\nu - \nu')c - \mathbf{u}\cdot(\nu\mathbf{n} - \mathbf{n}'\nu')]. \tag{8.93b}$$

Apply formula (8.93) to *Compton scattering*. Take $\mathbf{u} = \mathbf{0}$, the initial velocity of the electron. Now obtain the formula (4.124), which had derived in Chapter 4.

Ans. Here $\Gamma = 1$. Also $\mathbf{n} \cdot \mathbf{n}' = \cos\theta$, where θ is the angle between the initial and the final directions of the photon. Inserting these values, we now get

$$\begin{aligned} \frac{h^2\nu\nu'}{c^2}(1-\cos\theta) &= m_o h(\nu-\nu'), \\ \text{or} \quad \frac{m_o h(\nu-\nu')c^2}{h^2\nu\nu'} &= (1-\cos\theta). \end{aligned} \tag{8.94}$$

Now

$$\frac{c\,(\nu-\nu')}{\nu\nu'} = \frac{c}{\nu'} - \frac{c}{\nu} = \lambda' - \lambda. \tag{8.95}$$

Hence,

$$\lambda' - \lambda = \frac{h}{m_o c}(1-\cos\theta). \tag{8.96}$$

We get back the same Compton scattering formula (4.124), which had derived in Chapter 4 using a long route.

3. Apply formula (8.93) to *inverse Compton scattering*. An electron moving along the negative x-axis, with a high velocity $\mathbf{u} = -u\,\mathbf{e}_x$, collides with a photon of energy $h\nu$ proceeding along the positive X-axis. After the collision the photon bounces back to the negative X-axis with energy $h\nu'$. Find an expression for the energy of the outgoing photon

Ans. Write the three 4-vectors involved in this problem, keeping the (time, x) components only.

$$\vec{\mathcal{P}}_\gamma = (h\nu/c)(1,1);\ \vec{\mathcal{P}}'_\gamma = (h\nu'/c)(1,-1);\ \vec{\mathcal{P}}_e = m_o\Gamma c(1,-u/c). \tag{8.97}$$

We shall rewrite the formula (8.93) by setting $\mathbf{n} \cdot \mathbf{n}' = -1$, $\mathbf{u} \cdot \mathbf{n} = -u$, $\mathbf{u} \cdot \mathbf{n}' = u$.

$$\begin{aligned} \frac{h^2\nu\nu'}{c^2}(1-\mathbf{n}\cdot\mathbf{n}') &= \frac{m_o\Gamma h}{c}[(\nu-\nu')c - \mathbf{u}\cdot(\nu\mathbf{n} - \mathbf{n}'\nu')], \\ \text{or} \quad 2\frac{h^2\nu\nu'}{c^2} &= \frac{m_o\Gamma h}{c}[(\nu-\nu')c + (\nu+\nu')u] \\ &= \frac{m_o\Gamma h}{c}[(c+u)\nu - (c-u)\nu'] \\ &= (m_o\Gamma h)\left[(1+u/c)\nu - (1-u/c)\nu'\right]. \end{aligned} \tag{8.98}$$

Now, in inverse Compton scattering the velocity of the incoming particle is very high, almost equal to the speed of light. We can therefore make the following approximations:

$$(1 + u/c) \approx 2.$$

Also,
$$\frac{1}{\Gamma^2} = 1 - \frac{u^2}{c^2} = (1 - u/c)(1 + u/c) \approx 2(1 - u/c) \tag{8.99}$$

$$\therefore \quad 1 - u/c \approx \frac{1}{2\Gamma^2}.$$

We now go back to (8.98) and write:

$$\begin{aligned} 2\frac{h^2\nu\nu'}{c^2} &\approx 2(m_\circ\Gamma h)\left[\nu - \frac{1}{4\Gamma^2}\nu'\right] \\ \times \frac{c^2}{2(h\nu)^2} &\Rightarrow \frac{\nu'}{\nu} = \frac{\Gamma m_\circ c^2}{h\nu}\left[1 - \frac{1}{4\Gamma^2}\frac{\nu'}{\nu}\right], \\ \text{or } &\left(1 + \frac{m_\circ c^2}{4\Gamma\, h\nu}\right)\frac{\nu'}{\nu} = \frac{\Gamma m_\circ c^2}{h\nu}. \end{aligned} \tag{8.100}$$

Let us write the left-hand side as

$$\text{l.h.s} = \frac{m_\circ c^2}{4\Gamma\, h\nu}\left(\frac{4\Gamma\, h\nu}{m_\circ c^2} + 1\right). \tag{8.101}$$

Then

$$\frac{\nu'}{\nu} = \frac{4\Gamma^2}{\left(\frac{4\Gamma\, h\nu}{m_\circ c^2} + 1\right)}. \tag{8.102}$$

We can also write this in terms of the initial and final photon energies.

$$\frac{\mathcal{E}'}{\mathcal{E}} = \frac{4\Gamma^2}{\left(\frac{4\Gamma\,\mathcal{E}}{m_\circ c^2} + 1\right)}. \tag{8.103}$$

4. Consider the case in which the incoming electron energy is 10^3 MeV (about 2000 times its rest mass). The initial photon energy is almost the same as (or more than) the energy of the electron. Show that the incoming electron transfers almost all of its energy to the γ-ray. (Taylor and Wheeler [6, Example 8-23, p. 270])

Ans. In this case $\Gamma \simeq 2000$, so that "+1" in the denominator can be ignored. Using this approximation in (8.103) we get the energy of the back scattered X-ray:

$$\mathcal{E}' \approx \Gamma m_o c^2, \tag{8.104}$$

which is the energy of the incoming electron.

5. Take a photon of Cosmic Microwave Background, having photon energy $h\nu$ equal to 10^{-3} eV. A *proton* in the Cosmic ray of kinetic energy 10^{14} MeV collides with it. Proton rest mass energy is 939 MeV. For simplicity take it as 10^3 MeV. Find the energy of the photon after the collision (See [4, p.123]).

Ans. We shall use Eq. (8.103), but replace m_o with m_p, the mass of a proton. In this case

$$\begin{aligned} (\Gamma - 1)m_p c^2 &= 10^{14}, \\ \text{or,} \qquad \Gamma - 1 &\approx 10^{11}. \\ \text{Take} \qquad \Gamma &= 10^{11}. \end{aligned} \tag{8.105}$$

The denominator in (8.103) is

$$\text{denom} = \frac{4\Gamma\,\mathcal{E}}{m_p c^2} + 1 = \frac{4 \times 10^{11} \times 10^{-3}}{10^9} + 1 = 1.4. \tag{8.106}$$

Hence,

$$\mathcal{E}' = \frac{4 \times 10^{22}}{1.4} \times 10^{-3} = 2.9 \times 10^{19}\ \text{eV} = 2.9 \times 10^{13}\ \text{MeV}. \tag{8.107}$$

The extremely "weak" photon gains an enormous energy of $\simeq 10^{13}$ MeV from the impact with the cosmic ray proton.

8.9.5. *Example 5: Doppler effect*

Example 1. A fast moving space station is moving with speed βc along a straight line and is emitting light at a frequency f_0. An observer O is located at a distance h from the straight line shown in Fig. 8.6. We are required to calculate the frequency f of the same light as detected by the observer as function of time.

The first thing to remember is that when an observer O receives light from a *moving* source S at time t, this light comes not from the present location C of the source, but from the *retarded location* B of the source,

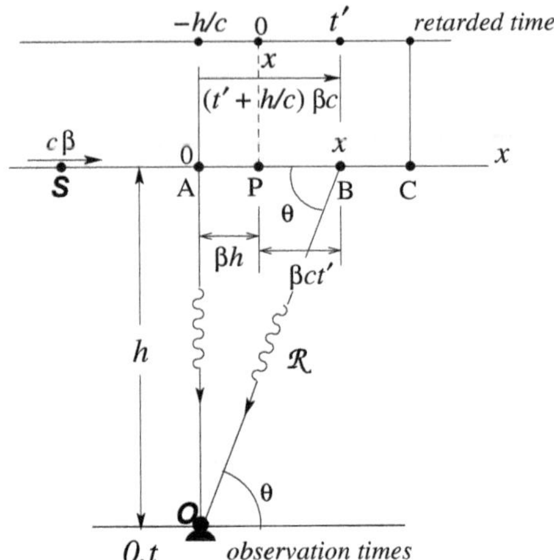

Fig. 8.6. Source moving at impact parameter b.

at the coordinate x, and emitted at the *retarded time* t', as explained in Fig. 8.6. These two times are related by the equation:

$$c\,(t - t') = \mathcal{R}, \tag{8.108}$$

where $\mathcal{R}$ is the distance between the source S and the observer O at the present time t. When we use the Doppler formula (3.31)

$$f = \frac{f_0}{\gamma(1 + \beta\cos\theta)}, \tag{8.109}$$

the angle θ is the angle at which the source was located at the retarded time t'. That is

$$\cos\theta = \frac{x}{R}. \tag{8.110}$$

Our first task is to find a relationship between x and t. It is assumed that the source is moving with speed $c\beta$ along the X-axis. We set the origin of the X-axis at A, vertically above O, and set time $t = 0$ when light emitted from A is received at O. We shall find the solution in several steps.

Step 1: Find t'_A, i.e. the time when the source was at A.

In formula (8.108) set $t = 0,\ t' = t'_A,\ \mathcal{R} = h$.

$$c\,t'_A = -h. \tag{8.111}$$

Step 2: Find a relation between the time ct' and the displacement x.

Referring to Fig. 8.6, S was at A, when $ct' = ct'_A = -h$, at P when time $ct' = 0$, at B when time $ct' = ct'$. S is moving with speed $c\beta$. Referring to Fig. 8.6

$$\begin{aligned} x &= \widehat{AB} = \widehat{AP} + \widehat{PB} = \beta h + \beta ct'. \\ \therefore ct' &= \frac{x}{\beta} - h. \end{aligned} \tag{8.112}$$

Step 3: Find a relation between t and x.

From (8.108a) and (8.112)

$$ct = r + ct' = r + \frac{x - h\beta}{\beta} = \sqrt{h^2 + x^2} + \frac{x - h\beta}{\beta}, \tag{8.113}$$

which is a quadratic equation. To get the roots we simplify the above equation by setting $h = 1$, $\chi \equiv ct$ to the form:

$$x^2(1 - \beta^2) - 2x\beta\,(\chi + 1) + \beta^2\,\chi\,(\chi + 2) = 0, \tag{8.114}$$

$$x = -\gamma^2\beta(\chi + 1) \times \left\{ -1 \pm \sqrt{1 - (1 - \beta^2)\left(1 - \frac{1}{(\chi + 1)^2}\right)} \right\}. \tag{8.115}$$

If we set $\chi = 0$, corresponding to $t = 0$, we get two solutions:

$$x = -\gamma^2\beta \times (-1 \pm 1) = 0, 2\gamma^2\beta. \tag{8.116}$$

The first solution $x = 0$ corresponds to light originating from the emitter at $t = -h/c$ (*retarded location*) and coming vertically down to the observer at $t = 0$. The second solution corresponds to light starting from the observer at a $t = 0$ and reaching the emitter at $x = 2\gamma^2\beta$ (*advanced location*) after time R/c. here $R^2 = h^2 + 2\gamma^2\beta^2$. We shall adopt first solution (i.e. with the + sign).

Now we go back to the Doppler formula (8.109) and set

$$\begin{aligned} \cos\theta &= \frac{x}{\sqrt{x^2 + h^2}}, \text{ with } h = 1; \\ \tilde{\mathrm{r}} &= \frac{f}{f_0}. \end{aligned} \tag{8.117}$$

Before we plot the frequency ratio as a function of time, we shall specialize Eq. (8.109) for two specific cases: $\theta = \pi/2$ corresponding to *transverse Doppler effect*, and $\theta = 0$, corresponding to *longitudinal Doppler effect.*

We collect the formulas for these two cases from (3.32)

$$\tilde{\mathrm{r}}_{\text{trans}} = \frac{1}{\gamma}, \qquad \tilde{\mathrm{r}}_{\text{long}} = \frac{1}{\gamma(1+\beta)} = \sqrt{\frac{1-\beta}{1+\beta}}. \tag{8.118}$$

Before plotting the frequency ratios for three values of the emitter velocity, namely $\beta = 0.8, 0.9, 0.95$, we shall obtain numerical values of the above ratios so that we can check the same values are obtained from the plots.

$$\tilde{\mathrm{r}}_{\text{trans}} = 0.6 \text{ for } \beta = 0.8, \quad 0.435 \text{ for } \beta = 0.9, 0.312 \text{ for } \beta = 0.95, \qquad \tilde{\mathrm{r}}_{\text{long}} = 0.333 \text{ for } \beta = 0.8, 0.229 \text{ for } \beta = 0.9, 0.160 \text{ for } \beta = 0.95. \tag{8.119}$$

In Fig. 8.7, we have plotted the frequency ratios as a function of the *observation time*. The above values are clearly reflected in the plots.

We shall add a numerical "feel" of the results obtained, assuming the light is emitted by a sodium gas vapour lamp of which the wavelength is

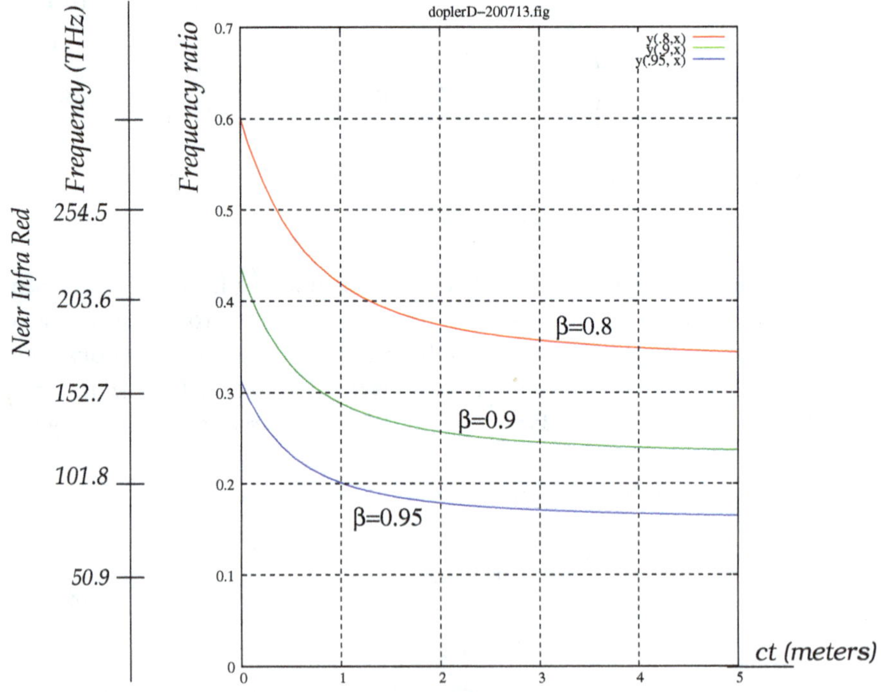

Fig. 8.7. Frequency ratio as a function of observation time.

5890 Å. The corresponding frequency is $f_0 = 5.09 \times 10^{14}$ Hz. The observed frequencies will then be $f = \tilde{\Gamma} \times 5.09 \times 10^{14}$ Hz $= \tilde{\Gamma} \times 509$ THz. The observed light falls in the Near Infra Red region, as shown in Fig. 8.7.

8.10. Exercises for the Reader II

R1 Consider elastic collision between two billiard balls A and B as in Fig. 8.3. The velocity of the incoming ball A is $v_{\text{in}} = \frac{4}{5}c$. Let the scattering angle (i.e. of the ball A after the collision) be $\theta = 30°$. Find the following quantities:

(a) The dynamic Lorentz factor of the incoming ball A.
(b) The boost velocity $c\beta$ (of the ZM frame with respect to the Lab frame), and the boost Lorentz factor γ.
(c) The bouncing angle ϕ of the ball B after the collision.
(d) The 3-momenta $\mathbf{p}_{\text{A-in}}$, $\mathbf{p}_{\text{A-out}}$, $\mathbf{p}_{\text{B-in}}$, $\mathbf{p}_{\text{B-out}}$ of the balls A and B before and after the collision.
(e) The total energies $\mathcal{E}_{\text{A-in}}$, $\mathcal{E}_{\text{A-out}}$, $\mathcal{E}_{\text{B-in}}$, $\mathcal{E}_{\text{B-out}}$ of the balls A and B before and after the collision.
(f) The En-Menta of the system of two balls before and after the collision.
(g) The rest mass M_0 of the system of two balls in the Lab frame S.

Ans. (a) $\Gamma_{\text{A-in}} = \frac{5}{3}$, (b) $\beta = \frac{1}{2}$, $\gamma = \frac{2}{\sqrt{3}}$, (c) $\phi = 52.4°$,

(d)

$$\begin{aligned}
\mathbf{p}_{\text{A-in}} &= m_o c\,(1.33,\ 0),\\
\mathbf{p}_{\text{B-in}} &= m_o c\,(0,\ 0),\\
\mathbf{p}_{\text{A-out}} &= m_o c\,(0.92,\ 0.36),\\
\mathbf{p}_{\text{B-out}} &= m_o c\,(0.41,\ -0.36).
\end{aligned} \tag{8.120}$$

(e)

$$\begin{aligned}
\mathcal{E}_{\text{A-in}} &= 1.67\, m_o c^2,\\
\mathcal{E}_{\text{B-in}} &= m_o c^2,\\
\mathcal{E}_{\text{A-out}} &= 1.46\, m_o c^2,\\
\mathcal{E}_{\text{B-out}} &= 1.20\, m_o c^2.
\end{aligned} \tag{8.121}$$

(f)

$$\vec{\mathcal{P}}_{\text{in}} = 2.66\, m_o c(1,\, 0.5,\, 0),$$
$$\vec{\mathcal{P}}_{\text{out}} = 2.66\, m_o c(1,\, 0.5\,, 0). \quad (8.122)$$

(g) $M_0 = 2.66\, m_o$.

R2 Consider the same scattering as in the previous exercise. Assume symmetric scattering of the two balls after collision, so that $\phi = \theta$. Let the dynamic Lorentz factor of the incoming particle be Γ. Show that the balls will bounce out with an angle Θ between them given by the formula:

$$\cos\Theta = \frac{K}{K + 4mc^2}, \quad (8.123)$$

where K represents the kinetic energy of the incoming particle A.

R3 A high-pressure sodium lamp contains vapourized sodium at a temperature of 2700 K. The sodium molecules are in random thermal motion with an average kinetic energy of $\frac{3}{2}k\text{T}$, where $k = 1.38 \times 10^{-23}$ J/K = Boltzmann constant. A young scientist analyzes the spectral lines of the light emitted by the gas. Take the mass of a sodium molecule as $m = 46$ u. The wavelength of the sodium line is 5890 Å:

(a) Find the average velocity of a sodium molecule.
(b) Find the broadening of a typical spectral line.

Note that you will get the same answer if you apply the non-relativistic formula.

Ans. (a) $5.7 \times 10^{-6}c$. (b) 0.065 Å.

R4 Consider the decay of a Λ-particle (cited earlier in the first problem in Sec. 4.11). Initially moving with a relativistic speed, it decays into particle #1 (proton) and # 2 (pion), leaving an angle θ between their tracks, as seen in a bubble chamber. Let P^μ, p_1^μ, p_2^μ represent the En-Menta of Λ, p and π^+, respectively. Using conservation of En-Mentum and invariance of its norm, show that

$$M^2c^4 = m_1^2c^4 + m_2^2c^4 + 2E_1E_2 - 2p_1p_2c^2\cos\theta, \quad (8.124)$$

where $p_i = |\mathbf{p}^i|$, $i = 1, 2$.

[Hint: Simplify both sides of the equation: $P^\mu P_\mu = (p_1^\mu + p_2^\mu)(p_{1\mu} + p_{2\mu})$.]

Chapter 9

Relativistic Rocket

9.1. Introduction

We shall present an example of how Minkowski's equation of motion works by demonstrating its application on a relativistic rocket. A relativistic rocket, in principle and for all theoretical calculations, is the same familiar rocket that the students have studied in their mechanics books,[a] with the difference that the exhaust gas is ejected with a "relativistic speed" u and, as a consequence, the rocket accelerates to a relativistic speed in due time. What we call relativistic speed is roughly the range: $c/3 \leq u \lesssim c$, where c is the speed of light. Because of the relativistic velocities involved in this case Newtonian mechanics breaks down, and we have to use Minkowskian mechanics, in particular Minkowskian equation of motion (EoM).

A relativistic rocket, i.e. a space-ship propelled by ejected gas to relativistic speed is not a reality [28]. However, one can still think of matter–antimatter annihilation rockets, and pion rockets for intellectual entertainment [29]. Even then the purpose behind our spending time on such an object is somewhat pedagogical. The exercises we are going to undertake are intended to sharpen ones understanding of Minkowskian equation of motion, employing 4-vectors.

We have derived the mass equation for a relativistic rocket (see Eq. (9.8)) using momentum–energy conservation principles [30–33].

[a]See [7, pp. 327–330]; [8, pp. 84–87, 144–147, 315–319, 328–332].

Some features of this chapter that may kindle a special interest in a student or a teacher of special relativity are as follows:

- We have subjected the two important equations derived in this chapter, namely, (a) the mass velocity equation (9.8), and (b) the EoM (9.29) to the *N.R. test*, by which we mean that all relativistic equations that have a non-relativistic (N.R.) analogue must converge to their N.R. counterparts when $v \ll c$.
- Taking u as the ejection velocity of the emitted gas/radiation, we have obtained two special solutions of the EoM, corresponding to (i) $u = c/3$ and (ii) $u = c$. We have plotted the velocity–time relation for both the cases, and shown that the plot for the case (i) closely follows the plot for the corresponding formula for $v = v(t)$ obtained using N.R. (Newtonian) mechanics, up to $v \approx 0.5c$.
- We have adopted a four-dimensional Minkowskian approach to obtain the EoM of the rocket, using 4-vectors, e.g. 4-velocity, 4-momentum, 4-acceleration, 4-force. For this purpose, we have adopted a mathematical formalism outlined by Moller [34].

Since the motion of the rocket will be one dimensional, confined to the x-direction, a typical 4-vector will have only t- and x- components and will be written as $\overrightarrow{\mathbf{A}} = (A^t, A^x)$.

9.2. The Rocket, Its Specifications

Let us now take a look at the rocket of our discussion. We have illustrated it in Fig. 9.1. It is moving along the X-axis with velocity $v(t)\,\text{m/s}$ with respect to an inertial frame S, which, for fixing the idea, we shall call the *ground frame* (GF). It is ejecting gas at a constant velocity $-u\,\text{m/s}$ and its *rest mass* at a constant rate $r = \frac{d\mu_o}{d\tau}$ kg/s, relative to its *instantaneous rest frame* (IRF) $S_o(\Theta)$, thereby generating a reaction force (in this case a thrust force) $\overrightarrow{\mathcal{R}}$. Our purpose is to find a formula for $\overrightarrow{\mathcal{R}}$, and then the equation of motion (EoM).

Note that we have labelled the IRF with the extra tag (Θ) to stress that it coincides with the rocket frame R at the event "Θ", which, for fixing the idea can be taken as "Θ: rocket passes a space station A". Every IRF has to be associated with one, and only one, event "Θ".

Three quantities are specified for the assessing the performance of the rocket: u, r and $M_i \equiv$ initial *rest* mass of the rocket at the instant $t = 0$, when it starts with zero velocity. In this chapter, $M = M(\Theta)$ will stand for

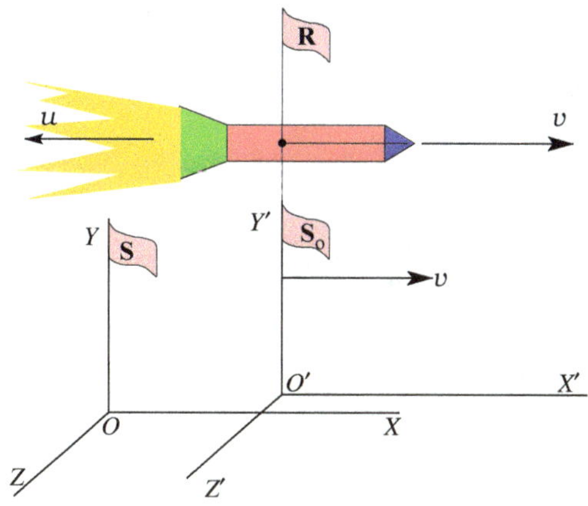

Fig. 9.1. The rocket.

the *instantaneous rest mass* of the rocket at the event Θ. When written as a function of the "ground time" t, it will appear as $M(t)$.

Let us consider two infinitely close events Θ_A and Θ_B (corresponding to the rocket passing two infinitely close space stations A and B on its path), the time–space coordinate differentials between them being $(c\,\delta t,\, \delta x)$ with respect to S, and $(c\,\delta\tau,\, 0)$ with respect to S_o. Between these events the rocket ejects a quantity of gas of *rest mass* $\delta\mu_{\text{o}}$. Consequently, its own velocity changes (i) from $v(t)$ to $v(t)+\delta v$ with respect to the GF S, (ii) from 0 to dv' with respect to S_o, and (iii) its *rest mass* changes from $M(t)$ to $M(t) + \delta M$. Note that the time differential between the events being infinitesimally small, $\delta\tau$ is the *proper time* between the events. Also, note that, the rate of emission of the *rest gas mass* with respect to the rocket frame is $r = \frac{d\mu_{\text{o}}}{d\tau} = \lim_{\delta\to 0} \frac{\delta\mu_{\text{o}}}{\delta\tau}$, which is taken as a constant.

In summary, the performance of the rocket is decided by three specifications: (i) its initial mass M_i, (ii) the rate $r = \frac{d\mu_{\text{o}}}{d\tau}$ at which *rest mass* is ejected from the rear end with respect to the IRF, (iii) the speed $-u$, with respect to the IRF, with which this rest mass is ejected. These quantities, being in the specification book supplied by the manufacturer, are frame independent and are to be taken as constants.

9.3. Review of the Non-relativistic Results

We shall briefly review the non-relativistic (N.R.) rocket formulas so that we can compare the relativistic results with their N.R. counterparts.

We shall drop the subscript "$_o$" from $d\mu_o$, because in the N.R. zone there is no such thing as proper mass. The N.R. formulas can be found in standard books on mechanics. We shall quote the following formulas [8]:

$$\delta v = -u\frac{\delta M}{M}, \tag{9.1a}$$

$$v = u\ln\left(\frac{M_i}{M(v)}\right); \quad \text{or,} \quad \frac{M(v)}{M_i} = e^{-\frac{v}{u}}, \tag{9.1b}$$

$$T = \text{thrust force} = ru, \tag{9.1c}$$

$$r = \frac{d\mu}{dt} = -\frac{dM}{dt}, \tag{9.1d}$$

$$M(v)\frac{dv}{dt} = T = ru, \quad \text{or, using (9.1b)} \quad M_i e^{-\frac{v}{u}}\frac{dv}{dt} = ru, \tag{9.1e}$$

$$v = -u\,\ln\left(1 - \frac{rt}{M_i}\right). \tag{9.1f}$$

In (9.1b), $M(v)$ is the same as $M(t)$, since $v = v(t)$. Equation (9.1d) in which $d\mu$ is the mass of the gas ejected in time dt, is a restatement of conservation of mass. The relationship between the velocity differential and mass differential shown in Eq. (9.1a) is a consequence of (i) conservation of mass, and (ii) conservation of linear momentum. The mass ratio equation (9.1b) is obtained by integrating the differentials in (9.1a). Equation (9.1e) represents the EoM of the rocket. Equation (9.1f) gives the solution of the EoM, subject to the initial condition: $v = 0$ when $t = 0$.

9.4. Relativistic Mass Equation

We now have the following Lorentz factors, corresponding to the velocities to be used:

$$g(u) = \frac{1}{\sqrt{1 - \frac{u^2}{c^2}}}, \quad \Gamma(v) = \frac{1}{\sqrt{1 - \frac{v^2}{c^2}}}, \quad \Gamma'(v') = \frac{1}{\sqrt{1 - \frac{v'^2}{c^2}}}. \tag{9.2}$$

As stated above, the rocket velocity changes from 0 to $\delta v'$, and it ejects a quantity of gas of *rest mass* $\delta\mu_o$ from the event Θ_A to the event Θ_B in the IRF S_o. We shall write the components of the 4-momentum of the rocket at Θ_A and Θ_B, and of the gas ejected between these events — all of them in S_o.

At this point, we emphasize once again that $M = M(\Theta) = M(t) = M(\tau)$ is the instantaneous rest mass of the rocket at the event Θ and hence, is a *4-scalar*.

The (t, x)-components of the momentum 4-vectors we shall write below will follow from (8.17), in which we shall set $m_{\circ} = M(t)$. Also, note that the dynamic Lorentz factors are: g for the ejected gas, and $\Gamma' = 1$ for the rocket, since $v' = 0$, i.e. the rocket is momentarily at rest in S_o. The 4-vectors written below have only (t, x)-components, and are valid in the IRF S_o:

$$\text{At } \Theta_A : \quad \overrightarrow{\mathcal{P}} = M(c, 0). \tag{9.3a}$$

$$\text{At } \Theta_B : \overrightarrow{\mathcal{P}} + \delta\overrightarrow{\mathcal{P}} = (M + \delta M)(c, \delta v'). \tag{9.3b}$$

$$\delta\overrightarrow{\mathcal{P}} = (\delta M c, M\delta v'). \tag{9.3c}$$

$$\delta\overrightarrow{\mathbf{p}} = \delta\mu_{\text{o}}\, g(c, -u). \tag{9.3d}$$

In the above $\delta\overrightarrow{\mathcal{P}}$ stands for the change in 4-momentum of the rocket, and $\delta\overrightarrow{\mathbf{p}}$ for the 4-momentum of the ejected gas, between the events Θ_A and Θ_B. We shall apply the conservation of 4-momentum in S_o, using the data in Eqs. (9.3c) and (9.3d).

$$\delta\overrightarrow{\mathcal{P}} = -\,\delta\overrightarrow{\mathbf{p}}. \tag{9.4a}$$

$$\text{From Eqs. (9.3c) and (9.3d) } t\text{-comp} : \delta M = -\delta\mu_{\text{o}}\, g. \tag{9.4b}$$

$$x\text{-comp} : M\delta v' = \delta\mu_{\text{o}}\, g\, u. \tag{9.4c}$$

$$\text{Hence, } M\delta v' = -\,\delta M\, u. \tag{9.4d}$$

Note that (i) the *rest mass* lost by the rocket equals the *relativistic mass* gained by the ejected gas, according to (9.4b), (ii) Eq. (9.4d) is valid in S_o. To validate it in S, we have to apply the velocity addition formula (4.6):

$$v + dv = \frac{dv' + v}{1 + \frac{dv'\,v}{c^2}} \approx (dv' + v)\left(1 - \frac{dv'\,v}{c^2}\right) \approx v + \left(1 - \frac{v^2}{c^2}\right) dv'. \tag{9.5}$$

Hence, (in the limit $dv' \to 0$),

$$dv = \left(1 - \frac{v^2}{c^2}\right) dv'. \tag{9.6}$$

This transforms Eq. (9.4d) to

$$\frac{dv}{\left(1 - \frac{v^2}{c^2}\right)} = -u\frac{dM}{M}. \tag{9.7}$$

Integrating (9.7) from $t = 0$ to $t = t$, setting $M = M_i$ (i for "initial"), and $v = 0$ at $t = 0$, we get

$$\frac{c}{2}\ln\frac{c+v}{c-v} = -u\ln\frac{M}{M_i},$$

$$\text{or } \frac{M}{M_i} = \left(\frac{c-v}{c+v}\right)^{\frac{c}{2u}} = \left(\frac{1-\beta}{1+\beta}\right)^{\frac{c}{2u}}. \tag{9.8}$$

One of the requirements of all relativistic formulas is that they must converge to the corresponding N.R. counterparts (if such counterparts exist) in the N.R. limit $v/c \to 0$. In this case, the N.R. mass formula is (9.1b). We shall show that this requirement is satisfied by the formula (9.8), using the definition of the Euler number:

$$e \overset{\text{def}}{=} \lim_{x\to 0}(1+x)^{1/x}. \tag{9.9}$$

Proof. We set $\beta = v/c$. Then in the limit $\beta \to 0$

$$\frac{1}{1-\beta} \overset{\beta\to 0}{\longrightarrow} (1+\beta). \quad \text{Hence, } (1-\beta) \overset{\beta\to 0}{\longrightarrow} \frac{1}{1+\beta}. \tag{9.10a}$$

$$\left(\frac{1-\beta}{1+\beta}\right)^{\frac{c}{2u}} \overset{\beta\to 0}{\longrightarrow} [(1+\beta)^{-2}]^{(\frac{c}{2v})(\frac{v}{u})} = [(1+\beta)^{\frac{1}{\beta}}]^{-(v/u)}. \tag{9.10b}$$

$$\text{Hence, } \left(\frac{1-\beta}{1+\beta}\right)^{\frac{c}{2u}} \overset{\beta\to 0}{\longrightarrow} e^{-(v/u)}, \tag{9.10c}$$

$$\text{or } \frac{M}{M_i} \overset{\beta\to 0}{\longrightarrow} e^{-(v/u)}, \tag{9.10d}$$

same as the N.R. formula (9.1b) (QED)

Is the formula (9.8) valid when $u = c$? To make sure, we shall retrace the steps from Eq. (9.3) downward, specializing them to $u = c$. Instead of assuming that the gas is ejected with velocity $-u$ with respect to IRF S_o, we shall assume that, between the events Θ_A and Θ_B, a beam of photons is emitted in the $-x$-direction with energy δE_o with respect to the IRF S_o. In this case, we use Eq. (8.26) for photon momentum:

$$\text{At } \Theta_A : \vec{\mathcal{P}} = M(c, 0). \tag{9.11a}$$

$$\text{At } \Theta_B : \vec{\mathcal{P}} + \delta\vec{\mathcal{P}} = (M + \delta M)(c, \delta v'). \tag{9.11b}$$

$$\delta\vec{\mathcal{P}} = (\delta M c, M\delta v'). \tag{9.11c}$$

$$\delta\vec{\mathbf{p}} = \left(\frac{\delta E_o}{c}, -\frac{\delta E_o}{c}\right). \tag{9.11d}$$

In the above $\delta\vec{\mathbf{p}}$ stands for the 4-momentum of the emitted photon.

We shall apply the conservation of 4-momentum in S_o, using the data in Eqs. (9.11c) and (9.11d).

$$\delta\vec{\mathcal{P}} = -\delta\vec{\mathbf{p}}. \tag{9.12a}$$

$$t\text{-comp: } \delta M = -\frac{\delta E_o}{c^2}. \tag{9.12b}$$

$$x\text{-comp: } M\delta v' = \frac{\delta E_o}{c} \tag{9.12c}$$

$$\text{Hence, } M\delta v' = -\delta M\, c. \tag{9.12d}$$

It follows that Eq. (9.4d) will be valid for $u = c$. As a consequence (9.8) is also valid for $u = c$. We shall rewrite this for this special case:

$$\frac{M}{M_i} = \sqrt{\frac{c-v}{c+v}} = \sqrt{\frac{1-\beta}{1+\beta}}, \quad \text{for a photon driven rocket.} \tag{9.13}$$

9.5. The Thrust 4-Force

The 4-momentum of the rocket at the event Θ_A can be written as $\vec{\mathcal{P}}(\Theta_A) = M(\Theta_A)\vec{\mathbf{V}}(\Theta_A)$. We have used Eq. (8.17), replaced m_o with $M(\Theta_A)$. The time difference between the events Θ_A and Θ_B is δt with respect to S, and $\delta\tau$ with respect to the IRF S_o. Differentiating $\vec{\mathcal{P}}$ with respect to τ we get

$$\frac{d\vec{\mathcal{P}}}{d\tau} = \frac{d}{d\tau}(M\vec{\mathbf{V}}) = M\frac{d\vec{\mathbf{V}}}{d\tau} + \frac{dM}{d\tau}\vec{\mathbf{V}}. \tag{9.14}$$

To evaluate $\frac{dM}{d\tau}$, we use (9.4b)

$$\frac{dM}{d\tau} = -g\frac{d\mu_o}{d\tau} = -gr,$$

$$\text{where } r = \frac{d\mu_o}{d\tau}. \tag{9.15}$$

To get a parallel formula for the photon-driven rocket, we refer to (9.12b), and get

$$\frac{dM}{d\tau} = -\frac{1}{c^2}\frac{dE_o}{d\tau}. \tag{9.16}$$

We can combine the two equations into one, *assuming* that the rocket is ejecting *relativistic mass*, either in the form of *matter* or in the form of *radiation* (we shall use the term radiation to mean photons), at the *constant rate* of ϱ kg/s in its *rest frame*.

$$\text{For matter emission:}\ \varrho = \lim_{\delta\tau\to 0} \frac{g\,\delta\mu_o}{\delta\tau} = g\frac{d\mu_o}{d\tau} = gr. \tag{9.17a}$$

$$\text{For photon emission:}\ \varrho = \lim_{\delta\tau\to 0} \frac{\delta E_o/c^2}{\delta\tau} = \frac{1}{c^2}\frac{dE_o}{d\tau}. \tag{9.17b}$$

Note that r is constant by assumption, and g is constant because u is so. Hence ϱ is constant in (9.17a). We now *assume* that if photons are ejected to generate the reaction force, then $\frac{dE_o}{d\tau}$ is also constant in (9.17b). Then by (9.15) – (9.17)

$$\frac{dM}{d\tau} = -\varrho = \text{constant} \tag{9.18}$$

for both matter and radiation.

We now go back to (9.14), and rewrite it as follows:

$$M\frac{d\vec{\mathbf{V}}}{d\tau} = \frac{d\vec{\boldsymbol{\mathcal{P}}}}{d\tau} - \frac{dM}{d\tau}\vec{\mathbf{V}} = \frac{d\vec{\boldsymbol{\mathcal{P}}}}{d\tau} + \varrho\vec{\mathbf{V}} \equiv \vec{\boldsymbol{\mathcal{R}}} \tag{9.19a}$$

$$\text{where}\ \vec{\boldsymbol{\mathcal{R}}} \stackrel{\text{def}}{=} \frac{d\vec{\boldsymbol{\mathcal{P}}}}{d\tau} + \varrho\vec{\mathbf{V}} \tag{9.19b}$$

is the "reaction 4-force", or better still the *thrust 4-force*. However, we are using the symbol R instead of T, because the latter symbol can be confused with time.

In the following equations, we write the (t, x) components of the 4-vectors in S_o:

$$\text{Using (9.4a), (9.3d), (9.17a):}\ \frac{d\vec{\boldsymbol{\mathcal{P}}}}{d\tau} = -\frac{d\vec{\mathbf{p}}}{d\tau} = -g\frac{d\mu_o}{d\tau}(c, -u)$$

$$= -\varrho(c, -u). \tag{9.20a}$$

$$\text{From (8.8):}\ \vec{\mathbf{V}} = (c, 0), \tag{9.20b}$$

$$\text{since}\ v' = 0,\ \Gamma' = 1.$$

$$\text{Hence, from (9.19b):}\ \vec{\boldsymbol{\mathcal{R}}} = -\varrho(c, -u) + \varrho(c, 0)$$

$$= \varrho(0, u) = (0, R). \tag{9.20c}$$

In other words, the reaction 4-vector has the following components with respect to S_o:

$$\vec{\mathcal{R}} = (\mathcal{R}'^t, \mathcal{R}'^x) = (0, R) \tag{9.21a}$$

$$\text{where } R = \varrho u = gru = \text{reaction 3-force, with respect to } S_o. \tag{9.21b}$$

Note that the (t,x)-components of the reaction 4-force $\vec{\mathcal{R}}$ in S_o are in agreement with (8.36).

We shall now find the (t,x)-components of the reaction 4-force: $\vec{\mathcal{R}} = (\mathcal{R}^0, \mathcal{R}^x)$, in the ground frame S, applying Lorentz transformation Eq. (7.77a), corresponding to the boost: $S_o(-c\beta, 0, 0)S$, to the (t,x)-components of $\vec{\mathcal{R}}$ in the IRF S_o, shown in (9.21).

$$\mathcal{R}^t = \Gamma(\mathcal{R}'^t + \beta\mathcal{R}'^x) = \Gamma\beta R. \tag{9.22a}$$

$$\mathcal{R}^x = \Gamma(\mathcal{R}'^x + \beta\mathcal{R}'^t) = \Gamma R. \tag{9.22b}$$

Note that the (t,x)-components of the reaction 4-force $\vec{\mathcal{R}}$ in S are in agreement with (8.36), which we rewrite in the present context as

$$\vec{\mathcal{R}} = (\mathcal{R}^t, \mathcal{R}^x) = \Gamma\left(\frac{Rv}{c}, R\right). \tag{9.23}$$

Referring back to Eq. (9.17)

- For radiation emission $R = \varrho c = \frac{1}{c}\frac{dE_o}{d\tau}$.
- For matter emission $R = \varrho u = gru = gT$, where $T = ru$ is the same thrust force of non-relativistic mechanics. See Eq. (9.1c). It changes to $R = gT$ as it enters the relativistic domain.

9.6. The Equation of Motion

We return to Eq. (9.19a) and write the equation of motion

$$M\frac{d\vec{\mathbf{V}}}{d\tau} = \vec{\mathcal{R}}, \tag{9.24}$$

where $M = M(\Theta)$ is the instantaneous rest mass of the rocket at the event Θ, and is a 4-scalar. All we now have to do is to write the x-component of the 4-vectors on either side of the equation, and simplify the same to obtain the acceleration $a = \frac{dv}{dt}$ of the rocket in the GF S. We shall, however, find it convenient to work out the acceleration $a_o = \frac{dv'}{d\tau}$ in the IRF S_o, and convert this acceleration to a using Eq. (8.46).

Consider the x-component of $\vec{\mathbf{V}}$ using (8.8). The kinematic quantities in S_o will be identified with "prime". Then,

$$\frac{dV'^x}{d\tau} = \frac{d(\Gamma' v')}{d\tau} = \Gamma' \frac{dv'}{d\tau} + \frac{d\Gamma'}{d\tau} v' = \frac{dv'}{d\tau} = a_o, \tag{9.25}$$

since $v'=$ instantaneous velocity of the rocket in $S_o = 0$. Consequently. $\Gamma' = 1$.

From (9.21), the x-component of $\vec{\mathcal{R}}$ is $R'^x = R = \varrho u$. We thus get a simple looking equation of motion, which is valid in S_o.

$$M(\Theta) a_o = \varrho u = \text{constant}. \tag{9.26}$$

Mass $\times$ acceleration is constant. But mass is not constant. Hence, the acceleration in IRF S_o is not constant.

We shall write the EoM in the ground frame S, by converting $a_o \to a$, the acceleration in the ground frame S, using (8.16), which gives $a_o = \Gamma^3 a$:

$$M(\Theta)\,\Gamma^3 a = \varrho u = \text{constant}. \tag{9.27a}$$

$$\text{or, } M(\Theta)\,\Gamma^3 \frac{dv}{dt} = \varrho u = \text{constant}. \tag{9.27b}$$

Now we rewrite Eq. (9.27b), using the mass equation (9.8):

$$\left(\frac{c-v}{c+v}\right)^{\frac{c}{2u}} \Gamma^3 \frac{dv}{dt} = \frac{\varrho u}{M_i} = \text{constant}, \tag{9.28}$$

where M_i is the initial (rest) mass of the rocket.

We shall set $\beta = v/c$, and $c/u = n$ in the index of the leftmost factor in Eq. (9.28). Here $n \geq 1$ is a positive real number greater than or equal to 1. $n = 1$ corresponds to $u = c$. On the other extreme $n \to \infty$ would converge to the N.R. EoM shown in (9.1e). We now simplify the left side:

$$\Gamma^3 = \frac{1}{(1-\beta^2)^{3/2}} = [(1+\beta)(1-\beta)]^{-3/2},$$

$$\left(\frac{c-v}{c+v}\right)^{\frac{c}{2u}} = \left(\frac{1-\beta}{1+\beta}\right)^{\frac{n}{2}},$$

$$\left(\frac{c-v}{c+v}\right)^{\frac{c}{2u}} \Gamma^3 \frac{dv}{dt} = c\left[\frac{(1-\beta)^{\frac{n-3}{2}}}{(1+\beta)^{\frac{n+3}{2}}}\right]\frac{d\beta}{dt}.$$

The EoM (9.28) now takes a simpler form

$$\left[\frac{(1-\beta)^{\frac{n-3}{2}}}{(1+\beta)^{\frac{n+3}{2}}}\right]\frac{d\beta}{dt} = \frac{\varrho u}{cM_i} = k = \text{constant.} \tag{9.29}$$

Let us rewrite the mass equation (9.8), setting $v = c\beta$; $c/u = n$:

$$\frac{M}{M_i} = \left(\frac{1-\beta}{1+\beta}\right)^{\frac{n}{2}}. \tag{9.30}$$

We assume that the rocket has no payload, all its mass will ultimately be ejected out to provide the thrust. In other words, the rocket operates until $M \to 0$, which happens when $\beta \to 1$.

We shall now show that the above EoM (9.29) will converge to the non-relativistic EoM as given in (9.1e). We shall set $\varrho = gr$ as per (9.17a), $\beta \to 0$ and use the definition of Euler's number e, as in (9.9).

Proof.

$$\begin{aligned}\left[\frac{(1-\beta)^{\frac{n-3}{2}}}{(1+\beta)^{\frac{n+3}{2}}}\right] &\xrightarrow{\beta\to 0} \left[\frac{1}{(1+\beta)^{\frac{n-3}{2}}}\right]\left[\frac{1}{(1+\beta)^{\frac{n+3}{2}}}\right] \\ &= \frac{1}{(1+\beta)^n} = \frac{1}{(1+\beta)^{c/u}} \\ &= \frac{1}{[(1+\beta)^{1/\beta}]^{v/u}} = \frac{1}{e^{v/u}}.\end{aligned} \tag{9.31}$$

Substituting this in (9.29) we get:

$$M_i e^{-v/u}\frac{dv}{dt} = \varrho u = gru = ru, \text{ since } g \to 1.$$

Thus, we get back (9.1e). (QED)

9.7. Solution of the EoM for Two Special Cases

We shall find solution of the EoM for two special cases, namely (i) $n = u/c = 3$, and (ii) $n = u/c = 1$. The first case corresponds to the transition zone from non-relativistic to relativistic domain; the latter corresponds to a photon rocket.

<u>*Example 1.*</u> Set $n = 3$, implying $u = c/3$.

The reason for choosing $n = 3$ is two-fold: (1) the EoM shown in (9.29) will assume the simplest form, the numerator within the square brackets becoming 1; (2) we are now at the threshold of transition from the non-relativistic (N.R.) to the relativistic domain, the Γ-factor is very close to 1, in fact $g = \frac{3}{2\sqrt{2}} = 1.06$. Our results obtained here should be close to the N.R. results, so that we may feel comfortable that we are on right track.

We specialize the EoM (9.29) for this special case:

$$\left[\frac{1}{(1+\beta)^3}\right]\frac{d\beta}{dt} = \frac{\varrho}{3M_i} \equiv k_3 = \text{constant}. \tag{9.32}$$

Integration, subject to the initial condition: $\beta = 0$ when $t = 0$ leads to the following solution:

$$\beta = \left[\frac{1}{\sqrt{1-2k_3 t}} - 1\right], \quad 0 \le t \le t_c. \tag{9.33}$$

The reader can verify the answer by differentiating β with respect to t.

We have written t_c to mean "critical time", when $M \to 0$, as explained below Eq. (9.30). In other words, t_c is the time at "burn out", assuming that the rocket has no payload, all its mass has been ultimately ejected out to provide the thrust. This happens when $\beta \to 1$. From Eq. (9.33):

$$\begin{aligned} &\left[\frac{1}{\sqrt{1-2k_3 t_c}} - 1\right] = 1, \\ &\text{or } 1 - 2k_3 t_c = \frac{1}{4} \quad \Rightarrow \quad k_3 t_c = \frac{3}{8}, \\ &\text{or } \frac{\varrho t_c}{M_i} = \frac{9}{8} \\ &\text{for the N.R. case: } \frac{\varrho t_c}{M_i} = 1. \end{aligned} \tag{9.34}$$

We have plotted the velocity–time relation (rather the β–t relation) in Fig. 9.2(a), using Gnuplot. On the same graph, we have also plotted the N.R. equation (9.1f).

We have set $M_i = 1\,\text{kg}$, $r = 1\,\text{kg/s}$. Setting $u/c = 1/3$ in the first of the equations in (9.2), we get $g = 3/\sqrt{8} = 1.06$. Hence ϱ (defined in Eq. (9.17a)) $= gr = 1.06\,\text{kg/m}$. Therefore $k_3 = \frac{\varrho}{3M_i} = 0.3535$. From Eq. (9.34) the critical time is $t_c = \frac{3}{8\times 0.3535} = 1.06\,\text{s}$ which has been set as the upper limit on the t axis.

The two plots, *relativistic* and *non-relativistic* almost coincide up to $t \approx 0.8\,\text{s}$, $\beta \approx 0.5$.

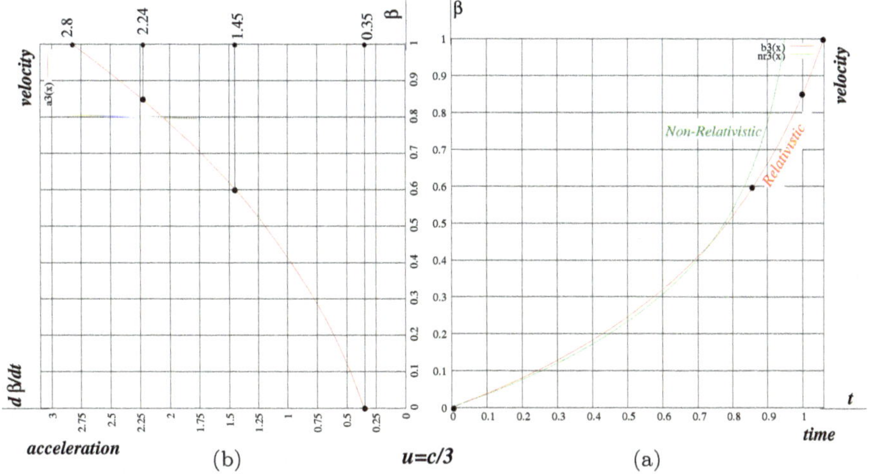

Fig. 9.2. Case I, $n = 3$. Plots for (a) velocity vs. time; (b) acceleration vs. velocity.

In Fig. 9.2(b), we have plotted $\frac{d\beta}{dt} = k_3(1+\beta)^3 = 0.3535(1+\beta)^3$. However, in this case we set the vertical axis to represent the independent variable β, matching it with the vertical β-axis of Fig. (9.2a). The horizontal axis, pointing to the left, represents the dependent variable $\frac{d\beta}{dt}$. We achieved this configuration by first plotting $\frac{d\beta}{dt}$ vs. β the usual way, then turning the plot anticlockwise by 90°. Our objective here has been to check whether the slope of the $\beta - t$-curve in Fig. 9.2(a) is corroborated by the measure of $d\beta/dt$ in Fig. 9.2(b). In order to judge the correspondence, we marked four selected points on the curve (a) and their corresponding points on (b), wrote the values of $\frac{d\beta}{dt}$ for these points on the upper horizontal axis of the plot box. Fair correspondence between these values in Fig (9.2b) and the corresponding slopes in 9.3(a) is discernible, suggesting that Eq. (9.33) is the solution of the EoM (9.32).

Example 2. Set $n = 1$, implying $u = c$.

This is the photon rocket mentioned in the Introduction. In this case, a jet of photons flowing out from the tail end of the rocket is serving as the propellant. We specialize the EoM (9.29) for this special case:

$$\left[\frac{1}{(1-\beta)(1+\beta)^2}\right]\frac{d\beta}{dt} = \frac{\varrho}{M_i} \equiv k_o = \text{constant}. \tag{9.35}$$

Integrating from $t = 0$ when $\beta = 0$ to $t = t$; $\beta = \beta$

$$\int_0^\beta \frac{d\beta}{((1-\beta)(1+\beta)^2} = k_o t, \tag{9.36}$$

we get

$$\frac{\beta}{2(1+\beta)} + \ln\sqrt{\frac{1+\beta}{1-\beta}} = k_o t = \left(\frac{\varrho}{M_i}\right) t,$$

or

$$t = \left(\frac{M_i}{\varrho}\right)\left[\frac{\beta}{2(1+\beta)} + \ln\sqrt{\frac{1+\beta}{1-\beta}}\right]. \tag{9.37}$$

We have plotted the velocity–time relation (rather the β–t relation) in Fig. 9.3(a). However, in Eq. (9.37) t is a function of β. Hence, using Gnuplot we first obtained β as the horizontal axis and t as the vertical axis. In order to reverse their roles, we transformed the plot by (i) a rotation through 90° in the anticlockwise direction, followed by (ii) a reflection about the vertical axis (i.e. about the new β-axis).

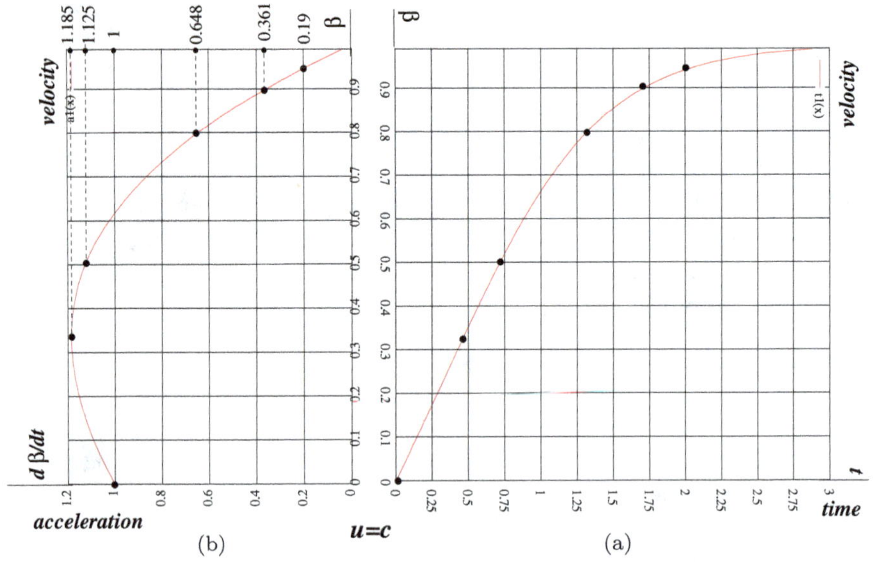

Fig. 9.3. Case II, $n = 1$. Plots for (a) velocity vs. time; (b) acceleration vs. velocity.

In Fig. 9.3(b), we have plotted $\frac{d\beta}{dt}$ (its axis pointing left) vs. β (its axis pointing upward). The procedure, objective, and explanations are the same as in Example 1.

It should be noted that in this case ϱ is given by (9.17b), which we rewrite and interpret as follows:

$$\varrho = \frac{1}{c^2}\frac{dE_o}{d\tau} = \frac{1}{c^2} \times \{\text{radiative power emitted from the tail of the rocket}\}. \tag{9.38}$$

For the plottings we have taken $M_i = 1\,\text{kg}$, $\varrho = 1\,\text{kg/s}$.

How long does the rocket operate? Until $\beta \to 1$, as mentioned below Eq. (9.30), and therefore, by (9.37), until $t \to \infty$.

It is seen from the plot in Fig. 9.3(a) that β approaches unity (or, v approaches c) asymptotically.

Chapter 10

Magnetism as a Relativistic Effect

10.1. Velocity-Dependent Force from a Velocity-Independent One under a Lorentz Transformation

We shall go back to Lorentz transformation of 4-force outlined in Sec. 8.7.4 and reconsider how a force $\mathbf{F}'$ on a particle at a particular event in S' will transform into a force $\mathbf{F}$ in S at the same event under the boost $S(\beta,0,0)S'$, so that $\boldsymbol{\beta} = \beta\mathbf{i}$. Specialize the force transformation formula (8.56) to this boost.

$$F_x = \frac{F'_x + \beta(\mathbf{F}' \cdot \boldsymbol{\nu}')}{1+\beta\nu'_x}; \quad F_y = \frac{F'_y}{\gamma(1+\beta\nu'_x)}; \quad F_z = \frac{F'_z}{\gamma(1+\beta\nu'_x)}. \tag{10.1}$$

Note that

$$F'_x + \beta(\mathbf{F}' \cdot \boldsymbol{\nu}') = F'_x + \frac{\beta\mathbf{v}' \cdot \mathbf{F}'}{c} = F'_x(1+\beta v'_x/c) + \frac{\beta}{c}(F'_y v'_y + F'_z v'_z).$$

$$\frac{F'_y}{\gamma} = \gamma(1-\beta^2)F'_y = \left[\gamma\left(1+\frac{\beta v'_x}{c}) - \frac{\gamma\beta}{c}(v'_x + c\beta\right)\right]F'_y. \tag{10.2}$$

Therefore,

$$\begin{aligned} F_x &= F'_x + \frac{\beta}{c}\left(\frac{F'_y v'_y + F'_z v'_z}{1+\beta\nu'_x}\right). \\ F_y &= \gamma F'_y - \frac{\gamma\beta}{c}\left(\frac{v'_x + c\beta}{1+\beta\nu'_x}F'_y\right). \\ F_z &= \gamma F'_z - \frac{\gamma\beta}{c}\left(\frac{v'_x + c\beta}{1+\beta\nu'_x}F'_z\right). \end{aligned} \tag{10.3}$$

Hence it follows from Eq. (10.4) and the velocity addition formula (4.23b) that

$$\begin{aligned} F_x &= F'_x + \frac{\gamma\beta}{c}(F'_y v_y + F'_z v_z). \\ F_y &= \gamma F'_y - \frac{\gamma\beta}{c} v_x F'_y. \\ F_z &= \gamma F'_z - \frac{\gamma\beta}{c} v_x F'_z. \end{aligned} \tag{10.4}$$

It will be a simple exercise to show that

$$\mathbf{v} \times (\mathbf{i} \times \mathbf{F}') = (v_y F'_y + v_z F'_z)\mathbf{i} - v_x(F'_y\mathbf{j} + F'_z\mathbf{k}). \tag{10.5}$$

We can now write the force whose components are given in Eq. (10.4) as

$$\mathbf{F} = F'_x\mathbf{i} + \gamma(F'_y\mathbf{j} + F'_z\mathbf{k}) + \frac{\gamma}{c}(\mathbf{v} \times (\boldsymbol{\beta} \times \mathbf{F}')). \tag{10.6}$$

We also break up the force $\mathbf{F}'$ into two components, viz., a component $\mathbf{F}'_{\|}$ which is parallel to the boost velocity $c\boldsymbol{\beta}$ and a component $\mathbf{F}_{\perp}$ which is perpendicular to $c\boldsymbol{\beta}$, so that, $\mathbf{F}' = \mathbf{F}'_{\|} + \mathbf{F}'_{\perp}$, and so that for the special boost $S(\beta, 0, 0)S'$,

$$\mathbf{F}'_{\|} = F'_x\,\mathbf{i};\ \mathbf{F}'_{\perp} = F'_y\mathbf{j} + F'_z\mathbf{k}. \tag{10.7}$$

The general form for the force transformation formula, valid for a general boost $S(\boldsymbol{\beta})S'$, now follows from Eq. (10.6):

$$\mathbf{F} = \mathbf{F}_0 + \mathbf{v} \times \mathbf{G}, \tag{10.8a}$$

$$\text{where, } \mathbf{F}_0 = \mathbf{F}'_{\|} + \gamma\mathbf{F}'_{\perp}, \tag{10.8b}$$

$$\text{and, } \mathbf{G} = \frac{\gamma}{c}(\boldsymbol{\beta} \times \mathbf{F}'). \tag{10.8c}$$

It is then seen from Eq. (10.8) that the transformed force $\mathbf{F}$ has a component $\mathbf{F}_0$ which is velocity independent if $\mathbf{F}'$ is so, and a part $\mathbf{v} \times \mathbf{G}$ which is explicitly velocity dependent provided $\mathbf{G}$ is not zero. We are therefore led to the following theorem:

Theorem 10.1. *Suppose a particle P moving in an arbitrary trajectory experiences a purely velocity-independent force $\mathbf{F}'$ as measured in an inertial frame of reference S'. Suppose this frame S' is moving with velocity $c\boldsymbol{\beta}$ with respect to another inertial frame S, and that the velocity of the particle at some instant t is $\mathbf{v}$ in S. Then the force $\mathbf{F}$, as measured in S at the time t, will be velocity dependent, being the sum of a velocity-independent component $\mathbf{F}_0$ and a purely velocity-dependent component $\mathbf{v} \times \mathbf{G}$ as given by Eq.* (10.8).

10.2. How Magnetic Force Originates from Lorentz Transformation

The starting point of the principles of electromagnetism has of two parts, namely (1) the *Lorentz force equation* which defines the electric field $\mathbf{E}$ and the magnetic field $\mathbf{B}$ in terms of the force $\mathbf{F}$ that a charged particle q will experience under the influence of a distribution of electric charges and currents, and (2) Maxwell's equations. Using these two sets of equations, one can understand and explain every phenomenon in the domain of electromagnetism, including attraction and repulsion between electric charges, electric currents, magnets, operations of motors and generators, as well as the propagation of electromagnetic waves.

Let us now consider a distribution of charge $\mathcal{Q}$ which is static in an inertial frame S'. Only an electrostatic field $\mathbf{E}'$ exists in this frame, so that the force experienced by a *moving* test particle of charge q is the *velocity-independent* force $\mathbf{F}' = q\mathbf{E}'$. If the distribution $\mathcal{Q}$ is in bulk motion with constant velocity $\mathbf{u} = c\boldsymbol{\beta}$ as seen from another frame S, then S and S' must be related to each other by the boost: $S(\boldsymbol{\beta})S'$. It now follows from the above theorem, in particular Eq. (10.8), that the same particle will be seen to experience a *velocity-dependent* force $\mathbf{F}$ in S, which is given by the formula:

$$\mathbf{F} = q(\mathbf{E} + \mathbf{v} \times \mathbf{B}) \tag{10.9a}$$

$$\text{where } \mathbf{E} = \mathbf{E}'_{\|} + \gamma\mathbf{E}'_{\perp}, \tag{10.9b}$$

$$\text{and } \mathbf{B} = \frac{\gamma}{c}(\boldsymbol{\beta} \times \mathbf{E}') = \frac{1}{c}\boldsymbol{\beta} \times \mathbf{E}, \tag{10.9c}$$

since $\boldsymbol{\beta} \times \mathbf{E}'_{\|} = \mathbf{0}$. Line (a) in the above equation is the *Lorentz force equation.* Lines (b) and (c) show how a pure electric field $\mathbf{E}'$ in the frame S' transforms into a combination of electric field $\mathbf{E}$ and magnetic field $\mathbf{B}$ in the frame S. Hence the main conclusion of this chapter.

Conclusion: *A distribution of charges $\mathcal{Q}$, when moving with uniform velocity $\mathbf{u} = c\boldsymbol{\beta}$, will create, in addition to an electric field $\mathbf{E}$, also a magnetic field $\mathbf{B}$. The emergence of the resulting magnetic force can be linked to the Lorentz transformation of contravariant 4-vectors which itself is a consequence of the postulates of special relativity. In this sense magnetism is a relativistic effect.*

Note that the above exercise does not shed any light on what happens when the charge distribution $\mathcal{Q}$ moves with an arbitrary (non-uniform) velocity. The field resulting from such a motion can be worked using the full set of Maxwell's equations.

We shall come back to Eq. (10.9) in Sec. 11.4 through Eq. (11.37), and in Sec. 11.6 through Eq. (11.63b) — by a different route, namely, Lorentz transformation of the electromagnetic field tensor.

It may be appropriate to close this discussion with a quotation from Leigh Page.[a] "The rotating armature of every generator and every motor in this age of electricity are steadily proclaiming the truth of the relativity theory to all those who have ears to hear."

[a]L. Page, Lecture at December 17, 1941 meeting of the American Institute of Electrical Engineers, New York. Quoted at the beginning of Chapter 3 of Ref. [5].

Chapter 11

Principle of Covariance with Application in Classical Electrodynamics

11.1. The Principle

Physics is geometry. Consider Newton's second law of motion:

$$\mathbf{F} = \frac{d\mathbf{p}}{dt}. \tag{11.1}$$

The left-hand side is a "prescribed vector" — implying thereby a straight line segment of pre-specified "length" and "direction". The right-hand side is a "constructed" vector, obtained through operations like "parallel transport", drawing a "vector triangle" and division of one of its side with a 3-scalar dt, as illustrated in Fig. 11.1.

A particle, while in motion, traces out a certain path Γ. Two nearby locations A and B on this path are reached by the particle at times t and $t + \delta t$. Its 3-momenta at these two instants are $\mathbf{p}(t)$ and $\mathbf{p}(t + \delta t)$, respectively. The average force $\mathbf{F}$ between these two instants, multiplied by the time δt is (approximately) equal to $\delta\mathbf{p}$, as illustrated in the upper box.

The ratio of the change in the momentum vector $\delta\mathbf{p}$ to the time difference δt is the average force $\mathbf{F}$ acting on the particle between A and B. This average force becomes the instantaneous force $\mathbf{F}$ at the instant t if we make the time difference infinitesimally small, i.e. when $\delta t \to dt$.

The process employed in this example epitomize the structure of most physical laws. The Lord of the universe conceived physical quantities of

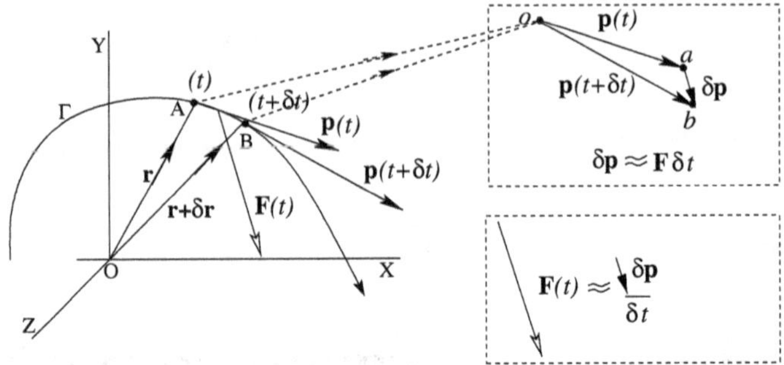

Fig. 11.1. Graphical construction of Newton's second Law of motion.

nature as "geometrical objects" and ordained them to shape through rules of geometrical constructions into different, but all the same, geometrical entities.

Even though the geometrical picture may not be overtly manifest in the equations of physics, geometrical concepts and constructions are subtly at work behind the structure of physical laws. This is because scalars, vectors and tensors are all geometrical objects, as we highlighted at the end of Sec. 5.1.1, and illustrated in Fig. 5.1. The shape of every physical equation — which either establishes a law — or proclaims a certain rule, convention, definition or relation — must be such that the geometrical objects on either side of the equation have come down to the same rank through processes of geometrical construction. This, in nutshell, is the *Principle of Covariance.*

Apparently then this principle is as old as Newtonian Mechanics. However, the advent of relativity adds extra dimension to this principle by reminding us that the physical quantities are no longer geometrical objects of the visible three-dimensional world, but are inhabitants of the four-dimensional space–time.

The Principle of Covariance declares that the mathematical expressions of all physical laws must be written in such a way that either side of the equation is a 4-tensor of the same rank and same sequence of contravariant and covariant indices. When an equation satisfies this demand, we say that it is covariant. By corollary, a law or a relationship which cannot be written covariantly, must have a limited application, spatially or temporally, and cannot be regarded as a universal law.

The law of motion written in the forms (8.28) and (8.33) are two examples of covariant equation.

In this chapter, we shall recast the familiar equations of classical electrodynamics, the equation of continuity, Maxwell's equations, the electromagnetic energy–momentum conservation laws in the covariant forms.

11.2. The Flux of a Vector Field in E^3

11.2.1. *2D surface embedded in 3D space, and the outward normal*

Figure 11.2(a) shows a two-dimensional surface S in the three-dimensional Euclidean space. The equation for such a surface is usually represented by a mathematical equation in the form

$$\begin{aligned} &\text{Implicit form } \Psi = \Psi(x, y, z) = C, \quad C = \text{constant.} \\ &\text{Parametric form } x = f(u, v),\ y = g(u, v),\ z = h(u, v). \end{aligned} \tag{11.2}$$

In the implicit form, it is not possible to tell from the equation which one is the independent variable, because all variables are treated as equal. In the parametric form, a point (x, y, z) on the surface S is determined by two parameters, suggesting that the surface is a two-dimensional object, embedded in the three-dimensional space spanned by the (X, Y, Z)-axes.

The simplest example of both can be provided by the equation of the surface of a sphere of radius R, and having the centre at the origin, shown

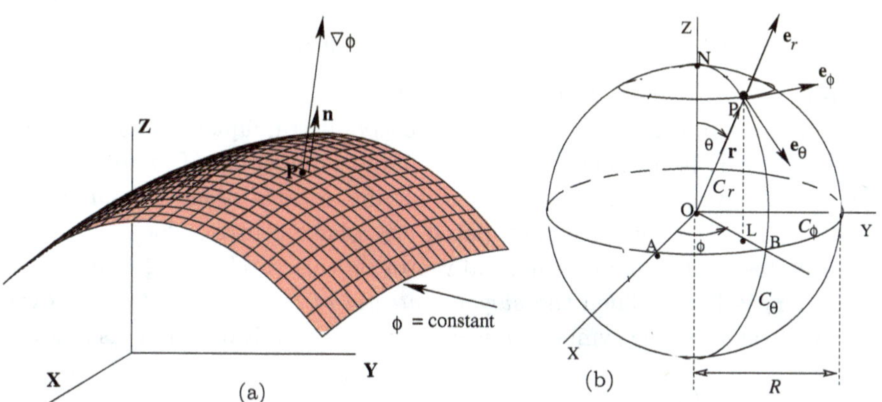

Fig. 11.2. (a) Surface and its normal; (b) a spherical surface shown with unit vectors in the spherical coordinate system.

in Fig. 11.2(b).

$$\begin{aligned} &\text{Implicit form } \Psi = x^2 + y^2 + z^2 = R^2. \\ &\text{Parametric form } x = R\sin\theta\cos\phi,\ y = R\sin\theta\sin\phi,\ z = R\cos\theta. \end{aligned} \tag{11.3}$$

In the second case, the spherical coordinates (θ, ϕ) serve as the parameters (u, v).

Surface integration of a vector field involves the *unit normal vector* $\mathbf{n}$ drawn at each point (x, y, z) on the surface S. The gradient vector $\boldsymbol{\nabla}\Psi$ points in the direction of the *outward normal* (i.e. in the direction in which it is increasing). Hence

$$\mathbf{n}(\mathbf{r}_0) = \left(\frac{\boldsymbol{\nabla}\Psi(\mathbf{r})}{|\boldsymbol{\nabla}\Psi(\mathbf{r})|}\right)_{\mathbf{r}=\mathbf{r}_0} \tag{11.4}$$

is the outward normal at a point $\mathbf{r} = \mathbf{r}_0$ on the surface. As an example, the *outward* normal drawn at a point $\mathbf{r}_0$ on the spherical surface S of Eq. (11.3) is given by the following expression:

$$\mathbf{n}(\mathbf{r}_0) = \left(\frac{x_0\mathbf{i} + y_0\mathbf{j} + z_0\mathbf{k}}{R}\right) = \mathbf{e}_r(\mathbf{r}_0), \tag{11.5}$$

where $(\mathbf{e}_r, \mathbf{e}_\theta, \mathbf{e}_\phi)$ are the three unit vectors on the surface, associated with the spherical coordinate system, being in the directions of increasing r, θ, ϕ, respectively (indicated by the curves $C_r.C_\theta, C_\phi$, respectively). Of these three, $\mathbf{e}_r$ is identified with $\mathbf{n}$, the normal vector, the other two being tangent vectors to the surface.

11.2.2. *Surface integral, flux of a vector field*

Consider a vector field $\mathbf{F}(\mathbf{r})$, "flowing" through an open surface S which is "oriented" (i.e. its unit normal at each point is uniquely defined). This surface S can be divided into a very large number N of small patches $\delta a_1, \delta a_2, \ldots, \delta a_N$, the largest of them having an area of δa; i.e. $\delta a_i \le \delta a$; $1 \le i \le N$. Consider one such patch δa_i whose centre is located at the coordinates $\mathbf{r}_i$. Let the unit normal vector on this patch be $\mathbf{n}_i$. The vector field at $\mathbf{r}_i$ is $\mathbf{F}(\mathbf{r}_i)$. Then the *surface integral of the vector field* $\mathbf{F}(\mathbf{r})$ over S, represented by the symbol φ_F is defined as the limit of a sum, as follows:

$$\varphi_F \equiv \iint_S \mathbf{F}(\mathbf{r}) \cdot d\boldsymbol{a} \stackrel{\text{def}}{=} \lim_{N\to\infty;\,\delta a\to 0} \sum_{i=1}^{N} \mathbf{F}(\mathbf{r}_i) \cdot \mathbf{n}_i\, \delta a_i, \tag{11.6}$$

and is called the *flux of the vector field* $\mathbf{F}(\mathbf{r})$ *across the surface* S.

We shall not elaborate on this further, but illustrate the surface integration with a familiar example, taking the vector field $\mathbf{F}(\mathbf{r})$ to be the electric field $\mathbf{E}(\mathbf{r})$ emanating from a point charge Q sitting at the origin. The surface S is the upper hemisphere with the centre at the origin, i.e. the surface of Eq. (11.3), expressed in the spherical coordinate system as follows: $r = R;\ 0 \leq \theta \leq \pi/2;\ 0 \leq \phi \leq 2\pi$.

In this case $\mathbf{F}(\mathbf{r}) = \mathbf{E}(\mathbf{r}) = \frac{Q}{4\pi\epsilon_0}\frac{1}{r^2}\mathbf{e}_r$, and $d\boldsymbol{a} = r^2 \sin\theta\, d\theta\, d\phi\, \mathbf{e}_r$, so that $\mathbf{E} \cdot d\boldsymbol{a} = \frac{Q}{4\pi\epsilon_0} \sin\theta\, d\theta\, d\phi$. Hence,

$$\varphi_E = \frac{Q}{4\pi\epsilon_0} \int_{\theta=0}^{\pi/2} \int_{\phi=0}^{2\pi} \sin\theta\, d\theta\, d\phi = \frac{Q}{2\pi\varepsilon_0} \tag{11.7}$$

is the flux of the $\mathbf{E}(\mathbf{r})$ field across the hemisphere.

The flux of a vector field is closely associated with Gauss's divergence theorem, which we state here as follows:

Theorem 11.1. *Let $\mathbf{F}(\mathbf{r})$ be a vector field which is continuous, along with continuous derivatives in a region of space $\mathcal{R}$. Let S be a closed piecewise smooth surface in $\mathcal{R}$, forming the boundary of the volume V within. Let $\mathbf{n}(\mathbf{r})$ be a unit outwardly directed normal vector on S at a point $\mathbf{r}$ on S. Then*

$$\iiint_V \boldsymbol{\nabla} \cdot \mathbf{F}(\mathbf{r}) \cdot d^3r = \iint_S \mathbf{F}(\mathbf{r}) \cdot \mathbf{n}\, da = \iint_S \mathbf{F}(\mathbf{r}) \cdot d\boldsymbol{a}. \tag{11.8}$$

11.2.3. *Continuity equation*

Let us consider a fluid in streamline motion[a] as in Fig. 11.3. This fluid is characterized by a velocity field $\mathbf{u}(\mathbf{r}, t)$ and a fluid mass density $\sigma(\mathbf{r}, t)$, both of which, in general, are unsteady fields, i.e. functions of t as well.

In pre-relativistic physics mass is conserved. There is an equation, sometimes called the *equation of continuity*, which expresses this conservation rule, namely,

$$\frac{\partial \sigma}{\partial t} + \boldsymbol{\nabla} \cdot (\sigma \mathbf{u}) = 0. \tag{11.9}$$

Note that the product $\sigma\mathbf{u}$ of fluid mass density and fluid velocity is the *fluid flux density*.

Conservation equitations of various physical quantities have a structure which is similar to Eq. (11.9). Since conservation equations play a very

[a]We *imagine* the fluid particles to be moving together without random thermal motion.

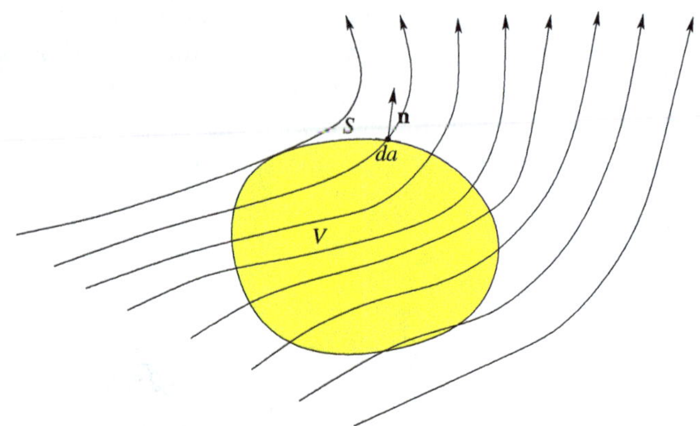

Fig. 11.3. Illustrating mass conservation.

important role in physics as well as in the remaining part of this chapter, we shall demonstrate how the above equation is derived.

Consider the imaginary closed surface S, *fixed* in space, and embedded inside a stream of fluid. Since fluid masses flow in and flow out of the space V inside S, the content of fluid mass in it is a function of time. We write it as follows:

$$M(t) = \iiint_V \sigma(\mathbf{r}, t) d^3 r. \tag{11.10}$$

The change of mass over time dt is

$$dM = \iiint_V \sigma(\mathbf{r}, t + dt) d^3 r - \iiint_V \sigma(\mathbf{r}, t) d^3 r = \left(\iiint_V \frac{\partial \sigma(\mathbf{r}, t)}{\partial t} d^3 r \right) dt. \tag{11.11}$$

Conservation of a physical quantity means that if the content of this quantity inside a fixed volume V has increased by a certain amount over a given time, the same amount must have flowed in through the boundary surface S in this time. Applied to mass conservation, it means:

$$\begin{aligned} dM &= -\left(\iint_S [\sigma(\mathbf{r}, t)\mathbf{u}(\mathbf{r}, t)] \cdot \mathbf{n}\, da \right) dt \\ &= -\left(\iiint_V \boldsymbol{\nabla} \cdot [\sigma(\mathbf{r}, t)\mathbf{u}(\mathbf{r}, t)] d^3 r \right) dt. \end{aligned} \tag{11.12}$$

We have used Gauss's divergence theorem to convert the first equality to the second one. The minus sign appears because the surface integral, if positive, would mean an outflow (the unit vector $\mathbf{n}$ is an outward normal). Equating the right-hand sides of the above two equations, we get

$$\left[\iiint_V \left(\frac{\partial\sigma(\mathbf{r},t)}{\partial t} + \boldsymbol{\nabla}\cdot\sigma(\mathbf{r},t)\mathbf{u}(\mathbf{r},t)\right) d^3r\right] dt = 0. \tag{11.13}$$

Since the result is valid for arbitrary V and arbitrary dt, the integrand is zero. Hence Eq. (11.9) follows. (QED)

All conservations laws have the same format as in Eq. (11.9), namely

$$\boxed{\frac{\partial}{\partial t}(\textit{volume density}) + \boldsymbol{\nabla}\cdot(\textbf{flux density}) = 0.} \tag{11.14}$$

We have written "volume density" in italics, and "flux density" in bold, to indicate that the former is a scalar, and the latter a vector.

How shall we write the continuity equation covariantly, i.e. in the context of relativity? By (1) making sure that the quantity in question is conserved, e.g. energy of a stream of particles flowing like a fluid, the electric charge contained in a charged fluid, as in Sec. 11.3; (2) multiplying the numerator and the denominator of the first term in (11.14) with c. Let ϱ (pronounced varrho) represent the volume density, and $\boldsymbol{\mathcal{J}}$ the corresponding flux density of this quantity (e.g. energy, charge). By definition, the flow of the quantity per unit time across the surface element $d\boldsymbol{a} = \mathbf{n}\, da$ is given by the relation

$$d\varphi = \boldsymbol{\mathcal{J}}\cdot d\boldsymbol{a}. \tag{11.15}$$

We now rewrite the continuity equation (11.14) in the following more precise language:

$$\frac{\partial\,(c\varrho)}{\partial(ct)} + \boldsymbol{\nabla}\cdot\boldsymbol{\mathcal{J}} = 0. \tag{11.16}$$

Written covariantly, as in Eq. (11.29)

$$\begin{aligned} \nabla_\mu \mathcal{J}^\mu(x) &= 0, \\ \text{where} \quad \mathcal{J}^\mu &= (c\varrho, \boldsymbol{\mathcal{J}}) \end{aligned} \tag{11.17}$$

represent the time and space components of the 4-vector density $\vec{\mathcal{J}}$ of the assumed conserved quantity.

11.3. Conservation of Electric Charge

There are only two types of force which can be understood in the "classical language", i.e. without using quantum mechanics. One of them is the force of gravity — the most commonly and universally experienced force of nature. This force, however, comes under the purview of General Relativity. Moreover, this force is too weak compared to the electromagnetic forces to have any effect at all on the motion of subatomic particles which can be in relativistic motion at the laboratories.

This leaves us with only one force, namely the electromagnetic force, which has a classical structure. We shall recast this classical structure of electromagnetic theory into a covariant form in order to illustrate the language of covariance. In the discussions to follow we shall not present any detailed discussion or derivation of the formulas, for which the reader has to look into standard books on electrodynamics.

We shall start with the basic postulates of electrodynamics and express them first in the "classical" language and then in the covariant format.

Postulate I. q is the electric charge of a particle, moving or stationary, then the measure of q is the same in all inertial frames. In other words, q is a 4-scalar.

Postulate II. Electric charge is conserved. This charge conservation law is customarily expressed in the form of the continuity equation (Sec. 11.2.3):

$$\frac{\partial \rho(\mathbf{r}, t)}{\partial t} + \boldsymbol{\nabla} \cdot \mathbf{J}(\mathbf{r}, t) = 0. \tag{11.18}$$

In the above equation, $\rho(\mathbf{r}, t)$ and $\mathbf{J}(\mathbf{r}, t)$ represent, respectively, the charge density and the charge current density at the event $\Theta = (ct, \mathbf{r}) = (x)$, as measured in a given frame of reference S, which for explicitness we shall call the Lab frame.

If the electric current distribution is due to a "streamline motion" of an electrically charge fluid, as shown in Fig. 11.4, and if $\mathbf{u}(\mathbf{r}, t)$ is the stream's 3-velocity (in the Lab frame) at the event (x), then

$$\mathbf{J}(\mathbf{r}, t) = \rho(\mathbf{r}, t)\, \mathbf{u}(\mathbf{r}, t). \tag{11.19}$$

We can think of a comoving frame of reference S_o moving with the charge stream at the event (x). The charge density $\rho_o(\mathbf{r}, t) = \rho_o(x)$ at (x), measured in the frame S_o, will be called the *proper charge density* of the fluid at (x), and will be treated as a *4-scalar* density. The dynamic Lorentz

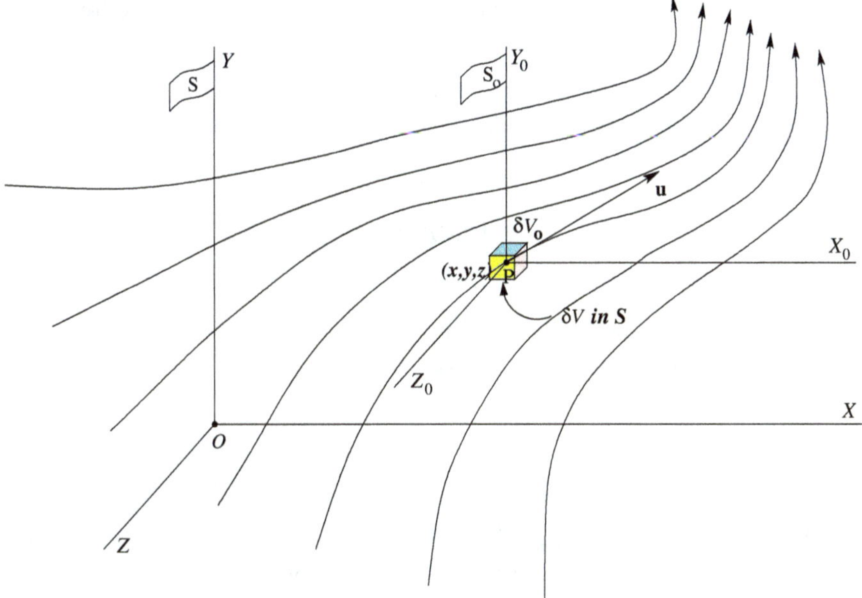

Fig. 11.4. Lab frame S and comoving frame S_o in a streamline flow of particles.

factor of the fluid at (x) is

$$\Gamma(x) = \frac{1}{\sqrt{1 - \frac{u^2(x)}{c^2}}}. \tag{11.20}$$

Imagine a collection of particles containing a quantity of charge δq, enclosed within the boundaries of a box whose volume is δV in S and δV_o in S_o, so that

$$\delta q = \rho_o(x)\delta V_o = \rho(x)\delta V. \tag{11.21}$$

Lorentz contraction of the dimension of this box along the direction of $\mathbf{u}$, changes its proper volume δV_o to the laboratory volume

$$\delta V = \frac{1}{\Gamma(x)}\delta V_o. \tag{11.22}$$

By (11.21) and (12.3)

$$\rho(x) = \Gamma(x)\rho_o(x). \tag{11.23}$$

The 4-velocity of the charged fluid at (x) is, according to (8.7),

$$U^\mu(x) = \Gamma(x)(c, \mathbf{u}(x)). \tag{11.24}$$

Therefore, we define the 4-current density of the electric charge fluid as

$$J^\mu(x) = \rho_o(x)U^\mu(x). \tag{11.25}$$

From (11.25), (11.24) and (11.23)

$$\begin{aligned} J^\mu(x) &= \rho_o(x)\Gamma(x)(c, \mathbf{u}(x)) \\ &= (\rho(x)c, \rho(x)\mathbf{u}(x)) \\ &= (\rho(x)c, \mathbf{J}(x)). \end{aligned} \tag{11.26}$$

Thus, $\rho(x)$ times c constitutes the time component, and $\mathbf{J}(x)$ constitutes the three space components of $J^\mu(x)$. Let us rewrite the continuity equation (11.18) as

$$\frac{\partial(c\rho)}{\partial(ct)} + \frac{\partial J_x}{\partial x} + \frac{\partial J_y}{\partial y} + \frac{\partial J_z}{\partial z} = 0. \tag{11.27}$$

The 4-gradient operators ∇_μ, ∇^μ were introduced and elaborated through Eqs. (7.72), (7.113). We rewrite the first one explicitly as follows:

$$\nabla_\mu \equiv \frac{\partial}{\partial x^\mu} = \left(\frac{\partial}{\partial x^0}, \frac{\partial}{\partial x^1}, \frac{\partial}{\partial x^2}, \frac{\partial}{\partial x^3}\right) = \left(\frac{\partial}{\partial ct}, \boldsymbol{\nabla}\right). \tag{11.28}$$

The charge continuity equation (11.18), rewritten in (11.27), now assumes the form:

$$\boxed{\nabla_\mu J^\mu(x) = 0.} \tag{11.29}$$

The left-hand side of the equation appears like a contraction, reducing it to a 4-scalar. The right-hand side is also a 4-scalar, having a single component 0. Hence, it is a covariant equation.

11.4. The Electromagnetic Field Tensor

Let us make the third postulate of electrodynamics.

Postulate III. The force experienced by a particle carrying an electric charge q is a velocity-dependent force, called *Lorentz force* (see Eq. 10.12), written as

$$\mathbf{F} = q(\mathbf{E} + \mathbf{v} \times \mathbf{B}), \tag{11.30}$$

where $\mathbf{v}$ is the velocity of the charged particle at the event point (x). The above equation also serves as the definition of the *electric field* $\mathbf{E}$ and the *magnetic field* **B** at the location of the particle. Let us write the dynamic Lorentz factor for the particle's velocity:

$$\Gamma = \frac{1}{\sqrt{1 - \frac{v^2}{c^2}}}. \tag{11.31}$$

Substituting the Lorentz force (11.30) in (8.36), the time and space components of the corresponding Minkowski force $\mathfrak{F}^\mu$ are now obtained compactly as

$$\vec{\boldsymbol{\mathcal{F}}} = \vec{\mathbf{e}}_\mu \mathfrak{F}^\mu = q\Gamma\left(\frac{1}{c}\mathbf{E}\cdot\mathbf{v},\ \mathbf{E} + \mathbf{v}\times\mathbf{B}\right), \tag{11.32}$$

and in an expanded form as

$$\begin{aligned}
\mathfrak{F}^0 &= \frac{\Gamma}{c}\mathbf{F}\cdot\mathbf{v} = \frac{q\Gamma}{c}(E_x v_x + E_y v_y + E_z v_z),\\
\mathfrak{F}^1 &= \Gamma F_x \quad = \frac{q\Gamma}{c}(E_x c + cB_z v_y - cB_y v_z),\\
\mathfrak{F}^2 &= \Gamma F_y \quad = \frac{q\Gamma}{c}(E_y c + cB_x v_z - cB_z v_x),\\
\mathfrak{F}^3 &= \Gamma F_z \quad = \frac{q\Gamma}{c}(E_z c + cB_y v_x - cB_x v_y).
\end{aligned} \tag{11.33}$$

The above equation tells us that the Minkowski 4-force acting on a charged particle q is a linear function of its 4-velocity, and therefore can be written as

$$\boxed{\mathfrak{F}^\mu = \frac{q}{c}F^{\mu\nu}V_\nu.} \tag{11.34}$$

In the above equation, V_ν is the covariant form of the 4-velocity vector V^μ, defined in (8.7) and is obtained from it by the lowering operation,

$$V_\mu = g_{\mu\nu}V^\mu = (\Gamma c, -\Gamma\mathbf{v}) = (\Gamma c, -\Gamma v_x, -\Gamma v_y, -\Gamma v_z). \tag{11.35}$$

$F^{\mu\nu}$, as defined by (11.34), is a very important contravariant tensor of rank 2, called *electromagnetic field tensor*, and must have the following components as suggested by Eq. (11.33):

$$F^{\mu\nu} = \begin{matrix} \\ 0 \\ 1 \\ 2 \\ 3 \end{matrix} \begin{matrix} \begin{matrix} 0 & \quad 1 & \quad 2 & \quad 3 \end{matrix} \\ \begin{pmatrix} 0 & -E_x & -E_y & -E_z \\ E_x & 0 & -cB_z & cB_y \\ E_y & cB_z & 0 & -cB_x \\ E_z & -cB_y & cB_x & 0 \end{pmatrix} \end{matrix}, \qquad F^{\nu\mu} = -F^{\mu\nu}. \tag{11.36}$$

The electromagnetic field tensor as written in (11.36) is an antisymmetric contravariant 4-tensor. Hence, its diagonal elements are all zero, and the off-diagonal elements on the upper side of the diagonal are equal and opposite to the corresponding off-diagonal elements on the lower side of the diagonal. Hence, it has only six independent components, and they are the six 3-scalar components of the $(\mathbf{E}, \mathbf{B})$ field.

Equation (11.34) is the covariant expression of the Lorentz force equation (11.30). Using the Lorentz transformation formula (7.80) for a contravariant tensor, the reader should be able to obtain the following transformation rule for the components of the $(\mathbf{E}, \mathbf{B})$ field under the boost: $S(\beta, 0, 0)S'$:

$$\begin{aligned} E'_x &= E_x, & cB'_x &= cB_x, \\ E'_y &= \gamma(E_y - \beta c B_z), & cB'_y &= \gamma(cB_y + \beta E_z), \\ E'_z &= \gamma(E_z + \beta c B_y), & cB'_z &= \gamma(cB_z - \beta E_y). \end{aligned} \tag{11.37}$$

To illustrate the procedure, we shall work out two of the above transformation formulas in details, namely for E'_x and B'_z, using the Lorentz matrix as given in (7.69), to transform the components of the tensor (11.36):

$$\begin{aligned} E'_x &= F'^{10} = \Omega^1_{.\alpha}\Omega^0_{.\beta}F^{\alpha\beta} = \Omega^1_{.0}\Omega^0_{.1}F^{01} + \Omega^1_{.1}\Omega^0_{.0}F^{10} \\ &= \gamma^2(1-\beta^2)E_x = E_x. \\ cB'_z &= F'^{21} = \Omega^2_{.\alpha}\Omega^1_{.\beta}F^{\alpha\beta} = \Omega^2_{.2}\Omega^1_{.0}F^{20} + \Omega^2_{.2}\Omega^1_{.1}F^{21} = \gamma(cB_z - \beta E_x). \end{aligned} \tag{11.38}$$

Note that even though each of the two lines above apparently involve a sum of 16 terms, corresponding to $\alpha, \beta = 0, 1, 2, 3$ we have accommodated only the non-zero terms, which happen to be only two in number in each case.

11.5. The Field Equations of Electrodynamics in the Covariant Language

We shall now establish covariance of the field equations of electrodynamics. These equations, expressed in their more familiar non-covariant form, are known as *Maxwell's equations*. They separate out into two sets of 1+3 equations, namely (1) the *inhomogeneous equations* containing the *source terms*, and (2) the *homogeneous equations* without the source terms:

Inhomogeneous part:
$$\nabla \cdot \mathbf{E}(\mathbf{r},t) = \frac{1}{\varepsilon_0 c} c\rho(\mathbf{r},t), \tag{11.39a}$$
$$\nabla \times c\mathbf{B}(\mathbf{r},t) - \frac{\partial \mathbf{E}(\mathbf{r},t)}{c\,\partial t} = \frac{1}{\varepsilon_0 c}\mathbf{J}(\mathbf{r},t), \tag{11.39b}$$

Homogeneous part:
$$\nabla \cdot c\mathbf{B}(\mathbf{r},t) = 0, \tag{11.39c}$$
$$\nabla \times \mathbf{E}(\mathbf{r},t) + \frac{\partial c\mathbf{B}(\mathbf{r},t)}{c\,\partial t} = \mathbf{0}. \tag{11.39d}$$

To convert the inhomogeneous part, given by (11.39a) and (11.39b), we first expand them into four separate equations in terms of the Cartesian components of $\mathbf{E}, c\mathbf{B}, \mathbf{J}$. Identifying these terms in the expanded expression as the components of $J^\mu(x)$ and $F^{\mu\nu}(x)$, with the help of (11.26) and (11.36) it should be easy to see that these two 1+3 equations fuse into a single 4-equation, i.e. one covariant equation:

$$\nabla_\alpha F^{\alpha\mu}(x) = \frac{1}{\varepsilon_0 c} J^\mu(x). \tag{11.40}$$

We shall demonstrate explicitly how (11.39a) shapes into the $\mu = 0$ component of (11.40):

$$\text{Left side} = \frac{\partial E_x}{\partial x} + \frac{\partial E_y}{\partial y} + \frac{\partial E_z}{\partial z} = \nabla_1 F^{10} + \nabla_2 F^{20} + \nabla_3 F^{30}$$
$$= \nabla_\mu F^{\mu 0}.$$

$$\text{Right side} = \frac{1}{\varepsilon_0 c} J^0. \qquad \text{Hence,} \quad \nabla_\alpha F^{\mu 0} = \frac{1}{\varepsilon_0 c} J^0. \qquad \text{(QED)}$$

With the above hint the reader should be able to demonstrate equivalence between the x, y, z components of (11.39b) and the $\mu = 1, 2, 3$ components of (11.40).

Now we shall convert Eqs. (11.39c) and (11.39d) into covariant form. For this purpose, we shall obtain from $F^{\mu\nu}$ its *dual* $\mathfrak{F}^{\mu\nu}$ by the following operation:

$$\mathfrak{F}^{\mu\nu}(x) = \frac{1}{2}\varepsilon^{\mu\nu\alpha\beta} F_{\alpha\beta}(x). \tag{11.41}$$

Here $F_{\alpha\beta}$ is the covariant tensor obtained from $F^{\mu\nu}$ by lowering both indices (Reader, confirm it):

$$F_{\alpha\beta} = g_{\alpha\mu}g_{\beta\nu}F^{\mu\nu} = \begin{pmatrix} 0 & E_x & E_y & E_z \\ -E_x & 0 & -cB_z & cB_y \\ -E_y & cB_z & 0 & -cB_x \\ -E_z & -cB_y & cB_x & 0 \end{pmatrix} \tag{11.42}$$

and $\varepsilon^{\mu\nu\alpha\beta}$ is the *Levi-Civita symbol.* We had defined this as a 3-symbol in Sec. 5.1.3, under Eq. (5.13). We now extend the same symbol to a 4-symbol, defined as

$$\varepsilon^{\mu\nu\alpha\beta} = \begin{cases} 0 & \text{if any two indices equal,} \\ 1 & \text{if } \mu\nu\alpha\beta = 0123\text{, or any even permutation of } 0123, \\ -1 & \text{if } \mu\nu\alpha\beta \text{ is any odd permutation of } 0123. \end{cases} \tag{11.43}$$

For example, 1023 and 1203 are obtained from a single (i.e. odd) permutation and from two (i.e. even) permutations of 0123. The reader should prove the following important identity:

$$\Omega^{\mu}_{\cdot\,\alpha}\Omega^{\nu}_{\cdot\,\beta}\Omega^{\kappa}_{\cdot\,\gamma}\Omega^{\lambda}_{\cdot\,\sigma}\,\varepsilon^{\alpha\beta\gamma\sigma} = \varepsilon^{\mu\nu\kappa\lambda}, \tag{11.44}$$

where $\hat{\Omega}$ represents a proper Lorentz transformation matrix. As a consequence, we can regard the *Levi-Civita symbol to be a contravariant tensor of rank 4, which transforms into itself under a proper Lorentz transformation.* There is only one other tensor, namely, the metric tensor $g_{\mu\nu}$ which has a similar property. Going back to (11.41) and (11.42), the reader should work out the components of $\mathfrak{F}^{\mu\nu}$, and show that

$$\mathfrak{F}^{\mu\nu} = \begin{matrix} \\ 0 \\ 1 \\ 2 \\ 3 \end{matrix} \begin{matrix} \;\;0 & \;\;1 & \;\;2 & \;\;3 \\ \end{matrix} \begin{pmatrix} 0 & -cB_x & -cB_y & -cB_z \\ cB_x & 0 & E_z & -E_y \\ cB_y & -E_z & 0 & E_x \\ cB_z & E_y & -E_x & 0 \end{pmatrix}. \tag{11.45}$$

It should now be a simple exercise to show that Eqs. (11.39c) and (11.39d) can be written covariantly as

$$\nabla_{\alpha}\mathfrak{F}^{\alpha\mu} = 0. \tag{11.46}$$

We shall demonstrate explicitly for one component, say $\mu = 3$.

$$\begin{aligned}
\nabla_\alpha \mathfrak{F}^{\alpha 3} &= \nabla_0 \mathfrak{F}^{03} + \nabla_1 \mathfrak{F}^{13} + \nabla_2 \mathfrak{F}^{23} + \nabla_3 \mathfrak{F}^{33} \\
&= \frac{\partial}{\partial\, ct}(-cB_x) + \frac{\partial}{\partial x}(-E_y) + \frac{\partial}{\partial y}(E_x) \\
&= -\left(\frac{\partial E_y}{\partial x} - \frac{\partial E_x}{\partial y} + \frac{\partial cB_z}{\partial\, ct}\right) = \left[\boldsymbol{\nabla} \times \mathbf{E} + \frac{\partial c\mathbf{B}}{c\,\partial t}\right]_x = 0. \qquad \text{(QED)}
\end{aligned}$$

In summary, Maxwell's equations written in 1+3 forms in (11.39) reduce to two covariant equations:

$$\boxed{\begin{aligned}
&\text{Inhomogeneous part: } \nabla_\alpha F^{\alpha\mu}(x) = \frac{1}{\varepsilon_0 c} J^\mu(x). && \text{(a)} \\
&\text{Homogeneous part: } \nabla_\alpha \mathfrak{F}^{\alpha\mu}(x) = 0. && \text{(b)}
\end{aligned}} \qquad (11.47)$$

There is one more way to express the homogeneous equation, namely,

$$\nabla^\mu F^{\nu\eta} + \nabla^\nu F^{\eta\mu} + \nabla^\eta F^{\mu\nu} = 0. \qquad (11.48)$$

Note that there is a cyclic permutation of the three indices μ, ν, η in the above equation. Also, due to antisymmetry of $F^{\mu\nu}$ the left-handside is identically zero, unless μ, ν, η are all different. This equation therefore represents only four equations, namely corresponding to $(\mu, \nu, \eta) = (0,1,2), (0,2,3), (0,3,1), (1,2,3)$. These four equations correspond to the 1+3 equations represented by (11.39c) and (11.39d). The four components of the operator ∇^μ are shown in (7.113b).

We shall verify equivalence between (11.48) and (11.39c) and (11.39d) for one combination, namely $\nu\mu\eta = 012$, leaving to the reader verification for the other three combinations:

$$\begin{aligned}
\nabla^0 F^{12} + \nabla^1 F^{20} + \nabla^2 F^{01} &= \frac{\partial(-cB_z)}{\partial\, ct} - \frac{\partial E_y}{\partial x} - \frac{\partial(-E_x)}{\partial y} \\
&= -\left[\boldsymbol{\nabla} \times \mathbf{E} + \frac{\partial\, c\mathbf{B}}{\partial\, ct}\right]_z = 0. \qquad \text{(QED)}
\end{aligned}$$

Therefore, Maxwell's equations written in 1+3 forms in (11.39) can be expressed covariantly in the second way:

$$\boxed{\begin{aligned}
&\text{Inhomogeneous part: } \nabla_\alpha F^{\alpha\mu}(x) = \frac{1}{\varepsilon_0 c} J^\mu(x). && \text{(a)} \\
&\text{Homogeneous part: } \nabla^\mu F^{\nu\eta} + \nabla^\nu F^{\eta\mu} + \nabla^\eta F^{\mu\nu} = 0. && \text{(b)}
\end{aligned}} \qquad (11.49)$$

There is a third way, perhaps a more powerful way, of expressing Maxwell's equations, namely, by writing them in terms of *potentials*. The homogeneous equations, i.e. Eqs. (11.39c) and (11.39d), define the scalar potential $\Phi(\mathbf{r}, t)$ and the vector potential $\mathbf{A}(\mathbf{r}, t)$ through the relations:

$$c\mathbf{B}(\mathbf{r}, t) = \boldsymbol{\nabla} \times c\mathbf{A}(\mathbf{r}, t), \tag{11.50a}$$

$$\mathbf{E}(\mathbf{r}, t) = -\boldsymbol{\nabla}\Phi(\mathbf{r}, t) - \frac{\partial c\mathbf{A}(\mathbf{r}, t)}{c\,\partial t}; \tag{11.50b}$$

because if we write $(\mathbf{E}, \mathbf{B})$ in the form of (11.50), the homogeneous part (11.39c) and (11.39d) of Maxwell's equations will be identically satisfied.

Adopting Lorentz gauge:

$$\boldsymbol{\nabla} \cdot \mathbf{A}(\mathbf{r}, t) + \frac{\partial \Phi(\mathbf{r}, t)/c}{c\,\partial t} = 0, \tag{11.51}$$

the source equations now reduce to the following inhomogeneous wave equations:

$$\nabla^2 \Phi - \frac{1}{c^2}\frac{\partial^2 \Phi}{\partial t^2} = \frac{1}{\varepsilon_0 c}\rho c, \tag{11.52a}$$

$$\nabla^2 \mathbf{A} - \frac{1}{c^2}\frac{\partial^2 \mathbf{A}}{\partial t^2} = \frac{1}{\varepsilon_0 c}\mathbf{J}. \tag{11.52b}$$

The set of Eqs. (11.50)–(11.52) constitute the *potential form* of Maxwell's equations (11.39). Together they are equivalent to the complete set of Maxwell's equations (11.39). We shall convert each of them into a covariant equation.

For this, we first define the 4-potential A^μ to be a contravariant 4-vector with space and time components:

$$A^\mu(x) \equiv \left(\frac{1}{c}\Phi(x), \mathbf{A}(x)\right), \tag{11.53}$$

and the d'Alembertian operator $\Box^2$, obtained by taking the "scalar product" of the operator ∇_μ with itself, as shown in (7.112c).

It is now obvious that the two 3-vector equations (11.50), defining the scalar and vector potentials, are equal to the following single "covariant equation", obtained by replacing the components of $(\mathbf{E}, c\mathbf{B})$ and $(\Phi/c, \mathbf{A})$ by the corresponding components of $F^{\mu\nu}$ and A^μ, identified from (11.36) and (11.53), respectively:

$$F^{\mu\nu}(x) = c[\nabla^{\mu} A^{\nu}(x) - \nabla^{\nu} A^{\mu}(x)]. \tag{11.54}$$

We shall demonstrate this equivalence for one specific example, namely, $\mu\nu = 10$.

$$\text{Right side} = c[\nabla^1 A^0 - \nabla^0 A^1] = -\frac{\partial \Phi}{\partial x} - \frac{\partial\, cA_x}{\partial\, ct} = E_x = F^{10}. \tag{QED}$$

Also, the Lorentz gauge condition (11.51) is clearly seen to be equivalent to the following covariant equation:

$$\nabla_{\mu} A^{\mu}(x) = 0. \tag{11.55}$$

Finally, the 1+3 inhomogeneous wave equations for the scalar and vector potentials, namely Eq. (11.52), now become one single covariant wave equation for the 4-potential:

$$\Box^2 A^{\mu}(x) = \frac{1}{\varepsilon_0 c^2} J^{\mu}(x). \tag{11.56}$$

In summary, we have the Potential form of Maxwell's equations:

$$\boxed{\begin{array}{l} \text{4-Pot defined:} \quad F^{\mu\nu}(x) = c[\nabla^{\mu} A^{\nu}(x) - \nabla^{\nu} A^{\mu}(x)]. \\ \text{Lorentz Gauge:} \ \nabla_{\mu} A^{\mu}(x) = 0. \\ \text{Wave equation:} \ \Box^2 A^{\mu}(x) = \dfrac{1}{\varepsilon_0 c^2} J^{\mu}(x). \end{array}} \tag{11.57}$$

We can supplement these equations with two more equations which represent the only invariants of the electromagnetic fields (i.e. they retain their values under a Lorentz transformation):

$$F^{\alpha\beta} F_{\alpha\beta} = 2(c^2 B^2 - E^2), \tag{11.58a}$$

$$F^{\alpha\beta} \mathfrak{F}_{\alpha\beta} = 2c\,\mathbf{B} \cdot \mathbf{E}. \tag{11.58b}$$

This exercise is best left to the reader. There are some interesting consequences of the above invariants, which the reader will explore through some examples in the problem set.

Before leaving this chapter, we remind the reader the covariant expressions of the equations of Classical Electrodynamics, laid down in the form of

(1) Continuity equation: (11.29);
(2) Minkowski force (corresponding the Lorentz force equation: (11.34);
(3) Maxwell's equations in terms of the field tensor in two ways: (11.47), (11.49);
(4) Maxwell's equations in terms of the 4-potential: (11.57).

We have framed each of these equations in a box to stress their importance.

11.6. EM Field of a Charged Particle in Uniform Motion

What we are doing in this section is partly a reflection of the statements we had made in Sec. 10.2. What we had done in that section using force transformation, will be done here for a very specific example, using field transformation. It is heartening to see that the force transformation and the field transformation, though they evolve from two entirely different offshoots of the special theory of relativity blend with each other, showing a larger homogeneity and consistency of the theory.

11.6.1. *Transformation from the rest frame to Lab frame*

The charge q is seated permanently at the origin of its *rest frame* S'. It is now viewed from the *Lab frame* S, moving in the *negative* X direction with velocity βc, relative to S', so that the charge is moving in the *positive* X direction with respect to the Lab, as explained in Fig. 11.5. The origins of S and S' coincide at $t = t' = 0$. Let us set $k = \frac{q}{4\pi\varepsilon_0}$. The EM field in the rest frame is

$$\mathbf{E}' = k\frac{\mathbf{r}}{r'^3} = \frac{k(x'\mathbf{i} + y'\mathbf{j} + z'\mathbf{k})}{r'^3}, \quad c\mathbf{B}' = \mathbf{0}, \tag{11.59}$$

where $r'^2 = x'^2 + y'^2 + z'^2$.

Let us now consider an event Θ having coordinates (x, y, z, ct) in S and (x', y', z', ct') in S'. Let the EM field at this event be $(\mathbf{E}, c\mathbf{B})$ in S and $(\mathbf{E}', c\mathbf{B}')$ in S'. The Lorentz transformation of the coordinates from S' to S is given by the formula (3.9), whereas the Lorentz transformation of the field components (*at the event* Θ) from S' to S is the inverse of the equations (11.37), i.e. with β replaced by $-\beta$.

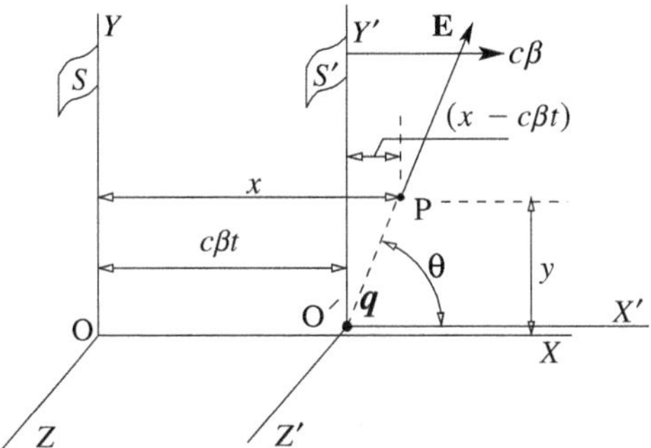

Fig. 11.5. The rest frame S of the point charge q moving with velocity $c\beta$ with respect to the Observer's frame S'.

Now consider the following event Θ_{P} = "the EM field due to the *above point charge* q is detected by a certain observer P" situated on the XY-plane, so that $z = z' = 0$. Ignoring the z coordinate, we transform the field given in (11.59) to the frame S', using (3.9) and (11.37). In this first conversion, the radial distance r' transforms into R, which we write in its relation to S

$$r'^2 = \gamma^2(x - \beta ct)^2 + y^2 + z^2 \equiv R^2. \tag{11.60}$$

The EM field $(\mathbf{E}, c\mathbf{B})$ in S transforms into the EM field $(\mathbf{E}', c\mathbf{B}')$ in S', whose components are given as

$$\begin{aligned} E_x &= E'_x = k\frac{x}{r'^3} = k\frac{\gamma(x - \beta ct)}{R^3}, \quad & cB_x &= cB'_x = 0, \\ E_y &= \gamma E'_y = k\frac{\gamma y'}{r'^3} = k\frac{\gamma y}{R^3}, & cB_y &= \gamma(cB'_y - \beta E'_z) = 0, \\ E_z &= \gamma E'_z = k\frac{\gamma z'}{r'^3} = k\frac{\gamma z}{R^3} = 0, & cB_z &= \gamma(cB'_z + \beta E'_y) = k\gamma\beta\frac{y}{R^3}. \end{aligned} \tag{11.61}$$

Let $\tilde{\mathbf{r}}$ be the radius vector stretching from the instantaneous location of charge q at the time t to the field point P, as seen from the Lab frame S. Then

$$\begin{gathered} \tilde{\mathbf{r}} = (x - \beta ct)\mathbf{i} + y\mathbf{j} + z\mathbf{k}, \\ \text{and note that: } \boldsymbol{\beta} \times \mathbf{E} = \beta\mathbf{i} \times (E_x\mathbf{i} + E_y\mathbf{j}) = \beta E_y\mathbf{k}. \end{gathered} \tag{11.62}$$

Now the six components of the transformed field written in (11.61) can be written compactly as

$$\mathbf{E} = \frac{q}{4\pi\varepsilon_0}\frac{\gamma\tilde{\mathbf{r}}}{R^3}, \tag{11.63a}$$

$$c\mathbf{B} = \boldsymbol{\beta} \times \mathbf{E}. \tag{11.63b}$$

Note that Eq. (11.63) can be obtained entirely from (10.9) of Sec. 10.2, if we write for the $\mathbf{E}'$ field the expression given in (11.59).

11.6.2. *Pictorial interpretation of the fields*

To get a picture of the **E** field, we shall obtain two sets of plot, showing (a) its angular distribution around the moving charge q; (b) its time variation at any given observation point P. For each case, we shall use "Gnuplot" to create exact plots corresponding to three values of β, namely $\beta \ll 1, = 0.4, 0.95$ in the first case, and $\beta = 0.4, 0.8, 0.95$ in the second case.

Angular distribution of the field around the moving charge

We shall obtain a pictorial interpretation of the **E** field described by Eq. (11.63a). Even though we had taken the field point P on the XY-plane for convenience, that restriction is withdrawn for writing the general expression for the field in terms of the radius vectors. The **E** field is *radial* in both frames of reference S' and S (i.e. emanating radially from the instantaneous location of the point charge). However, it is the *isotropic* Coulomb field in the rest frame S, whereas *angle dependent* in the Lab frame S'.

To picture this we have set up "displaced" Cartesian axes $\tilde{X}, Y, Z$ of the Lab frame, the origin A of which coincides with the instantaneous location of the charge q at the instant t, shown in Fig. 11.6. We shall denote the "displaced" Cartesian coordinates as $(\tilde{x}, y, z)$, and set up the spherical coordinates (r, θ, ϕ), as illustrated in Fig. 11.6. It is obvious that $\tilde{x} = x - \beta ct$ whereas the "displaced" y, z coordinates are the same as with respect to the original axes.

The relations (1) between the $(\tilde{x}, y, z)$ and (r, θ, ϕ), and (2) between the Cartesian unit vectors and the spherical unit vectors, are given by the

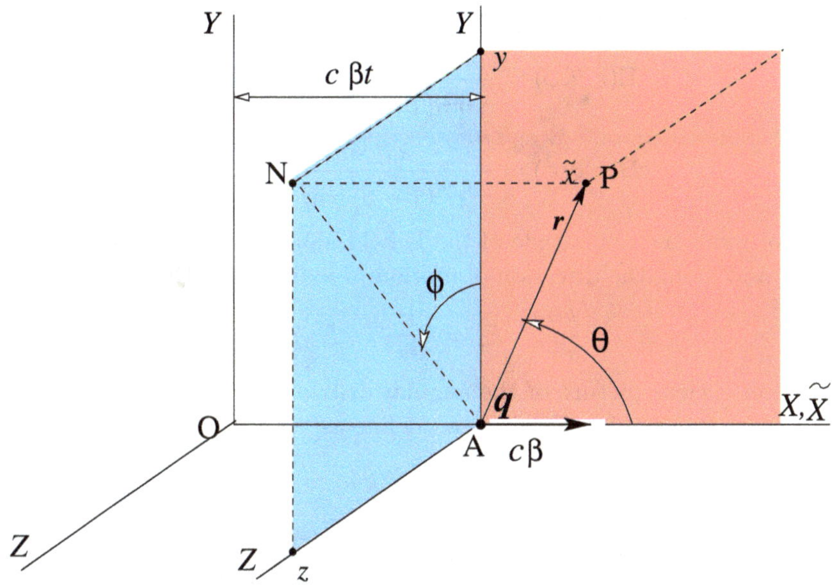

Fig. 11.6. Displaced axes with its origin at the instantaneous location of the moving particle at time t.

following formulas:

$$\begin{aligned} y &= r\, \sin\theta\, \cos\phi, & \mathbf{e}_r &= \sin\theta\,(\cos\phi\,\mathbf{j} + \sin\phi\,\mathbf{k}) + \cos\theta\,\mathbf{i}, \\ z &= r\, \sin\theta\, \sin\phi, & \mathbf{e}_\theta &= \cos\theta\,(\cos\phi\,\mathbf{j} + \sin\phi\,\mathbf{k}) - \sin\theta\,\mathbf{i}, \\ \tilde{x} &= r\, \cos\theta, & \mathbf{e}_\phi &= -\sin\phi\,\mathbf{j} + \cos\phi\,\mathbf{k}. \end{aligned} \tag{11.64}$$

The $(\mathbf{E}, \mathbf{B})$ field of Eq. (11.63) can be written as

$$\mathbf{E}(r, \theta, \phi) = k\frac{\gamma\, r}{R^3}\,\mathbf{e}_r, \tag{11.65a}$$

$$c\mathbf{B}(r, \theta, \phi) = \beta E \sin\theta\,\mathbf{e}_\phi. \tag{11.65b}$$

To get a make the picture complete, we need to express R in terms of r, θ and β, beginning with Eq. (11.60), exploiting the Cartesian $\to$ spherical conversion formulas (11.64), and using the relation (11.60):

$$R^2 = \gamma^2\tilde{x}^2 + y^2 + z^2 = \gamma^2 r^2\left(1 - \beta^2\sin^2\theta\right). \tag{11.66}$$

Therefore, we can rewrite Eq. (11.65) as

$$\mathbf{E}(r,\theta,\phi) = \frac{k}{\gamma^2 r^2 \left(1-\beta^2\sin^2\theta\right)^{3/2}}\,\mathbf{e}_r, \tag{11.67a}$$

$$c\mathbf{B}(r,\theta,\phi) = \frac{k\,\beta\,\sin\theta}{\gamma^2 r^2 \left(1-\beta^2\sin^2\theta\right)^{3/2}}\,\mathbf{e}_\phi. \tag{11.67b}$$

If we write the magnitude of the $\mathbf{E}$ field along the YZ-plane ($\theta = \pi/2$) as $E_\perp$, and along the direction of motion ($\theta = 0$) as $E_\parallel$, then

$$E_\perp = \frac{k\gamma}{r^2}, \quad E_\parallel = \frac{k}{\gamma^2 r^2}, \quad E_\perp/E_\parallel = \gamma^3. \tag{11.68}$$

To get a clear picture of the angular distribution of the field, we shall plot $\mathbf{E}$ on a "unit sphere", which we define as

$$r_{\mathrm{u}} = \sqrt{k} = \sqrt{\frac{q}{4\pi\varepsilon_0}},$$

$$\text{so that, from (11.67)}\quad \mathbf{E}(r_{\mathrm{u}},\theta,\phi) = \frac{1}{\gamma^2\left(1-\beta^2\sin^2\theta\right)^{3/2}}\,\mathbf{e}_r. \tag{11.69}$$

Figure 11.7 shows plots of this field surrounding the moving point charge, on a unit sphere, corresponding to three values of β, namely $\beta \ll 1$, $\beta = 0.8$, $\beta = 0.95$. The $E_\perp/E_\parallel$ ratios for the above three values of β are shown in Table 11.1. The length of the arrow in each figure is proportional to the strength of the field. We have plotted the field *exactly* (i.e. the lengths of the arrows exactly) using Gnuplot.

Table 11.1. $E_\perp/E_\parallel$ for 3 values of β.

β	γ	$E_\perp/E_\parallel$
$\ll 1$	≈ 1	≈ 1
0.8	1.66	4.63
0.95	3.2	32.84

Time variation at any given observation point P

The observation point is taken on the Y-axis, at $x = 0, y = 1$. Substituting these values in (11.63a), and setting $\frac{q}{4\pi\varepsilon_0} = 1, c = 1$ we get the x- and y-components of the $\mathbf{E}$ field (its z-component is zero).

$$E_x(t) = -\frac{\gamma\beta t}{[(\gamma\beta t)^2+1)]^{3/2}}, \quad E_y(t) = \frac{\gamma}{[(\gamma\beta t)^2+1)]^{3/2}}. \tag{11.70}$$

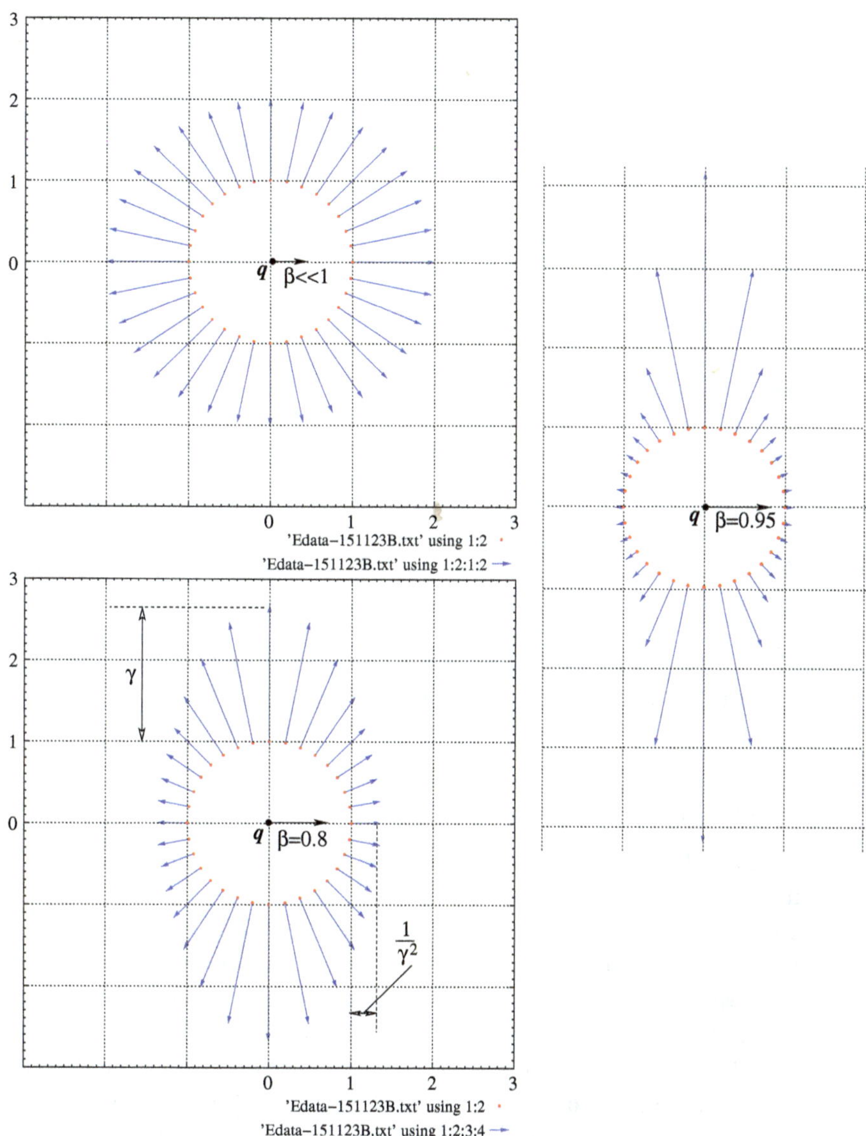

Fig. 11.7. Electric field lines on unit sphere corresponding to $\beta \ll 1, = 0.8, = 0.95$.

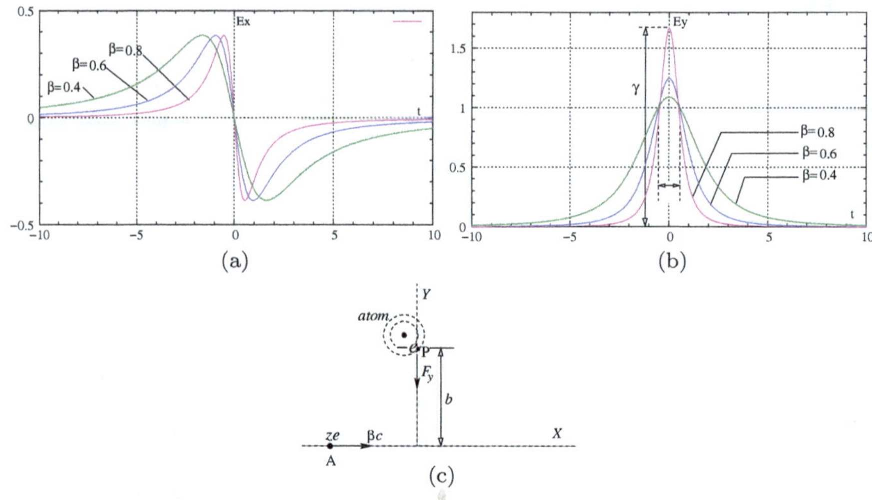

Fig. 11.8. (a) and (b) Plots of the E_y, E_x as functions of time, corresponding to $\beta = 0.4, 0.6, 0.8$. (c) Ion A passing an atomic electron with velocity $c\beta$ in its passage through matter.

We have plotted these two fields as functions of t, and corresponding to $\beta = 0.4, 0.8, 0.95$ in Figs. 11.8(a) and 11.8(b).

11.6.3. *Ionizing effect of a heavy ion in its passage through matter*

The formula (11.63a), and its time plot shown in Figs. 11.8(a) and 11.8(b) have important and interesting applications in nuclear physics. They are used to calculate the ionizing effect of a "heavy ion", e.g. a proton (hydrogen ion), α-particle (helium ion), as they pass through matter with high velocity, close to the speed of light, and the *Range* of such particles in matter (i.e. distance travelled before being stripped).

We shall get an appreciation of the "unit radius" defined in (11.69) with a realistic example, say a "heavy charged particle" A of positive electric charge ze (e.g. a proton with $z = 1$ or an α-particle for which $z = 2$, the corresponding "light particle" in this case is electron). As the ion moves through a piece of matter, e.g. gold foil, or an aluminium strip it ionizes the atoms, by pulling out electrons with its electric force. To make things simple we take the ionizing particle to be a proton, for which $q = e = 1.6 \times 10^{-19}$ C. Setting $\frac{1}{4\pi\epsilon_0} = 9 \times 10^9\,\text{N.m}^2/\text{C}^2$, we get $r_{\text{u}} = 3.8 \times 10^{-5}$ m, or, 38 microns. This "unit radius" is therefore quite large compared to the radius of the ionized atoms. We have used it only for the convenience of plotting, without attaching any further significance to it.

The phenomenon we are talking about is a *collision* of A with e^- (even though they do not physically touch each other), depicted in Fig. 11.8(c). The particle A of charge ze moves along the X-axis with velocity $v = c\beta$, encountering an atomic electron of charge $-e$, located at P, at an impact parameter b. Being heavy (infinitely heavy by assumption), it moves along without being deflected from its straight path. As it zips past by the atom it exerts a short burst of electric force $\mathbf{F}_e$ on all electrons in the atom, lasting for a very short time Δt, compared to the time period of the electronic orbits (pre-quantum classical picture). In this short burst, the movement of the electron is so small that it can be treated to be stationary. (Obviously, the magnetic force $\mathbf{F}_m$ will have practically no effect.) As a result, some of the electrons in the atom will receive enough energy to be ejected out of the atom, thereby ionizing it, while some others will be excited to higher energy levels.

Let us get back to some calculations. It is seen from the plots in Fig. 11.8 that E_x is antisymmetric, hence no effect when integrated over the t-axis. In contrast E_y is symmetric. When integrated over the t-axis, it will give a net area under the curve, $\tilde{a} \approx h\,\delta t$, where $h = \frac{k\gamma}{b^2}$ and $\delta t = 2t_{1/2} = \frac{2b}{\gamma v}$, the width of the curve at "half height", standing for the effective "duration of the impact". (Reader may verify the estimate.) Then $\tilde{a} \approx \frac{2k}{\gamma v}$. The passage of the heavy particle gives a sharp impulse, i.e. a momentum transfer $\delta p = e\tilde{a} = \frac{ze^2}{4\pi\varepsilon_0}\frac{2}{vb}$, in the negative Z-direction (net pulling effect of the traversing particles). The corresponding energy transfer is

$$\Delta\mathcal{E} = \frac{\Delta p^2}{2m_e} = \frac{1}{(4\pi\varepsilon_0)^2}\frac{2(ze^2)^2}{m_e v^2 b^2}. \tag{11.71}$$

The above formula has important application in nuclear physics [36–38].

A charged particle, like α-ray, while passing through matter, keeps losing its kinetic energy by transferring it to atomic electrons, thereby liberating them, and producing electron–ion pairs, and ultimately getting stopped after traversing an average distance R, called its *range* in the given material. For example, the 5.4 MeV α-particle [38] emitted by the polonium isotope P^{210}, carrying energy 5.3 MeV, has a range of about 10 cm in dry air. As it passes through, the material it keeps losing energy, by producing electron–ion pairs, which in this case is about 6600 pairs per mm.

We shall not pursue this matter further. For an exact calculation of these effects and an estimation of the Range of the particle, the interested reader can look up standard books in Nuclear Physics as cited in the footnote.

11.7. Exercises for the Reader III

Problem 11.1. The relative nature of the electric field **E** and the magnetic field **B** can be seen in an elementary manner as follows. Let a particle of electric charge q move along the X-axis with the velocity $c\beta$, as seen from a Lab frame S, with respect to which there is a uniform magnetic field $\mathbf{B} = B\,\mathbf{j}$ and no **E** field at all. According to the Lorentz force equation (11.30), the particle experiences a magnetic force $\mathbf{F} = q\,c\beta\,B\,\mathbf{k}$. With respect to the rest frame S' of the particle, however the particle is at rest, and hence, there is no magnetic force. The same magnetic force therefore now appears as an *electric force*. Therefore, there is an electric field equal to $\mathbf{E} = c\beta\,B\,\mathbf{k}$, in the Z-direction.

The answers suggested above are only approximate in view of the fact that we did not apply the correct force transformation equation (8.55) to obtain the force in S'. (a) Apply the suggested correction to obtain the **E** field in S'. (b) Check your answer using the LT equation (11.37) for the electromagnetic field under the boost:$S(c\beta, 0, 0)S'$. (c) Does the absence of "magnetic experience" by the particle in the rest frame imply vanishing **B** field in S'? Check your answer using (11.37) again.

Problem 11.2. The conclusions of Problem 11.2 raise many paradoxes. A "pure" magnetic field in S can be produced by a "neutral" wire carrying electric current, as is usually the case for a d.c. circuit. How can such a wire cause an electric field in another frame of reference?

To solve this paradox, consider a line current I flowing through a straight wire along the X-direction in a Lorentz frame S. This current can be considered to be due to a stream of negative charges (namely, the electrons) of charge density $\rho_{\text{neg}} = -\rho_o$ with a drift velocity $-c\nu$ along the X-axis, overlying a stationary distribution of positive charges (namely, positive ions at the lattice points of the conductor) of equal and opposite charge density $\rho_{\text{pos}} = \rho_o$. Assuming that the wire has a circular cross-section of radius a, the electric current density and the charge density ρ for this configuration are given as follows:

$$j_x = \begin{Bmatrix} \rho_o c\,\nu, & r \leq a \\ 0, & r > a \end{Bmatrix}; \; j_y = j_z = 0; \; \rho = \rho_{\text{neg}} + \rho_{\text{pos}} = 0. \tag{11.72}$$

This configuration produces only **B** field, but no **E** field.
(a) Find the **B** field at the point $P(0, y, 0)$ on the Y-axis produced by the above current.
Hint: The $(\mathbf{E}, \mathbf{B})$ fields at P due to a line charge density λ and a line current density I, both along the X-axis, are given as

$$\mathbf{E} = \frac{\lambda}{2\pi\varepsilon_0\, y}\mathbf{j}, \quad \mathbf{B} = \frac{I}{2\pi\varepsilon_0\, c\, y}\mathbf{k}. \tag{11.73}$$

(b) Using (11.37) show that the EM field at P, as seen by an observer S' who is moving along the X-axis with velocity $c\beta$, is

$$E'_x = E'_z = 0,\ E'_z = -\frac{\gamma\nu\beta\rho_o a^2}{2\varepsilon_0 y}; \quad cB'_x = cB'_y = 0;\ cB'_z = \frac{\gamma\nu\rho_o a^2}{2\varepsilon_0 y}. \tag{11.74}$$

(c) Using LT of the 4-current density shown in (11.72) show that the electric current density and charge density with respect to S' are

$$j'_x = \left\{\begin{matrix} \rho_o\gamma c\nu, & r \leq a \\ 0, & r > a \end{matrix}\right\};\ j'_y = j'_z = 0;\ \rho' = \rho'_{\text{neg}} + \rho'_{\text{pos}} = -\gamma\nu\beta\rho_o. \tag{11.75}$$

(d) With the help of (11.73), and using the current and charge densities shown in (11.75) obtain the $(\mathbf{E}, \mathbf{B})$ fields at P as measured by S'. Verify that you get the same answer as in (11.74).

Moral: A wire which is neutral with respect to a Lorentz frame S is not neutral with respect to another Lorentz frame S', which is why there is an **E** *field in S', in addition to the* **B** *field due to the electric current.*

Problem 11.3. We saw in the last two problems that a pure magnetic field in one Lorentz frame will appear as a combination of magnetic field and electric field in another. We shall examine the reverse of this case in this problem. Consider a frame S in which the field is purely **E**, there being no **B**-component. The force on a charge q moving with velocity $\mathbf{v} = c\boldsymbol{\nu}$ is therefore $\mathbf{F} = q\mathbf{E}$.
(a) Using the force and velocity transformation formulas (8.55) and (4.23) (and their inverses if necessary), show that the force components in the

frame S', under the boost: $S(c\beta, 0, 0)S'$ becomes

$$\begin{aligned} F'_x &= q[E_x - \beta\gamma(\nu'_y E_y + \nu'_z E_z)], \\ F'_y &= q\gamma(1 + \beta\nu'_x)E_y, \\ F'_z &= q\gamma(1 + \beta\nu'_x)E_z. \end{aligned} \tag{11.76}$$

(b) Hence, show with the help of (11.30) that there are both **E** and **B** fields in S', given as

$$\begin{aligned} &E'_x = E_x,\ E'_y = \gamma E_y, \quad E'_z, = \gamma E_z, \\ &cB'_x = 0,\ cB'_y = \gamma\beta E_z,\ cB'_z = -\gamma\beta E_y. \end{aligned} \tag{11.77}$$

(c) Check Eq. (11.77) against the general field transformation equation (11.37).

Part IV
4-Momentum Conservation in Continuous Media

Chapter 12

The Energy Tensor

12.1. Why Energy Tensor?

Forces of gravity originate from massive objects like the earth, the Sun, the stars — as Newton's theory of Universal Gravitation tells us. Mass loses the pristine purity enjoyed in Newtonian mechanics, because the relativistic mass is no longer constant (being a function of velocity), and the rest mass is no longer conserved. What replaces mass is energy, thanks to $\mathcal{E} = mc^2$. Energy, in turn, is the time component of a larger entity, namely 4-momentum (or, En-Mentum, a term coined in Sec. 8.4).

For an analytical study of gravitational field in Newtonian formalism, its source is expressed in terms of mass density. That density now needs to be replaced by some 4-momentum density. However, density being something per unit volume — and a unit volume in one frame being not so unit in another — the search for an appropriate density leads towards a 4-tensor which is generally known as energy–momentum–stress tensor. We would however prefer to give it a shorter name — energy tensor. Einstein's General Theory of Relativity traces the source of gravitation to this energy tensor. There is another reason for knowing the energy tensor. Even within the ambit of special relativity, it is a legitimate urge to write the conservation equations for energy and momentum in the appropriate language, namely covariantly. One would stumble upon the energy tensor in this attempt to write covariant conservation equations, which is also a prerequisite to the study of relativistic quantum mechanics and quantum field theory. The concept of energy tensor is built upon a more elementary three-dimensional base namely, the stress tensor. We had presented a detailed exposition of stress tensor in Sec. 5.2, and of Maxwell's stress tensor in particular in Chapter 6. The volume force density $\mathbf{f}_s(\mathbf{r})$ in matter (solid or fluid) at a point $\mathbf{r}$ was shown to be the divergence of the stress tensor field $\widehat{\mathcal{T}}(\mathbf{r})$, in Eq. (5.79), which we are rewriting here:

$$\mathbf{f}_s(\mathbf{r}) = \boldsymbol{\nabla} \cdot \widehat{\mathcal{T}}(\mathbf{r}). \tag{12.1}$$

12.2. Minkowski Volume Force Density

Passing from Euclidean space to Minkowski space–time, one might expect the four-dimensional generalization of stress tensor to be a 4-tensor field whose 4-divergence would yield Minkowski volume force density. However, density functions do not exhibit well-defined transformation property unless volume is measured in the *comoving frame*. We had touched on this aspect in Sec. 11.3. We transplant Fig. 11.4 from that section here and relabel it as Fig. 12.1. It shows a stream of particles constituting a fluid in motion. An infinitesimal volume δV (shown coloured in the figure), identified at the event point $(x) = (ct, \mathbf{r})$, contains a collection of fluid particles, which together possess a rest mass δm_o, a quantity of charge δq, and is moving

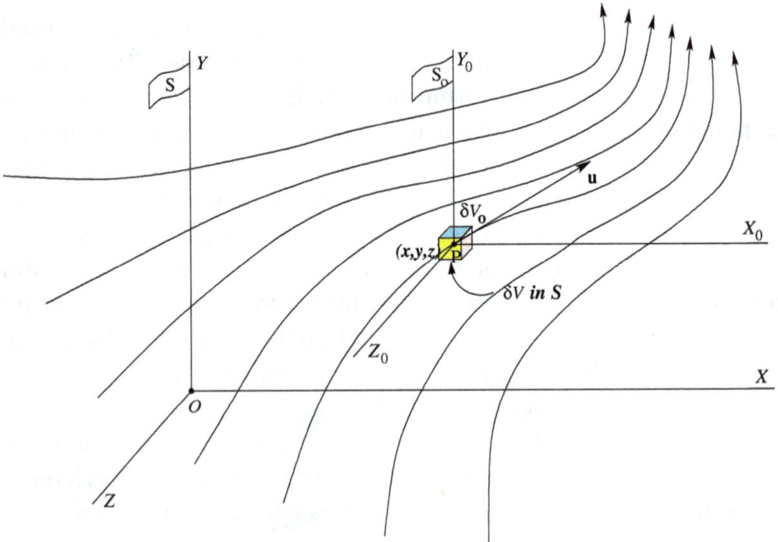

Fig. 12.1. Lab frame S and comoving frame S_o in a streamline flow of particles.

with the velocity $\mathbf{u}(\mathbf{r}, t)$ with respect to the Lab frame S. The corresponding 4-velocity U^μ and the Lorentz factor $\Gamma(x)$, copied from Eqs. (11.24) and (11.20) are

$$U^\mu(x) = \Gamma(x)(c, \mathbf{u}(x)) = \Gamma(c, u_1, u_2, u_3), \tag{12.2a}$$

$$\Gamma(x) = \frac{1}{\sqrt{1 - \frac{u^2(x)}{c^2}}}. \tag{12.2b}$$

Lorentz contraction of the dimension of this box along the direction of $\mathbf{u}$, changes its proper volume δV_o to the laboratory volume

$$\delta V = \frac{1}{\Gamma(x)} \delta V_o, \quad \text{or,} \ \delta V_o = \Gamma(x) \delta V. \tag{12.3}$$

Let the Minkowski force acting on these particles (inside the proper volume δV_o) be $\delta \vec{\boldsymbol{\mathcal{F}}}(x) = \vec{\mathbf{e}_\mu}\, \delta \mathcal{F}^\mu(x)$, the corresponding 3-force $\delta \mathbf{F}(x)$, and the power received $\delta \Pi(x)$. Then from (8.31)

$$\begin{aligned} \delta \vec{\boldsymbol{\mathcal{F}}}(x) &= \Gamma(x) \left(\frac{\delta \Pi(x)}{c}, \ \delta \mathbf{F}(x) \right) \\ &= \Gamma(x)\, \delta V \left(\frac{1}{c} \frac{\delta \Pi(x)}{\delta V}, \frac{\delta \mathbf{F}(x)}{\delta V} \right) \\ &= \delta V_o \left(\frac{1}{c} \frac{\delta \Pi(x)}{\delta V}, \frac{\delta \mathbf{F}(x)}{\delta V} \right). \end{aligned} \tag{12.4}$$

We define Minkowski volume 4-force density $\vec{\boldsymbol{f}}(x)$ to be the *Minkowski force per unit proper volume* — the 3-scalar density ϖ (to be pronounced as var-pi) as the power received per unit *lab volume*, or power density and 3-vector density $\mathbf{f}(x)$ as the 3-force per unit *lab volume*, as explained below:

$$\vec{\boldsymbol{f}}(x) \equiv \lim_{\delta V_o \to 0} \frac{\delta \vec{\boldsymbol{\mathcal{F}}}(x)}{\delta V_o}, \tag{12.5a}$$

$$\varpi(x) \equiv \lim_{\delta V \to 0} \frac{\delta \Pi(x)}{\delta V}, \tag{12.5b}$$

$$\mathbf{f}(x) \equiv \lim_{\delta V \to 0} \frac{\delta \mathbf{F}(x)}{\delta V}. \tag{12.5c}$$

It follows from (12.4) and (12.5) that

$$\delta \vec{\boldsymbol{\mathcal{F}}}(x) = \delta V_o \, \vec{\boldsymbol{f}}(x), \tag{12.6a}$$

$$\text{where} \quad \vec{\boldsymbol{f}}(x) = \left(\frac{\varpi}{c}, \ \mathbf{f}(x) \right). \tag{12.6b}$$

An example of Minkowski volume 4-force density $\overrightarrow{\boldsymbol{f}}(x)$ and its time and space components can be seen in Eq. (12.20a).

At this point, we shall remind the reader of the convention we adopted in Sec. 8.1 and onwards. A 4-vector $\overrightarrow{\mathbf{A}}$ can also be written in terms of its 4-components $\{A^\mu; (\mu = 0,1,2,3)\}$ as $\overrightarrow{\mathbf{A}} = \overrightarrow{\mathbf{e}_\mu} A^\mu$. A 4-tensor can also be written in terms of its 4×4 components $\{T^{\mu\nu}; (\mu,\nu = 0,1,2,3)\}$ as $\overrightarrow{\overrightarrow{\mathbf{T}}} = \overrightarrow{\mathbf{e}_\mu} T^{\mu\nu} \overrightarrow{\mathbf{e}_\nu}$. Sometimes we shall refer to a 4-vector or a 4-tensor as a whole "geometrical object" like $\overrightarrow{\mathbf{A}}$ and $\overrightarrow{\overrightarrow{\mathbf{T}}}$. Sometimes we shall write the same quantities in terms of its components as A^μ and $T^{\mu\nu}$.

The 4-stress tensor is now defined to be a symmetric tensor: $\overrightarrow{\overrightarrow{\mathcal{T}}}(x) = \overrightarrow{\mathbf{e}_\mu} \mathcal{T}^{\mu\nu}(x) \overrightarrow{\mathbf{e}_\nu}$ of rank 2, satisfying the requirement:

$$\boxed{\overrightarrow{\boldsymbol{f}} \equiv \overrightarrow{\boldsymbol{\nabla}} \cdot \overrightarrow{\overrightarrow{\mathcal{T}}} \Rightarrow f^\mu(x) \equiv \nabla_\alpha T^{\alpha\mu}(x).} \tag{12.7}$$

We may as well call it *Minkowski 4-stress tensor.*

Note from (7.113) the time and space components of the operator are given by $\overrightarrow{\boldsymbol{\nabla}} = (\frac{1}{c}\frac{\partial}{\partial t}, \boldsymbol{\nabla})$.

In the next few sections, we shall work towards such a tensor for a stream of incoherent dust, consisting of electrically charged particles, subjected to only electromagnetic forces.

12.3. Energy and Momentum Conservation in One Voice

Before proceeding further we shall apply Lorentz force (11.30) to a volume distribution of electric charges. Write the Lorentz force $\delta\mathbf{F}$ on the charge content δq inside an infinitesimal volume δV, by replacing q with $\rho\,\delta V$, where ρ is the charge density as defined in Sec. 11.3. Now divide each side with δV, and use the definition of volume current density $\mathbf{J}$, as defined in (11.19). The Lorentz force density at the event point (x) is now

$$\boldsymbol{f}_{\text{em}} = \lim_{\delta V \to 0} \frac{\delta\mathbf{F}}{\delta V} = \rho\mathbf{E} + \mathbf{J} \times \mathbf{B}. \tag{12.8}$$

There are two important theorems that are used to state the conservation laws involving the electromagnetic forces. We shall state them as two theorems, because they follow directly from Maxwell's equations.

Consider the same stream of charged particles subjected to electromagnetic forces of their own creation. We shall apply energy and momentum conservation theorems to this system of particles.

(A) *The energy theorem*, also called *Poynting's theorem*[a] is written as follows:

$$\mathbf{E}\cdot\mathbf{J}+\frac{\partial w}{\partial t}=-\boldsymbol{\nabla}\cdot\mathbf{S}. \tag{12.9}$$

We have proved the above theorem in Appendix A.1.

We interpret the terms appearing in the above equation as follows:

$$\begin{aligned}\mathbf{E}\cdot\mathbf{J} &= \text{work done by the field on the fluid particles per unit volume,}\\ &= \text{rate of change of kinetic energy per unit volume,}\end{aligned} \tag{12.10a}$$

$$w=\frac{\varepsilon_0}{2}(E^2+c^2B^2)=\text{field energy density,} \tag{12.10b}$$

$$\mathbf{S}=\varepsilon_0 c^2(\mathbf{E}\times\mathbf{B})=\text{field energy flux density}=\text{Poynting's vector.} \tag{12.10c}$$

All densities alluded to in the context of energy–momentum theorems (12.9) and (12.13) are *lab* densities. See comments after Eq. (12.4).

To justify the above interpretation we integrate over a volume V bounded by a surface S, and applying Gauss's theorem, we get

$$\begin{aligned}\iiint_V\left(\mathbf{E}\cdot\mathbf{J}+\frac{\partial w}{\partial t}\right)dv &= -\iiint_V \boldsymbol{\nabla}\cdot\mathbf{S}\,dv\\ &= -\iint_S \mathbf{S}\cdot d\boldsymbol{a}.\end{aligned} \tag{12.11}$$

LHS = rate of change of [mch energy + fld energy] inside V.
RHS = − *outflux* of fld energy across S = *influx* of fld energy across S.
Therefore,
rate of change of [mch energy + fld energy] per unit volume = influx density of fld energy per unit volume.
Our interpretation is justified.

(B) *The momentum theorem*:

We proved the following theorem, which follows from Maxwell's equations, as Eq. (6.50c).

$$(\rho\mathbf{E}+\mathbf{J}\times\mathbf{B})+\left\{\frac{\partial}{\partial t}(\varepsilon_0\mathbf{E}\times\mathbf{B})\right\}=\boldsymbol{\nabla}\cdot\widehat{\mathcal{T}}_{(\text{em})}. \tag{12.12}$$

---

[a]See [13, pp. 258–261].

We shall interpret the two terms on the LHS as follows. The $(\mathbf{E}, \mathbf{B})$ field exerts a force on the existing charge–current distribution according the Lorentz force equation. The first term represents this force $\mathbf{f}_{\text{em}}$, as in Eq. (12.8), equal to the rate of change of the momentum of the particles per unit volume represented by $\mathbf{P}$, which we shall refer to as *mechanical momentum density.*

However, when these fields start changing with time they create a propagating em field which carries away energy and momentum. The second term should represent the rate of change of this *field momentum density,* to be represented by the symbol $\boldsymbol{g}$:

$$\boldsymbol{g} \overset{\text{def}}{=} \varepsilon_0(\mathbf{E} \times \mathbf{B}) = \frac{\mathbf{S}}{c^2}. \tag{12.13}$$

Now we can rewrite Eq. (12.12) as

$$\frac{\partial \mathbf{P}}{\partial t} + \frac{\partial \boldsymbol{g}}{\partial t} = \boldsymbol{\nabla} \cdot \widehat{\mathcal{T}}_{(\text{em})}. \tag{12.14}$$

To justify the above interpretation, we shall integrate (12.14) over a volume V bounded by a surface S, and applying Gauss's theorem, we get:

$$\begin{aligned} \frac{d}{dt}\left(\iiint_V \mathbf{P}\, d^3r\right) + \frac{d}{dt}\left(\iiint_V \boldsymbol{g}\, d^3r\right) &= \iiint_V \left(\boldsymbol{\nabla} \cdot \widehat{\mathcal{T}}_{(\text{em})}\right) dv \\ &= \iint_S \widehat{\mathcal{T}}_{(\text{em})} \cdot \mathbf{n}(\mathbf{r})\, da. \end{aligned} \tag{12.15}$$

LHS = The rate of change of [Mch momentum + Fld momentum] inside V.
RHS = Total em force transmitted across S = Influx of fld momentum across S.
Hence, we interpret Eq. (12.15) as saying that
Rate of change of [Mch momentum + Fld momentum] per unit volume = Influx density of Fld momentum per unit volume.
Our interpretation is justified.

Equations (12.9) and (12.14) are two equations expressing conservation of energy and momentum, separately. The spirit of relativity will demand that they should be integrated into a single equation, unifying conservation of energy and momentum as a conservation of En-Mentum (a name for 4-momentum coined in Sec. 8.4). As a first step towards this we rewrite Eqs. (12.9) and (12.14) in such a way that the left-hand side will represent the charged particles and the right-hand side the em field:

$$\mathbf{E}\cdot\mathbf{J} = -\left[\frac{\partial w}{\partial t} + \boldsymbol{\nabla}\cdot\mathbf{S}\right], \tag{12.16a}$$

$$\rho\mathbf{E} + \mathbf{J}\times\mathbf{B} = -\frac{\partial \boldsymbol{g}}{\partial t} + \boldsymbol{\nabla}\cdot\widehat{\boldsymbol{\mathcal{T}}}_{(\mathrm{em})} \tag{12.16b}$$

$$= -\left[\frac{\partial \boldsymbol{g}}{\partial t} + \boldsymbol{\nabla}\cdot\widehat{\boldsymbol{\Phi}}_{(\mathrm{em})}\right], \tag{12.16c}$$

In the last equation, $\widehat{\boldsymbol{\Phi}}_{(\mathrm{em})}$ is the momentum "outflux density", *equal and opposite* to momentum "influx density" $\widehat{\boldsymbol{\mathcal{T}}}_{(\mathrm{em})}$. See Eq. (6.53).

Equations (12.16a) and (12.16c) represent the time component and the space components of one 4-vector equality.

The *right side terms* can be combined into a 4-vector, which we shall define to be the *negative* 4-divergence of a 4-tensor $\overrightarrow{\overrightarrow{\mathbf{M}}}$, namely the *Maxwell's energy 4-tensor*. This tensor is an upgradation of the Maxwell's stress $\widehat{\boldsymbol{\mathcal{T}}}_{(\mathrm{em})}$ defined in Eq. (6.46), except that $(-\widehat{\boldsymbol{\mathcal{T}}}_{(\mathrm{em})})$, defined as $\widehat{\boldsymbol{\Phi}}_{(\mathrm{em})}$, forms the 3×3 core of this upgradation. The 4×4 components of this tensor will be written as $M^{\mu\nu}$. The time and space components of the new 4-vector are as follows:

$$\textit{Time component:}\quad \frac{1}{c}\left(\frac{\partial w}{\partial t} + \boldsymbol{\nabla}\cdot\mathbf{S}\right) \stackrel{\mathrm{def}}{=} \nabla_\alpha M^{\alpha 0}. \tag{12.17a}$$

$$\textit{Space component:}\quad \left[\frac{\partial \boldsymbol{g}}{\partial t} + \boldsymbol{\nabla}\cdot\widehat{\boldsymbol{\Phi}}_{(\mathrm{em})}\right]_k \stackrel{\mathrm{def}}{=} \nabla_\alpha M^{\alpha k};\ k = 1,2,3. \tag{12.17b}$$

The subscript k on the left-hand side implies x, y, z components of the vector corresponding to $k = 1, 2, 3$, respectively.

It is now easy to identify the 16 components of the Maxwell's Energy 4-tensor $\overrightarrow{\overrightarrow{\mathbf{M}}}$ by taking a close look at Eq. (12.17), and recalling the components of the operator ∇_α shown in (7.113). Equation (12.17a) yields the components of the column 0, and Eq. (12.17b) the components of the columns $k = 1, 2, 3$. Remember that $c\boldsymbol{g} = \mathbf{S}/c$, according to (12.13). For further help, see Appendix A.3. Moreover,

$$M^{\mu\nu}(x) = \begin{matrix} \\ 0 \\ 1 \\ 2 \\ 3 \end{matrix} \begin{matrix} \begin{matrix} 0 & \quad 1 & \quad 2 & \quad 3 \end{matrix} \\ \begin{pmatrix} w & S_x/c & S_y/c & S_z/c \\ S_x/c & \Phi^{11}_{\mathrm{em}} & \Phi^{12}_{\mathrm{em}} & \Phi^{13}_{\mathrm{em}} \\ S_y/c & \Phi^{21}_{\mathrm{em}} & \Phi^{22}_{\mathrm{em}} & \Phi^{23}_{\mathrm{em}} \\ S_z/c & \Phi^{31}_{\mathrm{em}} & \Phi^{32}_{\mathrm{em}} & \Phi^{33}_{\mathrm{em}} \end{pmatrix} \end{matrix}. \tag{12.18}$$

We have written $\Phi^{11}_{\rm em}, \Phi^{12}_{\rm em}, \ldots$ to mean $\Phi^{xx}_{\rm em}, \Phi^{xy}_{\rm em}, \ldots$, respectively. Note that $M^{\mu\nu}$ is symmetric and traceless.

$$\begin{aligned} M^{\mu\nu} &= M^{\nu\mu}, \\ M^{\mu}_{\cdot\,\mu} &= 0. \end{aligned} \tag{12.19}$$

Both properties are Lorentz invariant, i.e. same in all inertial frames.

The *left side terms* of (12.16) can be combined into another 4-vector, namely $\vec{\boldsymbol{f}}_{\rm fld\to mat}$, the Minkwski force per unit proper volume *from* the em field *on* the charged particles in the fluid, thereby changing the En-Mentum of the particles of the charged fluid:

$$\vec{\boldsymbol{f}}_{\rm fld\to mat}(x) = \left(\frac{1}{c}\mathbf{E}\cdot\mathbf{J},\ \rho\mathbf{E} + \mathbf{J}\times\mathbf{B}\right) \tag{12.20a}$$

$$= \vec{\mathbf{e}}_{\mu}\,\frac{1}{c}F^{\mu\alpha}J_{\alpha}, \tag{12.20b}$$

where $F^{\mu\alpha}$ and J_{α} are, respectively, the electromagnetic field 4-tensor and the electric current density 4-vector, both of them defined in Chapter 11, as Eqs. (11.36) and (11.25) respectively.

Equations (12.20a) and (12.20b) are analogous to Eqs. (11.32) and (11.34) of Chapter 11. In fact (b) follows from (12.20a) in the same way as Eq. (11.25) follows from Eq. (11.32).

The conservation equations for 4-momentum, appearing disjointedly as (12.9) and (12.14), will now join into the following single 4-equation[b]:

$$\frac{1}{c}\,F^{\mu\alpha}J_{\alpha} = -\nabla_{\beta}\,M^{\beta\mu}(x). \tag{12.21}$$

12.4. Euler's (Non-relativistic) Equation of Motion for a Perfect Fluid

Our objective now is to construct the energy tensor of the simplest "closed system". The term "closed" in this context means that the system is self-contained in all its dynamical behaviour, i.e., all dynamical processes take place due to forces of interaction within the system, there being no scope for exchange of energy and momentum with anything outside. The total energy and the total momentum of a closed system are therefore fully conserved.

[b]See [33, p. 152].

A closed system contains both matter and forces. The only kind of classical forces that can receive relativistic treatment are electromagnetic forces. Before linking up matter with electromagnetic forces, we shall consider an oversimplified model which consists of matter in the form of perfect fluid — sometimes also called "classical fluid" — moving under the influence of internal and external forces whose origin we need not specify at this moment. We shall first lend a *non-relativistic* treatment to this fluid, so that transition to a relativistic formalism becomes smooth in the next section. The equation of motion of this perfect fluid is known as Euler's equation.

By perfect fluid, we mean a fluid which does not offer any viscous forces, which as the reader knows, causes shear stresses in the fluid. A perfect fluid, whether at rest or in motion, can sustain only normal compressive stresses inside, familiarly known as "pressure".

Let us consider a fluid in streamline motion as previously illustrated in Fig. 12.1. This fluid is characterized by a velocity field $\mathbf{u}(\mathbf{r}, t)$ and a fluid mass density $\sigma(\mathbf{r}, t)$, both of which, in general, are unsteady fields, i.e. functions of t as well. The divergence of $\mathbf{u}$ is called *dilatation*, a term we shall explain with the help of Figs. 12.2(a) and 12.2(b).

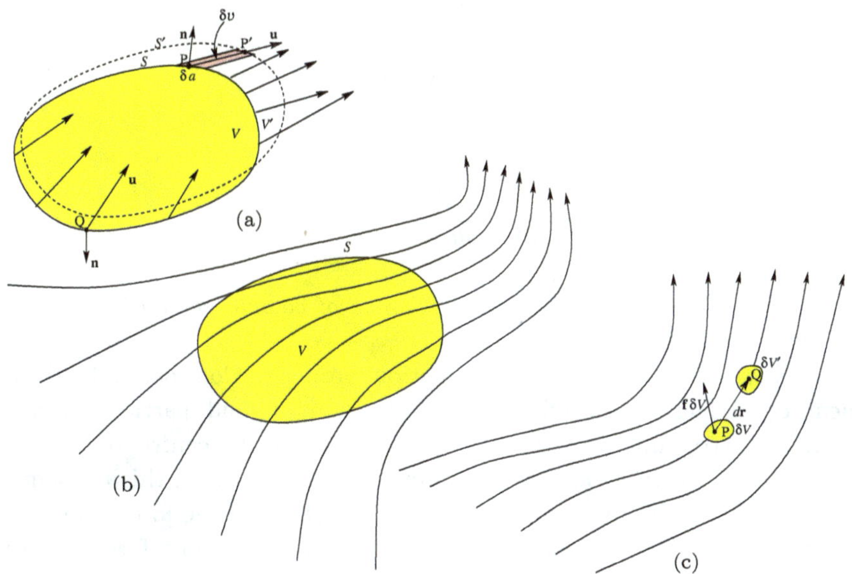

Fig. 12.2. Explaining fluid motion.

We have shown a stream of fluid in motion, inside of which we have marked out a volume V at time t. Since the fluid particles on the surface S of V have different velocities $\mathbf{u}(\mathbf{r},t)$, the boundary S not only moves with the particles lying on it, but also changes to a different shape S' (shown with broken line) at the time $t+dt$. Consequently, V will also change to a different volume, say V'.

Consider a film of fluid particles lying over a tiny area δa centred at the point P. These particles move a tiny distance $\mathbf{u}\,dt$ from P to P′ in time dt. In this time, a volume of fluid δv flows out from V, crossing the tiny surface area da. The volume that flows out is $\delta v = [\mathbf{u}\cdot\mathbf{n}]\delta a\,dt$.

There are certain regions of S, say at P, where $\mathbf{u}\cdot\mathbf{n}$ is positive, and the outflux (i.e. volume outflow) is positive. There are some other regions, say, at Q, where $\mathbf{u}\cdot\mathbf{n}$ is negative, and the outflux is negative. The net outflux of fluid volume is the surface integral of $\mathbf{u}$ over the boundary surface S. This can be written as

$$dV = V' - V = \left[\iint_S (\mathbf{u}\cdot\mathbf{n})\,da\right]dt = \left[\iiint_V (\boldsymbol{\nabla}\cdot\mathbf{u})\,d^3r\right]dt, \qquad (12.22)$$

where we have used Gauss's theorem to convert the surface integral to a volume integral. We reduce the finite volume V to an infinitesimal volume δV, thereby avoid integration, and get

$$d(\delta V) = [(\boldsymbol{\nabla}\cdot\mathbf{u})\,\delta V]\,dt. \qquad (12.23)$$

Therefore[c]

$$\boxed{\frac{1}{\delta V}\frac{d(\delta V)}{dt} = \boldsymbol{\nabla}\cdot\mathbf{u}.} \qquad (12.24)$$

In other words, $\boldsymbol{\nabla}\cdot\mathbf{u}$ is the *rate of change of volume per unit volume* — or, more compactly *dilatation.*

Now we take up equation of motion properly. Consider a fluid element consisting of an infinitesimal collection of fluid particles moving along the stream (Fig. 12.2(c)). At the time t its centre of mass is located at P where it occupies a volume δV. The mass of this element is $\delta m = \sigma(\mathbf{r},t)\delta V$, its momentum $\delta\mathbf{p} = \delta m\,\mathbf{u}((\mathbf{r},t))$ and the force impressed on it $\delta\mathbf{F} = \mathbf{f}(\mathbf{r},t)\,\delta V$, where $\mathbf{f}(\mathbf{r},t)$ represents the volume force density.

[c]See [34, p.136.]

Applying Newton's second law of motion to this fluid element,

$$\frac{d}{dt}(\delta \mathbf{p}) = \delta \mathbf{F}, \tag{12.25a}$$

$$\text{or, } \frac{d}{dt}[\delta m\, \mathbf{u}(\mathbf{r},t)] = \mathbf{f}(\mathbf{r},t)\, \delta V. \tag{12.25b}$$

Note that in the above equation $\frac{d}{dt}$ represents *convective derivative* whose meaning we shall explain with a more general example. Let there exist a certain field $f(x,y,z,t)$ in the fluid (e.g., temperature, fluid velocity, pressure). The value of this field at the location of the particle changes from $f(x,y,z,t)$ to $f(x+dx, y+dy, z+dz, t+dt)$ as the particle moves with velocity $\mathbf{u}$ from the location $\mathbf{r} = (x,y,z)$ at the time t to take up a new location $\mathbf{r} + d\mathbf{r} = (x+dx, y+dy, z+dz)$ at the time $t+dt$. The net change is

$$\begin{aligned} df &= \frac{\partial f}{\partial x}dx + \frac{\partial f}{\partial y}dy + \frac{\partial f}{\partial z}dz + \frac{\partial f}{\partial t}dt \\ &= \left[\frac{\partial f}{\partial x}u_x + \frac{\partial f}{\partial y}u_y + \frac{\partial f}{\partial z}u_z + \frac{\partial f}{\partial t}\right] dt \\ &= \left[\mathbf{u}\cdot\boldsymbol{\nabla} + \frac{\partial}{\partial t}\right] f(x,y,z,t)dt \\ &= \left(\frac{df}{dt}\right)_{\mathrm{c}} dt. \end{aligned} \tag{12.26}$$

We have attached a subscript "c" to stress that the time rates of the changes of physical quantities in motion are given by their *Convective Derivatives.*

$$\frac{df}{dt} \equiv \left(\frac{df}{dt}\right)_{\mathrm{c}} \stackrel{\text{def}}{=} \left(\mathbf{u}\cdot\boldsymbol{\nabla} + \frac{\partial}{\partial t}\right) f(x,y,z.t). \tag{12.27}$$

Using Eqs. (12.24) and (12.27), we establish a few relations for future reference:

$$\begin{aligned} \frac{d}{dt}[\sigma\, \delta V] &= \frac{d\sigma}{dt}\delta V + \sigma\frac{d\,\delta V}{dt} \\ &= \left[\left\{\frac{\partial \sigma}{\partial t} + (\mathbf{u}\cdot\boldsymbol{\nabla})\sigma\right\} + \sigma\,\{\boldsymbol{\nabla}\cdot\mathbf{u}\}\right]\delta V \qquad (12.28a) \\ &= \left[\frac{\partial \sigma}{\partial t} + \boldsymbol{\nabla}\cdot(\sigma\mathbf{u})\right]\delta V, \qquad (12.28b) \end{aligned}$$

where σ represents any *density function,* of which the mass density is particular example.

In N.R. physics mass is conserved. Consider the two terms in the first equality in (12.28a). The first term, if positive, means increase in mass in dV due to density fluctuation. The second term, if positive, means increase in mass due to volume fluctuation. However, both of them cannot be positive. Increase in one term is nullified by decrease in the other. Together they represent zero change. We get back the mass conservation equation, known as *continuity equation.*

$$\frac{d}{dt}[\sigma\,\delta V] = 0 \quad \Rightarrow \quad \frac{\partial\sigma}{\partial t} + \boldsymbol{\nabla}\cdot(\sigma\mathbf{u}) = 0. \tag{12.29}$$

We shall convert Eq. (12.28) to a momentum equation. Replace the scalar density σ with the density of the x-component of momentum σu_x in the above equation, and get

$$\frac{d}{dt}[(\sigma u_x)\,\delta V] = \left[\frac{\partial(\sigma u_x)}{\partial t} + \boldsymbol{\nabla}\cdot(\sigma u_x\mathbf{u})\right]\delta V. \tag{12.30}$$

The above relation holds for all the three components u_x, u_y, u_z. Multiplying the components with $\mathbf{e}_x, \mathbf{e}_y, \mathbf{e}_z$ and adding them together, we get

$$\frac{d}{dt}[\sigma\mathbf{u}\,\delta V] = \left[\frac{\partial}{\partial t}(\sigma\mathbf{u}) + \boldsymbol{\nabla}\cdot(\sigma\mathbf{u}\mathbf{u})\right]\delta V \tag{12.31}$$

Going back to Eq. (12.25), noting that $\delta m\,\mathbf{u}(\mathbf{r},t) = \sigma\mathbf{u}\,\delta V$ and using (12.31) we get the general equation of motion for the fluid:

$$\frac{\partial}{\partial t}(\sigma\mathbf{u}) + \boldsymbol{\nabla}\cdot(\sigma\mathbf{u}\mathbf{u}) = \mathbf{f}. \tag{12.32}$$

Equation (12.32) is the general equation of motion of a fluid, to be referred to as the *Euler's equation.*

12.5. Relativistic Equation of Motion for a Continuous Incoherent Media

We shall upgrade the E^3 version of the fluid equation of motion (12.32) to M^4. The starting point of the former was (12.25a). The starting point of the latter will be the M^4 version of this equation, i.e. Eq. (8.28) in which we set $\overrightarrow{\mathcal{P}} \to \delta\overrightarrow{\mathcal{P}}$ and $\overrightarrow{\mathcal{F}} \to \delta\overrightarrow{\mathcal{F}}$:

$$\frac{d(\delta\overrightarrow{\mathcal{P}})}{d\tau} = \delta\overrightarrow{\mathcal{F}}. \tag{12.33}$$

Here $\delta\overrightarrow{\mathcal{P}}$ is the 4-momentum of the mass content of the same fluid volume δV considered in Sec. 12.4, and $\delta\overrightarrow{\mathcal{F}}$ is the Minkowski force on this volume.

We shall write the left and the right side of the above equation

$$\delta\overrightarrow{\boldsymbol{\mathcal{P}}} = \overrightarrow{\mathbf{e}_\mu}\,\delta p^\mu; \quad \text{so that} \quad \frac{d(\delta\overrightarrow{\boldsymbol{\mathcal{P}}})}{d\tau} = \overrightarrow{\mathbf{e}_\mu}\left(\frac{d(\delta p^\mu)}{d\tau}\right), \tag{12.34a}$$

$$\delta\overrightarrow{\boldsymbol{\mathcal{F}}} = \overrightarrow{\boldsymbol{f}}(x)\,\delta V_o = \overrightarrow{\mathbf{e}_\mu}\,(f^\mu(x)\,\delta V_o)\,. \tag{12.34b}$$

Then the EoM (12.33) can be written in the form:

$$\frac{d\,\delta p^\mu}{d\tau} = f^\mu(x)\,\delta V_o, \tag{12.35a}$$

$$\text{or} \quad \frac{d\,\delta p^\mu}{dt} = f^\mu(x)\,\delta V. \tag{12.35b}$$

To go from (12.35a) to (12.35b), we divided each side with Γ, and recalled Eqs. (4.16) and (12.3).

If we make use of the expression for $\overrightarrow{\boldsymbol{f}}$ given in (12.6), the EoM takes the form:

$$\boxed{\frac{d\,\delta p^\mu(x)}{dt} = f^\mu(x)\,\delta V = \left(\frac{\varpi(x)}{c},\ \mathbf{f}(x)\right)\delta V.} \tag{12.36}$$

Equation (12.36) will be the backbone of our arguments. It states that the rate of change of 4-momentum of an *infinitesimal* volume δV of particles at the event point (x) is equal to the total 4-force acting on this volume. We shall convert it to a beautiful form in Eq. (12.44)

At this point let us be aware that mass is not conserved in relativistic mechanics. Mass conservation is violated, even if infinitesimally, in all real situations. Mass of a system changes when chemical reactions take place, when atoms absorb or emit light, when a gas expands or is compressed. Even for the perfect fluid, whose dynamics was given a relatively simple non-relativistic treatment in Sec. 12.4, its mass is continuously changing because of the work being done by fluid pressure. This effect has to be taken into consideration.

To make our task manageable, we shall think of a system of particles forming a *tenuous* fluid in motion. The constituent particles move along streamlines without a bond between them, i.e. no collision occurs. Also, the constituent particles — atoms, molecules, nuclei, electrons — whatever they may be, remain in their original ground states through the dynamical processes, and, hence, do not emit or absorb light, so that *their rest masses do not change*. The particles are charged, and the electromagnetic field created by their charges determine their equation of motion.

Let Fig. 12.1 represent a segment of this flowing fluid. An infinitesimal volume δV of this fluid, at the event point (x), possesses a rest mass δm_0, which is the sum of the rest masses of all the constituent particles inside δV. That is, $\delta m_0 = \sum_{i=1}^{\delta N} m_{oi}$, where δN is the number of particles inside δV and m_{oi} is the rest mass of the ith particle in this infinitesimal collection. Let σ_o stand for proper density of rest mass, which we define as

$$\sigma_o(x) = \lim_{\delta V_o \to 0} \frac{\delta m_0}{\delta V_o}, \tag{12.37}$$

where δV_o is the proper volume of the above collection, i.e. volume measured in the instantaneous rest frame. In contrast to σ_o, we use another symbol σ to denote density of relativistic mass in the observer's frame S. Seen from the observer's frame, the above collection of δN particles are now confined within a smaller volume $\delta V = \delta V_o/\Gamma$ and the relativistic mass of this collection is $\delta m = \Gamma \delta m_0$. Therefore,

$$\sigma(x) = \lim_{\delta V \to 0} \frac{\delta m}{\delta V} = \lim_{\delta V_o \to 0} \frac{\Gamma\, \delta m_0}{(\delta V_o/\Gamma)} = \Gamma^2 \sigma_o(x). \tag{12.38}$$

We shall work out the equations of motion of the energy and momentum content of the volume δV. The relativistic mass of this volume is

$$\delta m = \sigma(x)\delta V. \tag{12.39}$$

Therefore, according to formulas (8.19) and (8.18(b), 8.18(c)) of Sec. 8.4, the 4-momentum of the mass content within this volume is $\delta p^\mu = (\delta p^0, \delta \mathbf{p})$ where

$$\begin{aligned} \delta p^0 &= \delta m\, c = (\sigma\, \delta V)\, c, \\ \delta \mathbf{p} &= \delta m\, \mathbf{u} = (\sigma\, \delta V)\, \mathbf{u}. \end{aligned} \tag{12.40}$$

Let us now go back to the equation of motion (12.36). We shall expand the left-hand side corresponding to $\mu = 0$, using the time component of δp^μ as given in (12.40). With some help from (12.28):

$$\frac{d\, \delta p^0}{dt} = c \frac{d}{dt}(\sigma\, \delta V) = c \left[\frac{\partial \sigma}{\partial t} + \boldsymbol{\nabla} \cdot (\sigma \mathbf{u}) \right] \delta V, \tag{12.41}$$

and corresponding to $\mu = i = 1, 2, 3$ in a similar way with help from (12.31):

$$\frac{d\, \delta \mathbf{p}}{dt} = \frac{d}{dt}(\sigma \mathbf{u}\, \delta V) = \left[\frac{\partial (\sigma \mathbf{u})}{\partial t} + \boldsymbol{\nabla} \cdot (\sigma \mathbf{u}\mathbf{u}) \right] \delta V. \tag{12.42}$$

We shall combine the above two equations to obtain the following identity:

$$\boxed{\begin{aligned}\frac{d\,\delta p^{\mu}}{dt} &\equiv \left(\left[\frac{\partial(c^2\sigma)}{\partial\, ct} + \boldsymbol{\nabla}\cdot(c\,\sigma\mathbf{u})\right], \left[\frac{\partial(c\,\sigma\mathbf{u})}{\partial\, ct} + \boldsymbol{\nabla}\cdot(\sigma\mathbf{u}\mathbf{u})\right]\right)\delta V \\ &= \nabla_{\beta}[\sigma_o(U^{\beta}U^{\mu})]\,\delta V.\end{aligned}} \tag{12.43}$$

We have used Eq. (8.7) which gives the 4-velocity $U^{\mu} = \Gamma(c, \mathbf{u})$, in which $\mathbf{u}$ is the velocity of a particle at the event point (x).

Going back to (12.36) we rewrite the same EoM in the compact form:

$$\boxed{\nabla_{\beta}[\sigma_o(U^{\beta}U^{\mu})] = f^{\mu},} \tag{12.44}$$

which is the EoM for a system of incoherent dust subjected to 4-force $\vec{f} = \vec{\mathbf{e}}_{\mu} f^{\mu}$ per unit *proper* volume.

Much of our discussions to follow will be based on Eq. (12.44).

The compact four-dimensional EoM (12.44) has the following (time–space) components:

$$\left(\left[\frac{\partial(c^2\sigma)}{\partial\, ct} + \boldsymbol{\nabla}\cdot(c\,\sigma\mathbf{u})\right], \left[\frac{\partial(c\,\sigma\mathbf{u})}{\partial\, ct} + \boldsymbol{\nabla}\cdot(\sigma\mathbf{u}\mathbf{u})\right]\right) = \left(\frac{\varpi}{c},\ \mathbf{f}\right). \tag{12.45}$$

The *time part* is the energy equation, and the *space part* the momentum equation.

We now define the *energy tensor* of the incoherent fluid (also called incoherent dust) as[d]

$$\mathcal{D}^{\mu\nu} \stackrel{\text{def}}{=} \sigma_o U^{\mu}U^{\nu}. \tag{12.46}$$

It is now very easy to identify the 4×4 components of $\mathcal{D}^{\mu\nu}$:

$$\mathcal{D}^{\mu\nu} = \sigma_o\Gamma^2 \times \left[\begin{array}{c|cccc} & 0 & 1 & 2 & 3 \\ 0 & c^2 & cu_x & cu_y & cu_z \\ 1 & u_x c & u_x^2 & u_x u_y & u_x u_z \\ 2 & u_y c & u_y u_x & u_y^2 & u_y u_z \\ 3 & u_z c & u_z u_x & u_z u_y & u_z^2 \end{array}\right]. \tag{12.47}$$

[d] See [34, p. 141].

The EoM, written as (12.44), takes the beautiful comprehensive form:

$$\nabla_\mu \mathcal{D}^{\mu\nu} = f^\nu. \tag{12.48}$$

12.6. Energy Tensor for a System of Charged Incoherent Fluid

We prepared the groundwork for this section in Sec. 12.3, in particular through Eq. (12.21). Before proceeding further we shall recognize the following two volume 4-force densities.

$$\begin{aligned}\vec{\boldsymbol{f}}_{\text{fld}\to\text{mat}} &= (f^0_{\text{fld}\to\text{mat}}, \boldsymbol{f}_{\text{fld}\to\text{mat}}) \qquad (12.49\text{a})\\ &= \text{4-force per unit } \textit{proper} \text{ volume } \textit{from} \text{ the em fld}\\ &\quad\ \textit{on} \text{ the particles in the dust,}\\ \vec{\boldsymbol{f}}_{\text{mat}\to\text{fld}} &= (f^0_{\text{mat}\to\text{fld}}, \boldsymbol{f}_{\text{mat}\to\text{fld}}) \qquad (12.49\text{b})\\ &= \text{4-force per unit } \textit{proper} \text{ volume } \textit{from} \text{ the particles in the dust}\\ &\quad\ \textit{on} \text{ the em fld.}\end{aligned}$$

Let us now understand the *effect* of above two 4-force densities.

$$\begin{aligned}&\text{Rate of change of } \textit{matter} \text{ 4-momentum per u.p.v. at}(x)\\ &\quad= \left[\frac{1}{c}\mathbf{E}\cdot\mathbf{J}, \rho\mathbf{E} + \mathbf{J}\times\mathbf{B}\right] = \frac{1}{c}F^{\mu\alpha}J_\alpha\\ &\quad= \vec{\boldsymbol{f}}_{\text{fld}\to\text{mat}}(x), \qquad (12.50\text{a})\\ &\text{Rate of change of } \textit{field} \text{ 4-momentum per u.p.v. at}(x)\\ &\quad= \left[\frac{1}{c}\left(\frac{\partial w}{\partial t} + \boldsymbol{\nabla}\cdot\mathbf{S}\right), \left(\frac{\partial \boldsymbol{g}}{\partial t} + \boldsymbol{\nabla}\cdot\widehat{\boldsymbol{\Phi}}_{(\text{em})}\right)\right] = \nabla_\beta M^{\beta\mu}(x)\\ &\quad= \vec{\boldsymbol{f}}_{\text{mat}\to\text{fld}}(x) \qquad (12.50\text{b})\end{aligned}$$

We can now go back to (12.21) and rewrite the 4-Momentum conservation equation as[e]:

$$\vec{\boldsymbol{f}}_{\text{fld}\to\text{mat}} = -\vec{\boldsymbol{f}}_{\text{mat}\to\text{fld}}. \tag{12.51}$$

The above equation represents a generalization of Newton's third law of motion for the 3-forces of action and reaction to the 4-forces of action and reaction between a charged fluid media and its own electromagnetic field.

[e]See [33, p. 152].

The EoM of the charged dust is given by Eq. (12.48), in which the "force" f^μ is now the electromagnetic force on matter, i.e. $f^\mu_{\text{fld}\to\text{mat}} = f^\mu_{\text{em}}$, as given in (12.20), exerted by the em field originating from the charge–current density J^μ present in the matter itself.

The reader may re-read the statement following Eq. (12.36) to clear a possible doubt. The quantity $\overrightarrow{f}_{\text{fld}\to\text{mat}}(x)\,\delta V$ is the 4-force exerted on matter inside the elementary *lab* volume δV, by the charge–current distribution residing on the rest of the volume $V-\delta V$, outside δV. In this sense, this 4-force is an "external" 4-force (not a "self force") determining the fate of the matter inside δV. The quantity $(\nabla_\alpha \mathcal{D}^{\alpha\mu}(x))\,\delta V$ is the time rate of change of the 4-momentum of the particles in this volume, according to Eqs. (12.46) and (12.43).

The EoM (12.48) is now written as

$$\nabla_\alpha \mathcal{D}^{\alpha\mu}(x) = f^\mu_{\text{fld}\to\text{mat}}.$$

$$\text{But,}\quad f^\mu_{\text{fld}\to\text{mat}} = -f^\mu_{\text{mat}\to\text{fld}} = -\nabla_\alpha\, M^{\alpha\mu}(x), \quad \text{by (12.51) and (12.50b)}\,. \tag{12.52}$$

$$\text{Hence, } \nabla_\alpha \mathcal{D}^{\alpha\mu}(x) = -\nabla_\alpha\, M^{\alpha\mu}(x).$$

The system consisting of the matter (represented by $\mathcal{D}^{\alpha\mu}$) and its own em fld (represented by $M^{\alpha\mu}$) is now a *closed system.* Its *energy tensor* is

$$T^{\alpha\mu}_{\text{dust}}(x) = \mathcal{D}^{\alpha\mu}(x) + M^{\alpha\mu}(x), \tag{12.53}$$

satisfying

$$\nabla_\alpha T^{\alpha\mu}_{\text{dust}} = 0. \tag{12.54}$$

In Newton's theory of gravitation a massive star, or a massive planet, is the source of gravitation. In Einstein's General Theory of Relativity mass is replaced by energy. However, energy itself has no respectable status, because energy is the time component of 4-momentum. Hence, energy is replaced by 4-momentum, and energy density (analogous to mass density) by the energy tensor, which is loosely the density of 4-momentum. Since a star is an isolated object, its energy tensor must have zero 4-divergence. Equation (12.53) gives the simplest example of such an energy tensor, and Eq. (12.54) tells us the desirable property of such a source of gravitation.

12.7. Energy Tensor of a Closed System

By the defining property (see first paragraph of Sec. 12.4), the energy and momentum of a closed system are strictly conserved. This conservation statement is succinctly expressed in the mathematical equation:

$$\nabla_\alpha T^{\alpha\mu}(x) = 0, \tag{12.55}$$

where $T^{\alpha\mu}(x)$ is the energy tensor of the system at the event point (x). Equation (12.54) gives an example of such a tensor.

Equation (12.55) represents four equations, of which the $\mu = 0$ component expresses energy conservation, and the $\mu = 1, 2, 3$ components the momentum conservation, since energy divided by c, and the momentum 3-components, together, constitute the 4-momentum p^μ.

Before we proceed towards the construction of the energy tensor of a perfect fluid (which we take up in the next section), Eq. (12.55), and its full import will help us in our mission.

To see the energy part of (12.55) we set $\mu = 0$. Now multiplying either side with c and expanding we obtain:

$$\frac{\partial T^{00}}{\partial t} + \sum_{k=1}^{3} \frac{\partial (cT^{k0})}{\partial x^k} = 0. \tag{12.56}$$

This becomes the *energy conservation equation*, conforming to the format of the continuity equation shown way back in Eq. (11.14), by identifying:

$$T^{00} = w, \quad T^{k0} = \frac{S^k}{c}, \tag{12.57}$$

where w stands for the energy density, and the 3-vector $\mathbf{S}$ for the energy flux density. Equation (12.56) can now be rewritten in the standard form of the continuity equation:

$$\frac{\partial w}{\partial t} + \boldsymbol{\nabla} \cdot \mathbf{S} = 0. \tag{12.58}$$

In a similar manner we set $\mu = 1, 2, 3$ to view the *momentum conservation* part of the equation (12.55), which yields:

$$\frac{1}{c}\frac{\partial}{\partial t} T^{0i} + \sum_{k=1}^{3} \frac{\partial T^{ki}}{\partial x^k} = 0, \quad i = 1, 2, 3. \tag{12.59}$$

As before, the above equation becomes the conservation equation for the ith component of momentum by identifying:

$$T^{0i}(x) = cg^i(x); \quad T^{ki}(x) = k\text{th component of } \mathbf{T}^i(x), \tag{12.60}$$

where $\boldsymbol{g}(x) = (g^1(x), g^2(x), g^3(x))$ stands for the momentum density vector, and $\mathbf{T}^i(x) = (T^{1i}(x), T^{2i}(x), T^{3i}(x))$ for the flux density vector for the ith component of momentum. Hence, we can rewrite Eq. (12.59) as

$$\frac{\partial g^i}{\partial t} + \boldsymbol{\nabla} \cdot \mathbf{T}^i = 0, \quad i = 1, 2, 3. \tag{12.61}$$

Note that T^{ki}, which stands for the kth Cartesian component of the momentum flux density vector $\mathbf{T}^i$, is the (ki)-element of the energy tensor $T^{\mu\nu}(x)$. If we multiply either side of the above equation with $\mathbf{e}_i$ and sum over i, and make use of Eq. (5.68), the momentum conservation is expressed compactly in the form:

$$\frac{\partial \boldsymbol{g}}{\partial t} + \boldsymbol{\nabla} \cdot \widehat{\mathbf{T}} = 0, \tag{12.62}$$

where $\widehat{\mathbf{T}}$ is the *momentum flux density 3-tensor*, and comprises the *space–space components* of the energy tensor.

The energy tensor is strictly symmetric, i.e.

$$T^{\mu\nu}(x) = T^{\nu\mu}(x). \tag{12.63}$$

The symmetricity of the space–space components is related to the conservation of angular momentum (see [34], p. 170). The symmetricity between the space–time, and the time–space components is hidden in (12.57) and (12.60), as we shall now show.

The mass energy equivalence $E = mc^2$ associates with every energy density $w(x)$ an effective mass density $\sigma(x)$ such that

$$w(x) = \sigma(x)\, c^2. \tag{12.64}$$

The energy flux density $\mathbf{S}(x)$ suggests an equivalent transport density $\mathbf{v}(x)$ such that

$$\mathbf{S}(x) = w(x)\, \mathbf{v}(x). \tag{12.65}$$

In the case of the electromagnetic field, this transport velocity is the same as the velocity of light. In the case of a perfect fluid in which pressure exists, it is however different from the velocity $\mathbf{u}(x)$ of the fluid, as we shall see in the next section.

Also, momentum $\mathbf{p}$ and the effective mass m are connected by the relation $\mathbf{p} = m\mathbf{v}$. This is true for relativistic and non-relativistic systems. It is obviously valid for a particle if we take m as the relativistic mass. It holds for a photon if we interpret E/c^2 as the effective mass of the photon. Extending the above mass–momentum relationship to the corresponding densities, we get $\boldsymbol{g} = \sigma\mathbf{v} = \frac{w}{c^2}\mathbf{v}$. Use of (12.65) now leads to the following relationship between $\boldsymbol{g}$ and $\mathbf{S}$:

$$\boldsymbol{g}(x) = \frac{1}{c^2}\mathbf{S}(x). \tag{12.66}$$

It now follows from (12.57) and (12.60) that $T^{0k} = T^{k0}$.
To summarize:

$$\begin{aligned} T^{00} &= w; \quad T^{0k} = cg^k = \frac{S^k}{c} = T^{k0}; \\ T^{ki} &= T^{ik} = (ki), (ik) \\ &= \text{component of the momentum flux density tensor } \widehat{\mathbf{T}}. \end{aligned} \tag{12.67}$$

12.8. Energy Tensor of a Perfect Fluid

We shall take up another simplified model of a closed system to shed further light on the meaning of energy tensor. We shall consider a gas, as in the previous section, but drop the assumption of tenuousness, so that the gas particles can be imagined to interact through collision, thereby giving rise to fluid pressure.

When a fluid is compressed, its energy increases. Therefore, the fluid under pressure has extra elastic energy which must show up as extra mass density. In order to keep the discussion simple, we shall assume (a) that the fluid is *perfect*, so that the stress field inside has the simplest possible form, as given by Eq. (5.81), and (b) the heat exchange can be ignored so that the dynamical processes can be treated as adiabatic. We shall present a simplified treatment for this case. For a rigorous treatment, with a proper analysis of the stress tensor inside the fluid, the reader may look up [34, Sec. 6.6].

As in Sec. 12.5, consider the change in the energy–momentum content of a given number δN of particles confined within a variable volume δV, as these particles move in bulk as a stream. The rest mass of this volume is $\sigma_o(x)\delta V_o$, whereas its relativistic mass in the observer's frame (which includes the mass $\Gamma^2\sigma_o(x)\delta V$ as given in Eq. (12.38) plus some extra mass generated in this volume due to the work done by the pressure force) is $\sigma(x)\delta V$.

As the fluid moves, the energy content of δV changes due to the work done along its surface by the surface force. Consider an area element $\delta \boldsymbol{a}$ at a point P on the surface, as in Fig. 12.2(a). The force on this area is $-p\mathbf{n}\,\delta a$, where $\mathbf{n}$ is an outward normal. The surface element is moving with the velocity $\mathbf{u}$ under the pressure force. The work being done by this force is $-p\mathbf{n}\cdot\mathbf{u}\,\delta a$ per unit time. The net work on the closed surface per unit time is the surface integral of this elementary work, which we shall convert into volume integral using Gauss's theorem, to obtain the rate at which the energy of the volume V is increasing per unit time.

$$\Pi = \iint_S -p\mathbf{n}\cdot\mathbf{u}\,da = \iiint_V -\boldsymbol{\nabla}\cdot(p\mathbf{u})\,d^3r \tag{12.68}$$

Taking the above volume V to be very small, equal to δV, and dividing both sides of the equation by δV we get the rate of change of energy per unit volume per unit time (same as what we called power density in Sec. 12.2) as

$$\varpi = -\boldsymbol{\nabla}\cdot(p\mathbf{u}). \tag{12.69}$$

Interestingly, we can expand the above expression into two parts:

$$\varpi_1 = -\boldsymbol{\nabla}p\cdot\mathbf{u}, \tag{12.70a}$$

$$\varpi_2 = -p\,(\boldsymbol{\nabla}\cdot\mathbf{u}). \tag{12.70b}$$

Equation (12.70a) shows how the pressure volume force $-\boldsymbol{\nabla}p$, as given by Eq. (5.82), alters its kinetic energy per unit volume per unit time, and Eq. (12.70b) shows how the expansion of the gas against the external pressure contributes an elastic energy (same as the potential energy) per unit volume per unit time.

We now write the energy equation, noting that σc^2 is the energy density. The rate of change of energy inside the volume δV is equal to $\varpi\,\delta V$. Hence, by (12.69)

$$c^2\frac{d}{dt}[\sigma\,\delta V] = \varpi\delta V = -\boldsymbol{\nabla}\cdot(p\mathbf{u})\,\delta V. \tag{12.71}$$

By Eq. (12.28)

$$\frac{d}{dt}[\sigma\,\delta V] = \left[\frac{\partial\sigma}{\partial t} + \boldsymbol{\nabla}\cdot(\sigma\mathbf{u})\right]\delta V. \tag{12.72}$$

Hence, the time component of the EoM:

$$\frac{\partial(c^2\sigma)}{\partial t} + \boldsymbol{\nabla}\cdot[(c^2\,\sigma + p)\mathbf{u}] = 0. \tag{12.73}$$

Equation (12.73) represents the $\mu = 0$ component of the fundamental conservation equation (12.55) and is equivalent to (12.56), and then to (12.58). Comparing, we identify the $T^{\mu 0}$ components of the energy tensor:

$$T^{00}(x) = w(x) = \sigma c^2, \tag{12.74a}$$

$$T^{k0}(x) = \frac{1}{c}S^k(x) = \left(\sigma + \frac{p}{c^2}\right)u^k c. \tag{12.74b}$$

Before writing the momentum equation of motion, we shall need an expression for the momentum density $\boldsymbol{g}$ in the frame S. This follows straight from Eqs. (12.66) and (12.74b).

$$\boldsymbol{g}(x) = \frac{\mathbf{S}(x)}{c^2} = \left(\sigma + \frac{p}{c^2}\right)\mathbf{u}. \tag{12.75}$$

The momentum content of the volume δV is therefore,

$$\delta\mathbf{p} = \left(\sigma + \frac{p}{c^2}\right)\mathbf{u}\,\delta V. \tag{12.76}$$

To obtain the rate of change of this momentum we shall use Eq. (12.31), replace σ with $\left(\sigma + \frac{p}{c^2}\right)$:

$$\frac{d\,\delta\mathbf{p}}{dt} = \left[\frac{\partial}{\partial t}\left\{\left(\sigma + \frac{p}{c^2}\right)\mathbf{u}\right\} + \boldsymbol{\nabla}\cdot\left\{\left(\sigma + \frac{p}{c^2}\right)\mathbf{uu}\right\}\right]\delta V. \tag{12.77}$$

The only force acting on this fluid element, as already assumed, is the pressure force, namely,

$$\delta\mathbf{F} = -(\boldsymbol{\nabla}p)\,\delta V. \tag{12.78}$$

The EoM for momentum is

$$\frac{d\,\delta\mathbf{p}}{dt} = \delta\mathbf{F}. \tag{12.79}$$

Therefore, from (12.77)–(12.79)

$$\frac{\partial}{\partial t}\left\{\left(\sigma + \frac{p}{c^2}\right)\mathbf{u}\right\} + \boldsymbol{\nabla}\cdot\left\{\left(\sigma + \frac{p}{c^2}\right)\mathbf{uu} + \hat{\mathbf{1}}p\right\} = 0. \tag{12.80}$$

Equation (12.80) is to be identified with (12.59). Hence, the remaining components of the energy tensor:

$$T^{0k}(x) = \left(\sigma + \frac{p}{c^2}\right) u^k c, \tag{12.81a}$$

$$T^{ik}(x) = \left(\sigma + \frac{p}{c^2}\right) u^i u^k + \delta^{ik} p. \tag{12.81b}$$

The components of $T^{\mu\nu}$ shown in (12.74) and (12.81) do not present a covariant expression, because the quantities on the right-hand side are not 4-vectors or 4-scalars. We shall correct this defect.

Let us go back to the IRS S_o in which the components of $T^{\mu\nu}$ form the following diagonal components, by setting $\mathbf{u} = \mathbf{0}$ in (12.74) and (12.81):

$$T^{\mu\nu}{}_{\text{rest}}(x) = \begin{pmatrix} \sigma_o c^2 & 0 & 0 & 0 \\ 0 & p_o & 0 & 0 \\ 0 & 0 & p_o & 0 \\ 0 & 0 & 0 & p_o \end{pmatrix}, \tag{12.82}$$

where p_o is the pressure in S_o. According to Corollary #3 in Sec. 8.7.4, the pressure p is Lorentz invariant, i.e.

$$p = p_o. \tag{12.83}$$

The relativistic mass density $\sigma(x)$ has to be expressed in terms of $\sigma_o(x)$. For this purpose, note that

$$T^{00} = \Omega^0_{.\mu} \Omega^0_{.\nu} T^{\mu\nu}{}_{\text{rest}}, \tag{12.84}$$

where $\widehat{\boldsymbol{\Omega}}$ represents the Lorentz transformation matrix corresponding to the boost: $\{-\mathbf{u} = -\boldsymbol{\beta} c\}$ from the rest frame S_o to the observer's frame S. Replacing β by $-\beta$ in Eq. (3.20) of Sec. 3.2 we get the relevant components of the LT:

$$\Omega^0_{.0} = \Gamma;\ \Omega^0_{.k} = \Gamma \beta_k. \tag{12.85}$$

Substituting (12.85) in (12.84), and recognizing the T^{00} component from Eq. (12.74a) we get the following expression for $\sigma(x)$:

$$\begin{aligned} \sigma c^2 &= \Gamma^2 \sigma_o c^2 + \Gamma^2 (\beta_x^2 + \beta_y^2 + \beta_z^2) p_0 \\ &= \Gamma^2 \sigma_o c^2 + \Gamma^2 \beta^2 p_0 \\ &= \Gamma^2 \sigma_o c^2 + (\Gamma^2 - 1) p_0. \end{aligned}$$

Hence,

$$\sigma = \Gamma^2 \left(\sigma_o + \frac{p_0}{c^2}\right) - \frac{p_0}{c^2}. \tag{12.86}$$

Substituting (12.83) and (12.86) in (12.74) and (12.81), we transform $T^{\mu\nu}$ to the following form:

$$\begin{aligned} T^{00} &= \left\{\Gamma^2\left(\sigma_o + \frac{p_0}{c^2}\right) - \frac{p_0}{c^2}\right\}c^2, \\ T^{0k} &= T^{k0} = \Gamma^2\left(\sigma_o + \frac{p_0}{c^2}\right)cu^k, \\ T^{ik} &= \Gamma^2\left(\sigma_o + \frac{p_0}{c^2}\right)u^k u^i + p_o\delta^{ki}. \end{aligned} \tag{12.87}$$

It is now obvious from Eq. (12.87) that the energy tensor for a perfect fluid has the following compact covariant expression:

$$T^{\mu\nu} = \left(\sigma_o + \frac{p}{c^2}\right)U^\mu U^\nu - pg^{\mu\nu}, \tag{12.88}$$

where $g^{\mu\nu}$ is the metric tensor and $U^\mu(x)$ is the 4-velocity field of the fluid.

Note that we have dropped the subscript "$_o$" under p, because $p = p_o$.

Compare the energy tensor of the perfect fluid with that of the incoherent dust shown in (12.46), and note the difference between the two energy tensors:

$$\delta T^{\mu\nu} = T^{\mu\nu} - \mathcal{D}^{\mu\nu} = p\left(\frac{U^\mu U^\nu}{c^2} - g^{\mu\nu}\right). \tag{12.89}$$

The extra mass–energy–momentum represented by $\delta T^{\mu\nu}$ is entirely due to fluid pressure.

Appendices

A.1. Energy Conservation in Electromagnetic Field

We shall prove Poynting's theorem as given in Eq. (12.9) using Maxwell's equation (11.39). The electric current density appears in Eq. (11.39b).

$$\mathbf{E}\cdot\mathbf{J} = \mathbf{E}\cdot\varepsilon_0 c\left\{\boldsymbol{\nabla}\times c\mathbf{B}(\mathbf{r},t) - \frac{\partial\mathbf{E}(\mathbf{r},t)}{c\,\partial t}\right\}.$$

However, $\mathbf{E}\cdot\boldsymbol{\nabla}\times\mathbf{B} = \mathbf{B}\cdot\boldsymbol{\nabla}\times\mathbf{E} - \boldsymbol{\nabla}\cdot(\mathbf{E}\times\mathbf{B})$ (an identity).

$$\begin{aligned}
\text{Hence, } \mathbf{E}\cdot\mathbf{J} &= \varepsilon_0 c^2\{\mathbf{B}\cdot\boldsymbol{\nabla}\times\mathbf{E} - \boldsymbol{\nabla}\cdot(\mathbf{E}\times\mathbf{B})\} - \varepsilon_0\mathbf{E}\cdot\frac{\partial\mathbf{E}(\mathbf{r},t)}{\partial t}\\
&= \varepsilon_0 c\mathbf{B}\cdot\left(-\frac{\partial c\mathbf{B}}{\partial t}\right) - \boldsymbol{\nabla}\cdot\varepsilon_0 c^2(\mathbf{E}\times\mathbf{B}) - \varepsilon_0\mathbf{E}\cdot\frac{\partial\mathbf{E}(\mathbf{r},t)}{\partial t}\\
&= -\frac{\varepsilon_0}{2}\frac{\partial}{\partial t}(E^2 + c^2B^2) - \boldsymbol{\nabla}\cdot\varepsilon_0 c\,(\mathbf{E}\times c\mathbf{B})\\
&= -\frac{\partial w}{\partial t} - \boldsymbol{\nabla}\cdot\mathbf{S}. \qquad \text{(QED)}
\end{aligned}$$

A.2. Examples of Lowering and Raising an Index

Example 1.

$$V_\mu = V^\nu g_{\nu\mu} = \begin{pmatrix} V^0 \\ V^1 \\ V^2 \\ V^3 \end{pmatrix}\begin{pmatrix} 1 & 0 & 0 & 0 \\ 0 & -1 & 0 & 0 \\ 0 & 0 & -1 & 0 \\ 0 & 0 & 0 & -1 \end{pmatrix}$$

$$= \left(V^0,\ -V^1,\ -V^2,\ -V^3\right) \tag{A.1}$$

$$A^\mu = g^{\mu\nu} A_\nu = \begin{pmatrix} 1 & 0 & 0 & 0 \\ 0 & -1 & 0 & 0 \\ 0 & 0 & -1 & 0 \\ 0 & 0 & 0 & -1 \end{pmatrix} (A^0, \quad A^1, \quad A^2, \quad A^3)$$

$$= \begin{pmatrix} A^0 \\ -A^1 \\ -A^2 \\ -A^3 \end{pmatrix} \tag{A.2}$$

Lowering or Raising $\Rightarrow$ No change in the time component. Sign change in the space component.

Example 2. Let

$$F^{\mu\nu} = \begin{pmatrix} f^{00} & f^{01} & f^{02} & f^{03} \\ f^{10} & f^{11} & f^{12} & f^{13} \\ f^{20} & f^{21} & f^{22} & f^{23} \\ f^{30} & f^{31} & f^{32} & f^{33} \end{pmatrix} \tag{A.3}$$

be a contravariant 4-tensor. We shall lower only the first index μ, then only the second index ν, then both indices μ, ν.

$$F_\mu^{\cdot\nu} = g_{\mu\alpha} F^{\alpha\nu} = \begin{pmatrix} 1 & 0 & 0 & 0 \\ 0 & -1 & 0 & 0 \\ 0 & 0 & -1 & 0 \\ 0 & 0 & 0 & -1 \end{pmatrix} \begin{pmatrix} f^{00} & f^{01} & f^{02} & f^{03} \\ f^{10} & f^{11} & f^{12} & f^{13} \\ f^{20} & f^{21} & f^{22} & f^{23} \\ f^{30} & f^{31} & f^{32} & f^{33} \end{pmatrix}$$

$$= \begin{pmatrix} f^{00} & f^{01} & f^{02} & f^{03} \\ -f^{10} & -f^{11} & -f^{12} & -f^{13} \\ -f^{20} & -f^{21} & -f^{22} & -f^{23} \\ -f^{30} & -f^{31} & -f^{32} & f^{33} \end{pmatrix}. \tag{A.4}$$

First index lowered $\Rightarrow$ No change in row 0. Sign change in rows 1, 2, 3.

$$F^{\mu}_{\cdot\,\nu} = F^{\mu\alpha} g_{\alpha\nu} = \begin{pmatrix} f^{00} & f^{01} & f^{02} & f^{03} \\ f^{10} & f^{11} & f^{12} & f^{13} \\ f^{20} & f^{21} & f^{22} & f^{23} \\ f^{30} & f^{31} & f^{32} & f^{33} \end{pmatrix} \begin{pmatrix} 1 & 0 & 0 & 0 \\ 0 & -1 & 0 & 0 \\ 0 & 0 & -1 & 0 \\ 0 & 0 & 0 & -1 \end{pmatrix}$$

$$= \begin{pmatrix} f^{00} & -f^{01} & -f^{02} & -f^{03} \\ f^{10} & -f^{11} & -f^{12} & -f^{13} \\ f^{20} & -f^{21} & -f^{22} & -f^{23} \\ f^{30} & -f^{31} & -f^{32} & -f^{33} \end{pmatrix}. \tag{A.5}$$

Second index lowered $\Rightarrow$ No change in col 0. Sign change in cols 1, 2, 3.

$$F_{\mu\nu} = g_{\mu\alpha} F^{\alpha}_{\cdot\,\nu} = \begin{pmatrix} 1 & 0 & 0 & 0 \\ 0 & -1 & 0 & 0 \\ 0 & 0 & -1 & 0 \\ 0 & 0 & 0 & -1 \end{pmatrix} \begin{pmatrix} f^{00} & -f^{01} & -f^{02} & -f^{03} \\ f^{10} & -f^{11} & -f^{12} & -f^{13} \\ f^{20} & -f^{21} & -f^{22} & -f^{23} \\ f^{30} & -f^{31} & -f^{32} & f^{33} \end{pmatrix}$$

$$= \begin{pmatrix} f^{00} & -f^{01} & -f^{02} & -f^{03} \\ -f^{10} & f^{11} & f^{12} & f^{13} \\ -f^{20} & f^{21} & f^{22} & f^{23} \\ -f^{30} & f^{31} & f^{32} & f^{33} \end{pmatrix}. \tag{A.6}$$

Both indices lowered $\Rightarrow$ No change in $\{00, kj, jk\}$ components. Sign change in $\{0k, k0\}$ components.

Example 3. Trace of the contravariant tensor $F^{\mu\nu}$ is defined as $F^{\mu}_{\cdot\,\mu}$; sum over μ. Going back to (A.5),

$$\begin{aligned} \mathrm{Tr}\{F\} &= F^{\mu}_{\cdot\,\mu} = f^{00} - (f^{11} + f^{22} + f^{33}) \\ &= \text{sum of the diagonal elements of } F^{\mu}_{\cdot\,\nu}. \end{aligned} \tag{A.7}$$

A.3. Components of Maxwell's Stress 3-Tensor and Maxwell's 4-Tensor, and Their Traces

Maxwell's 3-tensor was written in a short form in Eq. (6.46). We shall now write down the 3 × 3 components of this tensor. The reader should verify them.

$$
\begin{aligned}
\mathcal{T}_{\text{em}}^{xx} &= \frac{\varepsilon_0}{2}[(E_x^2 - E_y^2 - E_z^2) + c^2(B_x^2 - B_y^2 - B_z^2)],\\
\mathcal{T}_{\text{em}}^{yy} &= \frac{\varepsilon_0}{2}[(E_y^2 - E_z^2 - E_x^2) + c^2(B_y^2 - B_z^2 - B_x^2)],\\
\mathcal{T}_{\text{em}}^{zz} &= \frac{\varepsilon_0}{2}[(E_z^2 - E_x^2 - E_y^2) + c^2(B_z^2 - B_x^2 - B_y^2)],\\
\mathcal{T}_{\text{em}}^{xy} &= \mathcal{T}_{\text{em}}^{yx} = \varepsilon_0[E_xE_y + c^2B_xB_y],\\
\mathcal{T}_{\text{em}}^{yz} &= \mathcal{T}_{\text{em}}^{zy} = \varepsilon_0[E_yE_z + c^2B_yB_z],\\
\mathcal{T}_{\text{em}}^{zx} &= \mathcal{T}_{\text{em}}^{xz} = \varepsilon_0[E_zE_x + c^2B_zB_x].
\end{aligned} \tag{A.8}
$$

We can now write the trace of the Maxwell 3-tensor:

$$\text{Tr}\{\mathcal{T}_{\text{em}}\} = \mathcal{T}_{\text{em}}^{xx} + \mathcal{T}_{\text{em}}^{yy} + \mathcal{T}_{\text{em}}^{zz} = -\frac{\varepsilon_0}{2}(E^2 + c^2B^2). \tag{A.9}$$

Maxwell's 4-tensor was defined by Eq. (12.17). We shall use the same equation to identify all the components of $M^{\mu\nu}(x)$.

From (12.17a): $$\frac{1}{c}\left(\frac{\partial w}{\partial t} + \boldsymbol{\nabla}\cdot\mathbf{S}\right) = \nabla_\alpha M^{\alpha 0},$$

or $$\frac{\partial w}{c\partial t} + \frac{\partial}{\partial x^j}(S_j/c) = \frac{\partial M^{00}}{c\partial t} + \frac{\partial}{\partial x^j}M^{j0} \quad \text{(sum over } j\text{)}.$$

Hence, $$M^{00} = w;\; M^{j0} = S_j/c. \tag{A.10}$$

In the following, we shall write $\Phi_{\text{em}}^{11}, \Phi_{\text{em}}^{12}, \ldots$, to mean $\Phi_{\text{em}}^{xx}, \Phi_{\text{em}}^{xy}, \ldots$ respectively.

From (12.17b): $$\left[\frac{\partial \mathbf{g}}{\partial t} + \boldsymbol{\nabla}\cdot\widehat{\boldsymbol{\Phi}}_{(\text{em})}\right]_k = \nabla_\alpha M^{\alpha k};\; k = 1, 2, 3,$$

or $$\frac{\partial c g_k}{c\partial t} + \frac{\partial}{\partial x^j}(\Phi_{\text{em}}^{jk}) = \frac{\partial M^{0k}}{c\partial t} + \frac{\partial}{\partial x^j}(M^{jk}); \begin{cases}\text{sum over } j\\ k = 1, 2, 3.\end{cases}$$

Hence, $$M^{0k} = cg_k;\; M^{jk} = \Phi_{\text{em}}^{jk}. \tag{A.11}$$

We can now write all the 4×4 components of $M^{\mu\nu}(x)$.

$$M^{\mu\nu}(x) = \begin{matrix} \\ 0 \\ 1 \\ 2 \\ 3 \end{matrix} \begin{matrix} \begin{matrix} 0 & 1 & 2 & 3 \end{matrix} \\ \begin{pmatrix} w & cg_x & cg_y & cg_z \\ S_x/c & \Phi^{11}_{\text{em}} & \Phi^{12}_{\text{em}} & \Phi^{13}_{\text{em}} \\ S_y/c & \Phi^{21}_{\text{em}} & \Phi^{22}_{\text{em}} & \Phi^{23}_{\text{em}} \\ S_z/c & \Phi^{31}_{\text{em}} & \Phi^{32}_{\text{em}} & \Phi^{33}_{\text{em}} \end{pmatrix} \end{matrix}. \tag{A.12}$$

Because of Eq. (12.13), $cg_k = \frac{S_k}{c}$, and the tensor is symmetric. The trace of the Maxwell's 4-tensor follows from (A.7) and (A.9).

$$\begin{aligned} \text{Tr}\{M\} &= M^{\mu}_{\cdot\,\mu} = M^{00} - (M^{11} + M^{22} + M^{33}) \\ &= w - (\Phi^{11}_{\text{em}} + \Phi^{11}_{\text{em}} + \Phi^{11}_{\text{em}}) \\ &= w + (\mathcal{T}^{xx}_{\text{em}} + \mathcal{T}^{yy}_{\text{em}} + \mathcal{T}^{zz}_{\text{em}}) = w - \frac{\varepsilon_0}{2}(E^2 + c^2 B^2) = 0. \end{aligned} \tag{A.13}$$

B.1. Useful Integrals

We shall write derive the values of some integrals required in this book. The integrands of all the integrals will have in their denominators integer/half-integer powers of the expression $(r^2+a^2-2ra\cos\theta)$, the integration variable will be θ, and the range of integration $[0, \pi]$. We shall do some preliminary work by changing the variable of integration from θ to η, accompanied by the change of the limits of integration, and conversion of the numerators for the first two cases:

$$\eta^2 = r^2 + a^2 - 2ra\cos\theta, \tag{B.1a}$$

$$\eta\, d\eta = ar\sin\theta\, d\theta, \tag{B.1b}$$

$$a - r\cos\theta = \frac{a^2 - r^2 + \eta^2}{2a}, \tag{B.1c}$$

$$(r^2 + a^2)\cos\theta - 2ra = \frac{(a^2 - r^2) - (a^2 + r^2)\eta^2}{2ra}. \tag{B.1d}$$

Lower limit: $\theta = 0 \Rightarrow \eta = \{(a - r),\ \text{if } a > r\};\quad \{(r - a),\ \text{if; } r > a\}.$ (B.1e)

Upper limit: $\theta = \pi \Rightarrow \eta = a + r.$ (B.1f)

Direct Evaluation

Using the above conversions hints it should not be difficult for the reader to establish the following integrals:

Integral #1

$$\Psi_1(a,r) \equiv \int_0^\pi \left[\frac{(a - r\cos\theta)}{(r^2 + a^2 - 2ra\,\cos\theta)^{\frac{3}{2}}}\right] \sin\theta\, d\theta = \begin{cases} \dfrac{2}{a^2} & \text{if } a > r, \\ 0 & \text{if } a < r. \end{cases} \tag{B.2}$$

Integral #2

$$\Psi_2(r,a) \equiv \int_0^\pi \left[\frac{(r^2 + a^2)\cos\theta - 2ra}{(r^2 + a^2 - 2ra\,\cos\theta)^3}\right] \sin\theta\, d\theta = 0. \tag{B.3}$$

Evaluation using Maxima

We have evaluated the following three integrals, using Maxima (version 5.13.0). We shall first write down the values of the integrals, and then show the commands used in Maxima to obtain these results. Let us write

$$\begin{aligned} \alpha &= 2\cos\theta; \quad \beta = 2(r^2 + a^2)\cos\theta - (3 + \cos^2\theta)ar. \\ \gamma &= \sin\theta; \quad \delta = (r^2 - 2a^2 + ar\cos\theta)\sin\theta. \end{aligned} \tag{B.4}$$

Integral #3

$$\Psi_3(r,a) = \int_0^\pi [(\alpha^2 - \gamma^2)\cos\theta - 2\alpha\gamma\sin\theta]\, \sin\theta\, d\theta = 0. \tag{B.5}$$

Integral #4

$$\begin{aligned} \Psi_4(r,a) &= \int_0^\pi \left[\frac{(\alpha\beta - \gamma\delta)\cos\theta - (\alpha\delta + \beta\gamma)\sin\theta}{(r^2 + a^2 - 2ra\,\cos\theta)^{5/2}}\right] \sin\theta\, d\theta \\ &= \begin{cases} \dfrac{12r}{a^4} & (a > r) \\ 0 & (a < r). \end{cases} \end{aligned} \tag{B.6}$$

Integral #5

$$\Psi_5(r,a) = \int_0^\pi \left[\frac{(\beta^2 - \delta^2)\cos\theta - 2\beta\delta\sin\theta}{(r^2 + a^2 - 2ra\,\cos\theta)^5}\right] \sin\theta\, d\theta = 0. \tag{B.7}$$

Maxima Commands, Inputs and Outputs

We shall write the interactive commands and prompts between the user and the Maxima so that the reader can verify the values of the integrals #4 and #5. Note the following:

1. Some output lines (e.g. %o5, %o6 in Ex. #4) are spread over two lines in which the first line contains the "indices", e.g. "to the power 2". These indices get displaced and detached from the base when the output is copied into any text file. To avoid this anomaly, we have brought them to one line using mathematical mode.
2. If the output is an expression of a definite integral, it is spread over seven lines (e.g. as in %o9 in Example #4), and the integral sign becomes unintelligible when copied. We have replaced these outputs and other outputs that appear too long and complicated with "...". All outputs except the final one are non-essential.

Input/Output for Integral #4

(%i1) aa: 2*cos(x);
(%o1) 2 cos(x)
(%i2) bb: 2 *(r^2 + a^2) *cos(x) – ((cos(x))^2 +3)* a* r;
(%o2) $2(\mathrm{r}^2 + \mathrm{a}^2)\cos(x) - \mathrm{ar}(\cos^2(x) + 3)$
(%i3) cc: sin(x);
(%o3) sin(x)
(%i4) dd: (r^2 –2* a^2 + a* r *cos(x))*sin(x);
(%o4) $(\mathrm{ar}\cos(\mathrm{x}) + \mathrm{r}^2 - 2\mathrm{a}^2)\sin(\mathrm{x})$
(%i5) f: (aa*bb – cc*dd)* cos(x) – (aa*dd+bb*cc)* sin(x);
(%o5) $\cos(\mathrm{x})\ (2\cos(\mathrm{x})\ (2(\mathrm{r}^2 + \mathrm{a}^2)\cos(x) - \mathrm{ar}(\cos^2(x) + 3))$
$- (\mathrm{a\ r}\cos(\mathrm{x}) + \mathrm{r}^2 - 2\mathrm{a}^2)\sin^2(x)) - \sin(\mathrm{x})$
$((2(\mathrm{r}^2 + \mathrm{a}^2)\cos(x) - \mathrm{ar}(\cos^2(x) + 3))\sin(x)$
$+ 2\cos(\mathrm{x})\ (\mathrm{a\ r}\cos(\mathrm{x}) + \mathrm{r}^2 - 2\mathrm{a}^2)\sin(\mathrm{x}))$
(%i6) et: abs(sqrt(r^2+a^2 – 2*r*a*cos(x)));
(%o6) $\mathrm{sqrt}(-2\ \mathrm{a\ r}\cos(\mathrm{x}) + \mathrm{r}^2 + \mathrm{a}^2)$
(%i7) h: (f/(et^5))*sin(x);
(%o7) ....
(%i8) assume (a – r > 0);
(%o8) [a > r]
(%i9) 'integrate (h, x);
(%o9) ...

(%i10) changevar (%, et − y, y, x);
Is y positive, negative, or zero?
pos;
solve: using arc-trig functions to get a solution.
Some solutions will be lost.
(%o10) ...
(%i11) %,nouns;
Is sqrt(r + 2 a r + a) − sqrt(r − 2 a r + a) positive, negative, or zero?
pos;
Is r + a zero or nonzero?
nonzero;

(%o11)
$$-\Big(\frac{\mathrm{sqrt}(r^2-2ar+a^2)(48a^2r^7+36a^3r^6+8a^4r^5+4a^5r^4)}{r-a}$$
$$-\frac{\mathrm{sqrt}(r^2+2ar+a^2)(48a^2r^7-36a^3r^6+8a^4r^5-4a^5r^4)}{r+a}$$
$$+\mathrm{sqrt}(r^2+2ar+a^2)(-48a^2r^6+12a^3r^5-4a^4r^4)$$
$$-\mathrm{sqrt}(r^2-2ar+a^2)(-48a^2r^6-12a^3r^5-4a^4r^4)\Big)/(16a^6r^6)$$

To simplify the last output (%o11), set

$$\mathrm{sqrt}(r^2-2ar+a^2)=\begin{cases}(a-r) & \text{if, } (a>r)\\ (r-a) & \text{if, } (a<r)\end{cases} \tag{B.8}$$

and get $\frac{4\times 48a^2r^7}{16a^6r^6}=\frac{12r}{a^4}$ for the first case and 0 for the second.

Input/Output for Integral #5

(%i1) bb: 2*(r^2 + a^2)*cos(x) − ((cos(x))^2 +3)*a*r;
(%o1) $2\ (r^2+a^2)\cos(x) - a\ r\ (\cos^2(x)+3)$
(%i2) dd: (r^2 −2*a^2 + a*r*cos(x))*sin(x);
(%o2) $(a\ r\cos(x)+r^2-2\ a^2)\sin(x)$
(%i3) f: (bb^2−dd^2)*cos(x) − 2*bb*dd*sin(x);
(%o3) $\cos(x)\ ((2\ (r^2+a^2)\cos(x)-a\ r\ (\cos^2(x)+3))^2$
$-\ (a\ r\cos(x)+r^2-2\ a^2)^2\sin^2(x)) - 2\ (a\ r\cos(x)+r^2-2\ a^2)$
$(2\ (r^2+a^2)\cos(x)-a\ r\ (\cos^2(x)+3))\sin^2(x)$
(%i4) ets: r^2+a^2 − 2*r*a*cos(x);
(%o4) $-\ 2\ a\ r\cos(x)+r^2+a^2$

```
(%i5)  h: (f/(ets^5))*sin(x);
(%o5)  (sin(x) (cos(x) ((2 (r²+ a²) cos(x) − a r (cos²(x) + 3))²
       − (a r cos(x) + r²− 2 a²)²sin²(x)) − 2 (a r cos(x) + r²− 2 a²)
       (2 (r²+ a²) cos(x) − a r (cos²(x) + 3)) sin²(x)))/(− 2 a r cos(x)
       + r²+ a²)⁵
(%i6)  assume (a−r > 0);
(%o6)  [a > r]
(%i7)  'integrate (h, x, 0, %pi );
(%o7)  ...
(%i8)  changevar (%, abs(sqrt(ets)) − y, y, x);
       Is y positive, negative, or zero?
       pos;
       solve: using arc-trig functions to get a solution.
       Some solutions will be lost.
(%o8)  ...
(%i9)  %,nouns;
       Is sqrt(r²+ 2 a r + a²) − sqrt(r²− 2 a r + a²) positive, negative,
       or zero?
       pos;
       Is r + a zero or nonzero?
       nonzero;
(%o9)  0
```

Epilogue

In the high reaches of the Himalayas, the ice of the Gangotri glacier melts, and as it descends down the mountains into the plains, is swelled by the tributaries and groundwaters to form a mighty river, the Ganga. But Ganga is not just a river. It is also a concept, one of the pillars of a Faith that has moulded a civilization. It could be an edifying experience, intellectually and physically, to journey down the course of the river, from the mountain to the sea, as it meanders past holy cities and historical monuments.

Nestled in the lofty heights of scientific analysis and philosophical ruminations, occasionally great and epoch making theories are born. Some of them cause deluge of such a magnitude as to shake everything on their way, altering the course of history, uprooting the foundations of conventional notions and concepts, and building on the ruins of destruction another edifice of much greater vigour and beauty. It can be an edifying experience to make an intellectual voyage downstream of the deluge and thrill at the profound upheavals that one intellectual feat of a human mind could bring about.

Sometimes the impact is noticeable to all sections of society, as for example in the case of Faraday's discovery of Electromagnetic Induction, leading to widespread use of electricity. More often, however, the impact is of a more subtle and esoteric nature, comprehensible to the avowed practitioners of the discipline. Newton's formulation of Classical Mechanics and Gravitation falls in the latter category. It replaced the decrepit Aristotelian beliefs with a system of analysis whose mind-boggling profundity, universality, power and simplicity laid the foundations of physics. Much of our studies in classical physics is a journey through the course of a river that finds its source in the laws of motion and gravitation as conceived by Newton, and streams through the macroworld of planets and satellites, the earthly world of missiles and locomotives, down to the microworld of atoms and molecules, as if to unify the three worlds in a single grand design. Even though the advent of Quantum Mechanics curbed the role of Newton in the microworld, and General Theory of Relativity redefined cosmological concepts, they are not to be regarded as an abandonment of Newtonian ideas, but a refinement of the same. The unifying spirit of Newton has been the driving force behind the pursuits of physics among all succeeding generations.

In this book, we tried to trace the course of another great river, the Theory of Relativity, which finds its origin in the belief that all frames of reference are equal. Seen in isolation, this principle is just an extrapolation of the egalitarian value system from society to reference frames, without any visible impact. However, in juxtaposition with the laws of electrodynamics, this innocuous hypothesis makes startling revelations about the relative nature of space and time, leading to paradoxes of *time dilation* and *length contraction* (Chapter 2). An immediate fallout is *Lorentz transformation*, providing uniform prescription for the conversion of the time and space coordinates of an event between frames of reference (Chapter 3).

The same Lorentz transformation does not seem to fit in within the scheme of Newtonian Mechanics unless *momentum* is redefined and the *energy* expression is modified. But this does not seem possible without recognizing *mass–energy equivalence*. The innocuous hypothesis then opens up an entirely new world where matter is annihilated to liberate energy. A new branch of physics is born — the physics of *Atomic Nuclei* and the physics of *Elementary Particles* with awesome forebodings of a nuclear holocaust (Chapter 4).

The mathematical language in which the laws of physics, in particular, the conservation of Energy and Momentum are expressed involve an exposition to the world of *tensors*. Maxwell's stress tensor gives a beautiful illustration of the stress that exists in empty space, in the vicinity of electric charges and currents, or anywhere else where electromagnetic field exists (Part II).

The vision of relativity is incomplete unless physical quantities are looked upon as four-dimensional geometrical objects having one time and three space components. Realization of this four-dimensional *world*, called *space–time*, begins with the construction of the *Minkowski metric* giving the expression for a *line element* stretching between two *events*. The vision of this four-dimensional scheme in the workings of the universe is then consummated through the *Principle of Covariance* that declares that the mathematical expressions of the laws of physics must be an equation between two 4-tensors of the same rank and type. This leads to the unification of energy and momentum within a 4-vector, the *4-momentum* (also called or En-Mentum in this book), and *Minkowski's equation of motion* (EoM) in terms of another 4-vector, the *4-force* (also called Pow-Force in this book). An illustration of this new scheme is provided by the *relativistic rocket*, and the *covariant equations of classical electrodynamics* (Part III).

Further example of the Principle of Covariance is provided by *energy tensor*, which not only illustrates the succinctness, elegance and the power of the energy–momentum (i.e. 4-momentum) conservation equations but also builds a bridge to the General Theory of Relativity, which is also the *Relativistic Theory of Gravitation* (Part IV).

Within the limited scope of this book we have thus taken the reader on a guided tour of the Relativity Valley. Originating from a modest creed, nurtured by the Laws of Classical Electrodynamics, swelled by the conservation laws of energy–momentum, embellished by the demands of *Covariance*, the relativity theory turns into a mighty river providing nourishment to all branches of modern physics. Midway downstream of our journey we had to call it a day as the river was about to enter a deep canyon of exquisite beauty — the domain of the General Theory of Relativity. Some of the readers who might have received an inspiration from their visit to the temple of energy tensor for a rafting expedition down these turbulent rapids may now feel dejected at this last moment betrayal of their captain. However, great expedition requires not only great courage, but also necessary equipments and above all, arduous preparations and training to face unexpected challenges in a land of adventure. It is now time to undergo this training from a master of tensor calculus. We shall resume our voyage, hopefully in future, after period of interlude to give some sobering time for the impatient expeditioner.

Bibliography

[1] J. H. Smith, *Introduction to Special Relativity*, W.A. Benjamin Inc., New York (1965).
[2] A. Einstein and L. Infeld, *Evolution of Physics*, Cambridge University Press, London (1938).
[3] W. Rindler, *Essential Relativity*, Van Nostrand Reinhold Co., New York (1969).
[4] W. Rindler, *Relativity, Special, General, and Cosmological*, 2nd. edn., Oxford University Press, New York (2006).
[5] C. W. Misner, K. S. Thorne and J. A. Wheeler, *Gravitation*, W.H. Freeman & Co. San Francsisco (1973).
[6] E. F. Taylor and J. A. Wheeler, *Spacetime Physics*, W.H. Freeman & Co., New York (1992).
[7] A. P. French, *Newtonian Mechanics: The MIT Introductory Course Series*, W.W. Norton & Co. (1971).
[8] S. Datta, *Mechanics*, Pearson Education, New Delhi (2013).
[9] G. Gamow, *Gravity, Classical and Modern Views*, Doubleday and Co., New York (1961). Indian publication by Vakils, Effer and Simons, Pvt. Ltd., Bombay.
[10] S. Chandrashekhar, *Truth and Beauty: Aesthetics and Motivations in Science*, The University of Chicago Press, Chicago (1987).
[11] A. P. French, *Vibrations and Waves*, CBS Publisher, New Delhi (1987).
[12] S. Glasstone, *Source book of Atomic Energy*, 3rd. Edition., Affiliated East West Press Pvt Ltd, New Delhi (1967).
[13] J. D. Jackson, *Classical Electrodynamics*, 3rd. edn., John Wiley & Sons, Singapore (1999).
[14] D. J. Griffiths, *Introduction to Electrodynamics*, 3rd edn., Pearson Education, New Delhi (2006).
[15] R. D. Evans, *The Atomic Nucleus*, Tata McGraw-Hill, Bombay Delhi (1955).
[16] B. Cohen, *Concepts of Nuclear Physics*, McGraw-Hill, New York (1971).

[17] S. Datta, Maxwell's stress tensor and conservation of momentum in electromagnetic field, *Physics Education*, **30**(3): 1 (2014), www.physedu.in.
[18] K. R. Simon, *Mechanics*, 2nd edn., Addison-Wesley, Reading (1960).
[19] Sir Edmund Whittaker, *Aether and Electricity, The Classical Theories,* , Thomas Nelson and Sons Ltd, London (1951).
[20] J. Clerk Maxwell, A dynamical theory of the electromagnetic field, *Philosophical Transactions of the Royal Society of London*, **155**: 459–512 (1864).
[21] J. Clerk Maxwell, *A Treatise on Electricity and Magnetism*, Vols. 1 and 2. 3rd edn., Dover (1954). http://archive.org/details/ATreatiseonElectricity Magnetism-Volume1.
[22] J. Schwinger (compiled by his students), *Classical Electrodynamics*, Westview (A member of the Perseus Books Group) (1998).
[23] H. Goldstein, *Classical Mechanics*, 2nd edn., Addison-Wesley Indian Edition, New Delhi (1989), p. 146.
[24] C. Moller, *The Theory of Relativity*, 2nd edn., Oxford University Press, London (1972).
[25] A. P. French, *Special Relativity: The M.I.T. Introductory Physics Series*, W.W. Norton & Co., New York (1968).
[26] R. D. Evans, *The Atomic Nucleus*, Tata McGraw-Hill Publishing Co., New Delhi (1955).
[27] S. Datta, Relativistic rocket, its equation of motion and solution for two special cases, *Physics Education*, **34**(4): 6 (2018).
[28] T. Singal and A. K. Singal, Is interstellar travel possible? *Physics Education*, **35**(4): (2019).
[29] Wikipedia, *Relativistic Rocket*, https://en.wikipedia.org/wiki/Relativistic_rocket.
[30] J. Ackeret, On the theory of rockets, *Helvetica Physica Acta*, **19**: 103 (1946).
[31] J. W. Rhee, Relativistic rocket motion, *American Journal of Physics* **33**(10): 587 (1965).
[32] K. B. Pomeranz, *American Journal of Physics*, **32**:955 (1964); **34**: 565, (1966); **37**:741, (1969).
[33] W. Rindler, *Relativity*, 2nd edn., Oxford University Press, USA (2006).
[34] C. Moller, *The Theory of Relativity*, 2nd edn., Oxford: Clarendon Press (1972).
[35] S. Datta, Magnetism as a relativistic effect, *Physics Education*, **21**(2) (2004).
[36] B. L. Cohen, *Concepts of Nuclear Physics*, McGraw-Hill, New York (1971).
[37] E. Segre, *Nuclei and Particles*, W.A. Benjamin, Inc (1965).
[38] I. Kaplan, *Nuclear Physics*, Addison-Wesley, Reading (1964).
[39] S. Datta, Energy tensor for charged incoherent dust in its own electromagnetic field, *Physics Education*, **35**(3): (2019).
[40] S. Datta, 1905 relativity papers of Einstein, *Physics Education*, **22**(1) (2005).
[41] S. Datta, Minkowski's space-time, *Physics Education*, **29**(2) (2013).
[42] D. Halliday, R. Resnick, J. Walker, *Fundamentals of Physics*, 4th edn., Asian Books Pvt. Ltd, New Delhi/John Wiley & Sons, Singapore (1998).

Index

www.ingramcontent.com/pod-product-compliance
Lightning Source LLC
Chambersburg PA
CBHW050536190326
41458CB00007B/1808